Dating the Passion

Time, Astronomy, and Calendars

Texts and Studies

Editors

Charles Burnett
Sacha Stern

VOLUME 1

The titles published in this series are listed at www.brill.nl/tac

Dating the Passion

The Life of Jesus and the Emergence of Scientific Chronology (200–1600)

By

C. Philipp E. Nothaft

BRILL

LEIDEN • BOSTON
2012

This book is printed on acid-free paper.

Library of Congress Cataloging-in-Publication Data

Nothaft, C. Philipp E.
Dating the Passion : the life of Jesus and the emergence of scientific chronology (200–1600) / by C. Philipp E. Nothaft.
p. cm. — (Time, astronomy, and calendars, ISSN 2211-632X ; v. 1)
Substantially the revised English version of the author's German dissertation, entitled Das Leben Jesu und die Entstehung der wissenschaftlichen Chronologie: Eine problemgeschichtliche Studie (200–1600), which was submitted to the Ludwig Maximilian University of Munich in October 2010.
Includes bibliographical references.
ISBN 978-90-04-21219-0 (hardback : alk. paper) 1. Chronology—History. 2. History—Methodology. 3. Chronology—Historiography. 4. Calendar—History. I. Title.

CE6.N68 2011
529—dc23

2011029856

ISSN 2211-632X
ISBN 978 90 04 21219 0

CONTENTS

PREFACE

The present book is the substantially revised English version of my German dissertation, entitled *Das Leben Jesu und die Entstehung der wissenschaftlichen Chronologie: Eine problemgeschichtliche Studie (200–1600)*, which was submitted to the Ludwig Maximilian University of Munich in October 2010. My *Doktorvater* Helmut Zedelmaier kindly agreed to supervise this somewhat unusual project and provided much-needed help and encouragement along the way. Anthony Grafton was willing to join as a co-supervisor and supported me in crucial ways besides lending his unsurpassed expertise in the history of historical chronology. My deepest thanks are due to both of them. A doctoral stipend from the Studienstiftung des deutschen Volkes ensured that I could spend most of 2009 and the first half of 2010 with research and writing, profiting from Munich's many excellent libraries. The Martin Buber Society of Fellows in the Humanities offered me a generous fellowship, thanks to which I could rework my dissertation into English whilst at the Hebrew University of Jerusalem during the academic year 2010/11. I would like to express my most heartfelt gratitude to everyone who made a contribution to my project during these years. Anthony Grafton, Leofranc Holford-Strevens, Sacha Stern, and Immo Warntjes have diligently read through earlier drafts of this book and improved it through many helpful suggestions. Sacha Stern and Charles Burnett have kindly agreed to make it the first volume of their new series on time, astronomy, and calendars. I also owe special thanks to the good people at the Martin Buber Society, in particular to David Shulman and his excellent staff, Yael Baron and Amitai Jacobsen, who have done their utmost to ensure a pleasant and productive stay in Jerusalem. Further thanks are due to Michael Allen, Ari Belenkiy, Jonathan Ben-Dov, Menso Folkerts, Yehoshua Granat, Eva Haverkamp, Magnus Quirin Löfflmann, Daniel McCarthy, Martin Mulsow, Dáibhí Ó Cróinín, Gerhard Schmitz, Christof Schuler, Israel Yuval, and my parents.

ABBREVIATIONS

AASS	*Acta Sanctorum*. 68 vols. Antwerp, 1643–1940.
CCCM	*Corpus Christianorum continuatio medievalis*. Turnhout, 1966–.
CCSL	*Corpus Christianorum series Latina*. Turnhout, 1953–.
CSCO	*Corpus Scriptorum Christianorum Orientalium*. Louvain, 1903–.
CSEL	*Corpus Scriptorum Ecclesiasticorum Latinorum*. Vienna, 1866–.
DHEE	*Diccionario de Historia Eclesiástica de España*. 5 vols. Madrid, 1972–87.
GCS	*Die griechischen christlichen Schriftsteller der ersten [drei] Jahrhunderte*. 60 vols. Leipzig, 1897–1989.
GCS-NF	*Die griechischen christlichen Schriftsteller der ersten Jahrhunderte. Neue Folge*. Berlin, 1995–.
KGW	Johannes Kepler. *Gesammelte Werke*. 21 vols. Munich, 1937–2009.
Krusch, *Studien* (I)	Bruno Krusch. *Studien zur christlich-mittelalterlichen Chronologie: Der 84-jährige Osterzyklus und seine Quellen*. Leipzig, 1880.
Krusch, *Studien* (II)	Bruno Krusch. *Studien zur christlich-mittelalterlichen Chronologie: Die Entstehung unserer heutigen Zeitrechnung*. Abhandlungen der Preußischen Akademie der Wissenschaften, Jg. 1937, phil.-hist. Kl., 8. Berlin, 1938.
MGH	Monumenta Germaniae Historica
Oppolzer	Theodor Oppolzer. *Canon der Finsternisse*. Denkschriften der kaiserlichen Akademie der Wissenschaften, math.-naturw. Cl., 52. Vienna, 1887.
PG	*Patrologiae cursus completus, series Graeca*. 161 vols. Paris, 1857–66.
PL	*Patrologiae cursus completus, series Latina*. 221 vols. Paris, 1844–65.
PLS	*Patrologiae cursus completus, series Latina. Supplementum*. 5 vols. Paris, 1958–74.

INTRODUCTION

According to the Russian mathematician Anatoly T. Fomenko, the history of Western historiography since the Renaissance has been characterized by a struggle of epic proportions between a brave gang of outlaws against an all-powerful syndicate. Among the members of the former group, Fomenko counts early modern characters such as the Jesuit antiquary Jean Hardouin (1646–1729) and Isaac Newton (1643–1727) as well as a string of more recent authors leading up to himself, who have spent the past few centuries trying to alarm the world that they have discovered nasty cracks in the basement of history. Sadly, their Cassandra-like calls are consistently ignored by a malicious cabal of historians and chronologers, who have been scheming to defend the traditional timeline that had been forged by the legendary Huguenot philologist Joseph Justus Scaliger (1583–1609). After several decades of arduous work on the plumbing of chronology, Fomenko believes he has found strong evidence that 'Scaligerian chronology', which is still slavishly followed by historians, is nothing but a gigantic hoax, a grand concoction, built on forgeries and historical wishful thinking. If Fomenko and his followers are right, our history textbooks may soon have to be rewritten to match his insights that—to mention only two of his more startling claims—Jesus Christ was born in AD 1152 and crucified in AD 1182 and that the book of Revelation was written after AD 1486.[1]

Abstruse as these and other theories proposed by the Russian school of 'New Chronology', which has found an equally dazzling German counterpart in Heribert Illig's 'Phantomzeittheorie', may be, they serve to highlight the fundamental importance of technical chronology for our understanding of human history. Chronology's role as an indispensable item in the tool kit of historical research stands in striking contrast to the relative neglect it has suffered at the hands of

[1] Anatoly T. Fomenko, *History: Fiction or Science?*, 2nd ed., 7 vols. (Paris: Delamere, 2003–6); Fomenko, *Empirico-Statistical Analysis of Narrative Material and its Applications to Historical Dating*, 2 vols. (Dordrecht: Kluwer, 1994). For an introduction to Fomenko's theories, see Florin Diacu, *The Lost Millennium* (Toronto: Vintage Canada, 2006).

twentieth-century historians. Although many dates in our textbooks are the result of intricate technical arguments, hingeing on the successful reconstruction of exotic calendars and the often precarious interpretation of ancient eras, the working knowledge necessary to assess these arguments is rarely taught in seminar rooms, while scholars interested in the subject are too often forced to work their way through long out-of-print publications in foreign languages. As the recent Russian and German onslaughts against history's very foundations go to show, such continuous lack of interest is easily exploited by those who inhabit the fringes of academic scholarship and may even serve to undermine the discipline's credibility in the public eye.[2]

If chronology has been an unfashionable and marginalized pursuit for the past hundred years, this holds true to an even greater extent for its history as a discipline. For all the obvious distortions inherent to his account, Fomenko's schematic view of the roots of technical chronology is in fact largely shared by mainstream historians, who have frequently singled out Scaliger, the arch-villain of Fomenko's tale, not as the culprit in a conspiracy theory, but as one of the main heroes in the development of modern historiography. With his two groundbreaking works, the *Opus de emendatione temporum* (1583^1, 1598^2, 1629^3) and the even more ambitious *Thesaurus temporum* (1606^1, 1658^2), in which he attempted to reconstruct the world chronicle of Eusebius of Caesarea, Scaliger is said to have singlehandedly invented historical chronology as a discipline.[3] Following the lead of his first modern biographer Jacob Bernays, subsequent scholars have painted a picture according to which the French polymath fertilized the grey and unscientific pastures of medieval chronography by combining philology with the new results of sixteenth-century astronomy, thereby

[2] For the German context, see Heribert Illig, *Das erfundene Mittelalter* (Düsseldorf: Econ, 1996); Uwe Topper, *Kalender-Sprung* (Tübingen: Grabert, 2006), and the refutations by Franz Krojer, *Die Präzision der Präzession* (Munich: Differenz-Verlag, 2003); Ronald Starke, *Niemand hat an der Uhr gedreht!* (Munich: Differenz-Verlag, 2009).

[3] The standard monograph on Scaliger's oeuvre is Anthony Grafton, *Joseph Scaliger*, 2 vols. (Oxford: Clarendon Press, 1983–93), the second volume of which deals with his chronological work. For a comprehensive bibliography up to 1993, see Anthony Grafton and Henk Jan de Jonge, "Joseph Scaliger: A Bibliography 1850–1993," in *The Scaliger Collection*, ed. Rijk Smitskamp (Leiden: Smitskamp Oriental Antiquarium, 1993), i–xxx.

creating a successful merger of two disciplines that had hitherto lain unconnected.[4]

Owing to his reputation as the founding father of historical chronology, Scaliger has also been invoked as an agent in the process of 'secularization' that characterizes much of early modern European historiography. In his influential study *Die Säkularisierung der universalhistorischen Auffassung*, Adalbert Klempt showed how the chronological contradictions between different versions of the Old Testament gradually moved scholars to abandon traditional 'biblicist' chronologies in favour of more neutral modes of time reckoning. In order to bypass the chaos of different creation eras and other ancient systems of dating, Joseph Scaliger created his Julian Period, an artificial 7980-year cycle beginning on 1 January 4173 BC that was mathematically derived from three common calendar cycles (the 19-year lunisolar cycle, the 28-year solar cycle, and the 15-year indictional cycle; 19 × 28 × 15 = 7980). Klempt singled out the introduction of the Julian Period, which is still used by astronomers as a basis for the Julian day count, as an important step towards the dissolution of the old regime of biblical world eras, which eventually led to the common adoption of our current BC/AD-system of counting years on an abstract timeline that could be extended *ad infinitum* in both directions.[5]

Other scholars have followed Klempt's lead by portraying Scaligerian chronology as a project bent on undermining the biblical framework of world history. In its most extreme version, this view has amounted to the claim that Scaliger himself had abandoned belief in a divine plan that endowed history with a definite beginning and an equally well-defined ending.[6] Such an assessment of Scaliger's approach, however, is

---

[4] Jacob Bernays, *Joseph Justus Scaliger* (Berlin: Hertz, 1855). See also Mark Pattison, "Joseph Scaliger," in *Essays*, ed. Henry Nettleship, 2 vols. (Oxford: Clarendon Press, 1889), 1:162–63: "Of the mathematical principles on which the calculation of periods rests, the philologians understood nothing. The astronomers, on their side, had not yet undertaken to apply their data to the records of ancient times. Scaliger was the first of the philologians who made use of the improved astronomy of the sixteenth century to get a scientific basis for historical chronology."

[5] Adalbert Klempt, *Die Säkularisierung der universalgeschichtlichen Auffassung* (Göttingen: Musterschmidt, 1960), 85. On Scaliger's Julian Period, see Anthony Grafton, "Joseph Scaliger and Historical Chronology: The Rise and Fall of a Discipline," *History and Theory* 14 (1975): 162–64, 181–85.

[6] See Arno Borst, *Schriften zur Komputistik im Frankenreich von 721 bis 818* (Hannover: Hahn, 2006), 1:111: "Er betrachtete die geschichtlichen Zeiten nicht mehr wie die Gläubigen als Schöpfung Gottes mit punktuellem Ursprung und abruptem Ende."

open to serious objections. Throughout the sixteenth and seventeenth centuries, chronologers, including Scaliger himself, continued to use the Old Testament scriptures as their principal source for estimating the age of the world and reconstructing the earliest stages of human history.[7] In the fifth book of his *Opus de emendatione temporum*, he appealed to the main biblical intervals from the creation down to the fall of Zedekiah in an effort to establish Wednesday, 21 April, 3949 BC, as the day on which the sun and the moon had been created according to Genesis 1:14. In Scaliger's eyes, the fact that astronomical calculations showed this day to have been a full moon and the seat of the vernal equinox lent additional credibility to his result, which in many ways followed traditional moulds of thinking and writing about the beginning of the world. Simply to conceive of him as the destructor of 'biblicist' schemes of chronology would clearly mean to miss out on an important element of his own work.[8]

In a similar vein, the originality of Scaliger's methods has oftentimes been overestimated. As Anthony Grafton has shown in his brilliant intellectual biography of the man and a number of important articles, Scaliger's work was in many places indebted to the work of earlier chronologers, in particular to German Protestant scholars such as Paul Crusius, whose unimposing *Liber de epochis* (1578) arguably comes much closer than Scaliger's *De emendatione temporum* (1583) to being the pioneering text of technical chronology.[9] Yet instead of

---

[7] On the background, see C. A. Patrides, "Renaissance Estimates of the Year of Creation," *Huntington Library Quarterly* 26 (1963): 315–22; G. J. R. Parry, "Trinity College Dublin MS 165: The Study of Time in the Sixteenth Century," *Historical Research* 62 (1989): 15–33; James Barr, "Why the World was Created in 4004 BC: Archbishop Ussher and Biblical Chronology," *Bulletin of the John Rylands University Library of Manchester* 67 (Spring 1985): 575–608; Barr, "Luther and Biblical Chronology," *Bulletin of the John Rylands University Library of Manchester* 72 (1990): 51–69; Anthony Grafton, "Kircher's Chronology," in *Athanasius Kircher*, ed. Paula Findlen (New York: Routledge, 2004), 171–87; Grafton, "The Chronology of the Flood," in *Sintflut und Gedächtnis*, ed. Martin Mulsow and Jan Assmann (Munich: Fink, 2006), 65–82.

[8] Joseph Justus Scaliger, *Opus novum de emendatione temporum* (Paris, 1583), 200: "Nam sine ullo dubio ex verbis Mosis a conditu mundi ad exodum colliguntur anni 2453.... Ab exodo ad fundamenta templi Solomonici scriptura ponit annos 480. A conditu ergo mundi ad templi Solomonici fundamenta colliguntur anni 2933. Hinc ad casum Sedekiae regis, & Templi vastationem anni 427. Summa annorum a conditu mundi ad casum Sedekiae, 3360. Hactenus verus epilogismus annorum a conditu mundi ad casum Sedekiae ex scriptura asseritur. Quem neque augere, neque imminuere possumus." See also Grafton, *Scaliger*, 2:262–70.

[9] Grafton, *Scaliger*, 2:109–39, 270–79. See also Grafton, "From *De die natali* to *De emendatione temporum*: The Origins and Settings of Scaliger's Chronology," *Journal*

paying due credit to his immediate predecessors, who had won much of the ground that he himself ventured to tread, Scaliger preferred to portray his project as the continuation of an ancient tradition. In his view, the methods of reconstructing lost dates and identifying historical years according to their astronomical 'character' had already been in active use among Greek and Roman scholars.[10]

There is indeed no reason to deny that the roots of chronological scholarship ultimately trace back to classical antiquity. Pioneers of Greek chronography such as Timaeus of Tauromenium (died ca. 250 BC) and Eratosthenes of Cyrene (died ca. 195 BC) tried to impose an orderly temporal grid on the chaotic torrent of historical traditions, both diachronically, by establishing the intervals between important past events, and synchronically, by linking the histories of *poleis* and kingdoms in different parts of the Mediterranean to one another.[11] By the late fifth century BC, Greek scholars had also begun to discuss the calendar dates of certain events, one noteworthy example being the fall of Troy, which Dionysius of Halicarnassus (*Roman Antiquities* 1.63), with artificial precision, assigned to the eighth day from the end of the Athenian month of Thargelion.[12] Chronological arguments of an even more advanced kind crop up in the first century BC in the work of the Roman antiquary Varro, who had asked an astrologer, Lucius Tarrutius of Firmum, to calculate the calendar dates of Romulus's conception and birth.[13] Writing in the third century AD, Censorinus would later praise Varro for his method of "counting back through intervals of eclipses" (*defectus eorumque intervalla retro dinumerans*), which indicates that the Roman polymath embraced the use of astronomy in his search for historical dates.[14]

---

of the Warburg and Courtauld Institutes 48 (1985): 100–43. The work of Paul Crusius will be treated in chapter eight below.

[10] Grafton, *Scaliger*, 2:247–62.

[11] See most recently Astrid Möller, "Epoch-Making Eratosthenes," *Greek, Roman, and Byzantine Studies* 45 (2005): 245–60; Denis Feeney, *Caesar's Calendar* (Berkeley: University of California Press, 2007), 7–67.

[12] Anthony Grafton and Noel Swerdlow, "Greek Chronography in Roman Epic: The Calendrical Date of the Fall of Troy in the *Aeneid*," *Classical Quarterly*, n.s., 36 (1986): 212–18.

[13] Anthony Grafton and Noel Swerdlow, "The Horoscope of the Foundation of Rome," *Classical Philology* 81 (1986): 148–53.

[14] Censorinus, *De die natali* (21.5), ed. Nicolaus Sallmann (Leipzig: Teubner, 1983), 51–52: "Sed hoc quodcumque caliginis Varro discussit, et pro cetera sua sagacitate nunc diversarum civitatium conferens tempora, nunc defectus eorumque intervalla

It remains doubtful, however, whether the activities of Varro and other ancient scholars can be fully assimilated to early modern methods of technical chronology. Anthony Grafton and Noel Swerdlow have carefully assessed the remaining evidence, but their conclusions show that Scaliger and his successors seriously overestimated the ancients when they styled themselves as heirs to a Greco-Roman tradition. In no concrete example that has been preserved from pagan antiquity did a historian try to establish a year on the basis of a previously known calendar date or use an actually observed and recorded eclipse to retrospectively date an event. What our sources reveal instead, is a widespread ancient concern with the 'ominous' character of certain calendar dates or with astrological considerations—where dates were simply chosen by looking for a significant celestial constellation in the proximity. As Grafton concisely puts it, "ancient scholars searched the skies not for tools but for portents."[15]

This result, helpful as it may be in providing an understanding of the ancient tradition, leaves a curious explanatory gap when it comes to the origins of early modern technical chronology. From the second half of the sixteenth century onwards, learned men on both sides of the confessional divide produced an impressive number of tomes and manuals, in which chronographic problems were discussed and past dates re-arranged. From Italy to England and on either bank of the Rhine, scholars were busy hunting for rare manuscripts and learning exotic languages, so they could spend their nights under dim candle light, brooding over the reconstruction of ancient calendars and lacunose lists of ancient kings. The results they achieved were often impressive and sometimes left little room for further refinement: "No Renaissance science, perhaps, reached more dramatic or more lasting results."[16]

retro dinumerans eruit verum lucemque ostendit, per quam numerous certus non annorum modo, sed et dierum perspici possit."

[15] Grafton, *Scaliger*, 2:261. See also Anthony Grafton and Noel Swerdlow, "Technical Chronology and Astrological History in Varro, Censorinus and Others," *Classical Quarterly*, n.s., 35 (1985): 454–65; Grafton and Swerdlow, "Calendar Dates and Ominous Days in Ancient Historiography," *Journal of the Warburg and Courtauld Institutes* 51 (1988): 14–42.

[16] Grafton, *Scaliger*, 2:5. See also Anthony Grafton, "Chronology and Its Discontents in Renaissance Europe: The Vicissitudes of a Tradition," in *Time: Histories and Ethnologies*, ed. Diane Owen Hughes and Thomas R. Trautmann (Ann Arbor: University of Michigan Press, 1995), 139–66; Grafton, "Tradition and Technique in

The accuracy of their findings was underwritten by the use of astronomical techniques, especially cyclical lunisolar calendars and calculated eclipses. As early as 1540, the German mathematician Petrus Apianus adorned his magnificent *Astronomicum Caesareum* with illustrated diagrams of historical lunar eclipses by which he hoped to correct the mistakes that Eusebius of Caesarea had supposedly made in marking the years for crucial events such as the capture of Nicias during the Sicilian expedition (413 BC) or Alexander's victory at Gaugamela (331 BC). Apianus's astronomical demonstrations showed that many of the chronological errors which marred the historical record thanks to the ignorance of chronographers or the carelessness of scribes could be remedied with the help of the latest astronomical tables. "By eclipses, after all, all events can be fixed to precise years, before Christ no less than after him."[17] A scholar mindful of this advice had little reason to despair of history and fall back into the kind of paranoid pyrrhonism which later caused authors such as the Jesuit Jean Hardouin (1646–1729), one of Fomenko's early modern predecessors, to proclaim that the bulk of the classical tradition was the brainchild of medieval forgers.[18] To us, however, Apianus's confidence in the use of astronomy as history's Ariadne's thread also raises some vexing new question. If Renaissance scholars could not directly emulate ancient authorities in their application of chronological techniques, where are the roots of their practices to be found?

A medieval tradition of technical chronology, however tenuous and primitive, might well have provided the proper breeding ground for the surge of chronological activity during the sixteenth century. At the same time, however, an assessment of the influence the roughly fifteen centuries of pre-modern Christianity may have had on the development of Western chronological ideas and techniques is not made easy by the relative dearth of existing literature. True, there is the ubiquitous presence of the computus, the characteristically medieval study of all things concerned with the calculation of the date of Easter. Recent studies have greatly expanded our knowledge of the computistical

Historical Chronology," in *Ancient History and the Antiquarian*, ed. M. H. Crawford and C. R. Ligota (London: The Warburg Institute, 1995), 15–31.

[17] Petrus Apianus, *Astronomicum Caesareum* (Ingolstadt, 1540), I3v: "Per ecleipses enim omnia certos in annos reduci possunt, Christum praecedentes non minus quam sequentes." For more on Apianus, see chapter eight below.

[18] See Anthony Grafton, "The Antiquary as Pariah: The Strange Case of Jean Hardouin," *Journal of the Warburg and Courtauld Institutes* 62 (1999): 241–67.

tradition, its manifold problems and sources, especially with regard to the earliest medieval centuries, when the field was dominated by Irish monks.[19] Yet the question remains if and how the calculations of medieval computists can be linked to the achievements of Renaissance chronologers. To judge from Jacob Bernays's characterization of pre-Scaligerian chronology as little more than an assembly of simple computistical procedures "or, at the most, an occasional aid for biblical exegesis and the understanding of the most common Latin classics," the prospect of finding such pathways would seem to be rather dim.[20] As a matter of fact, even the late Arno Borst, who devoted much of his scholarly life to the study of medieval computistical sources, was more interested in stressing the divide which separated the humble and pious monks who copied computistical manuscripts from Scaliger's chronological science than in looking for the latter's premodern roots.[21]

Those, however, who have closely studied the manifold spoils of medieval historiography have occasionally come to strikingly different conclusions. In his important 1980-study of medieval historiographic culture, Bernard Guenée expressed his admiration for the successful medieval "conquest of time," which he claims had a fruitful influence on modern historical scholarship.[22] This assessment finds some justification in the fact that medieval scholars were evidently able to juggle a confusing variety of dating techniques and year counts. As Laura Smoller aptly puts it, they had to

---

[19] See most recently Immo Warntjes, *The Munich Computus* (Stuttgart: Steiner, 2010), and Warntjes, "Irische Komputistik zwischen Isidor von Sevilla und Beda Venerabilis: Ursprung, karolingische Rezeption und Forschungsperspektiven," *Viator* 42 (forthcoming).

[20] Bernays, *Scaliger*, 90: "Vom Beginn des Mittelalters bis auf Scaliger war die Chronologie nur ein Gemenge von Handgriffen zur Osterberechnung und zum Kalendermachen, oder höchstens ein gelegentliches Hilfsmittel für biblische Exegese und zum Verständnis der gangbarsten lateinischen Klassiker. Er hat ihr eine wissenschaftliche Grundlage gegeben durch Benutzung der gerade zu seiner Zeit, welche den Copernicus und Tycho Brahe sah, so sehr erweiterten Astronomie." All translations are my own, except where stated.

[21] Arno Borst, *The Ordering of Time* (Chicago: University of Chicago Press, 1993). An earlier version of this book was published as Borst, "Computus: Zeit und Zahl im Mittelalter," *Deutsches Archiv* 44 (1988): 1–82.

[22] Bernard Guenée, *Histoire et culture historique dans l'occident medieval* (Paris: Éditions Aubier-Montaigne, 1980), 147–48: "Son plus grand mérite est assurément la conquête du temps, et, tout en disant les murs auxquels celle-ci s'est heurtée, il convient de ne pas sous-estimer les succès auxquels elle a abouti et dont notre moderne érudition est souvent l'heureuse héritère." See also ibid., 147–65.

> correlate dates based on regnal years, dates from the founding of Rome (*ab urbe condita*), dates expressed in terms of the fifteen-year Indiction cycle, and dates in the more recent anno domini form, all the while encountering inaccuracies, confusion, and omissions.[23]

There is reason to suspect that this medieval drive to know and understand the dates of important events, especially when they directly concerned the history of mankind's salvation, was much more firmly ingrained in Christian tradition than is sometimes admitted. An important hint in this direction comes from the title-page of Scaliger's very own *Opus de emendatione temporum*, which quotes Tatian's second-century dictum that "people whose chronological records are inconsistent cannot write true history either."[24] In Tatian's case, these words were part of his project of defending the Christian worldview against the claims of cultural superiority made by Greek philosophy and paganism. By comparing various chronographic traditions with the chronology of the Hebrew scriptures, Tatian was able to show that Moses lived four centuries before the Trojan War and was thus a writer and teacher of greater antiquity (and hence credibility) than Homer, Plato, or any other Greek cultural hero.[25] Other early Christian apologists, such as Theophilus of Antioch and Clement of Alexandria shared this sentiment and drew on the synchronisms between events, rulers, and famous men in their attempts to demonstrate the superiority of Judeo-Christian tradition over pagan philosophy. Their project was one of the roots for the emergence of Christian chronography in the third century, which saw writers such as Julius Africanus (died ca. 240) fuse biblical and profane history into an integrated chronological overview of the entire history of salvation, from the beginning of the world to the present (and sometimes beyond). The genre reached its apex with the chronicle of Eusebius of Caesarea (died 339), who made ample use of the possibilities provided by the bound codex, creating a striking new way of displaying the history of different nations and kingdoms in parallel columns—a visual approach to history that was to remain popular until modern times.[26]

---

[23] Laura Ackerman Smoller, *History, Prophecy, and the Stars* (Princeton, NJ: Princeton University Press, 1994), 82.

[24] Tatian, *Oratio ad Graecos* (31), ed. Molly Whittaker (Oxford: Clarendon Press, 1982), 59.

[25] Ibid. (39), ed. Whittaker, 71.

[26] On the tradition of Christian chronography, see Anna-Dorothee von den Brincken, *Studien zur lateinischen Weltchronistik bis in das Zeitalter Ottos von Freising*

Unlike Eusebius's work, in which Greek Olympiads competed with regnal eras and a supplementary count of years from Abraham, modern visual representations of historical time have long been used to adapt all events to a single era, a timeline for all of world history, based on counting the years both before and after the 'incarnation of the Lord'.[27] This practice of relying on a single continuous era that can both designate the present year and put it into immediate relation with any past event, is by no means a cultural universal. In China and other parts of East Asia, the dominant method of telling the year was based not on a single era, but on a frequently changing succession of 'era names', which were combined with a sexagesimal cycle in order to facilitate the precise identification of a given year. The 'era name'-system is still in use in Japan, where the accession of Emperor Ahikito to the throne in 1989 saw the beginning of the current era *Heisei* ('peace everywhere'), one of four successive eras used during the twentieth century. East Asia's preference for changing 'era names' led to a compartmentalization of historical time, which in turn reinforced the need for chronological tables as a means to comprehend the flow and synchronicity of historical events. In Chinese historiography, the tradition of using such tables as an aid to students of history can be traced back to the *Chronological Tables of the Twelve Dukes*, composed around 100 BC as part of Sima Qian's *Shiji* ('Records of the Great Historian').[28]

Single eras with a significance comparable to that of the 'incarnation era' in Western Christianity are also known in Islam, where the *Hijra* marks the year of the prophet Muhammad's migration from Mecca to Medina (AD 622), and in Judaism, where the *minyan ha-olam* or 'era of the world', beginning with 3761/60 BC as the assumed year of the world's creation, gained prominence during

(Düsseldorf: Triltsch, 1957); Alden A. Mosshammer, *The Chronicle of Eusebius and Greek Chronographic Tradition* (Lewisburg, PA: Bucknell University Press, 1979); William Adler, *Time Immemorial* (Washington, DC: Dumbarton Oaks, 1989); Martin Wallraff, ed., *Julius Africanus und die christliche Weltchronistik* (Berlin: de Gruyter, 2006).

[27] On the origins of this practice, see Anna-Dorothee von den Brincken, "Beobachtungen zum Aufkommen der retrospektiven Inkarnationsära," *Archiv für Dipolmatik* 25 (1979): 1–20. On the visualization of history, see now Daniel Rosenberg and Anthony Grafton, *Cartographies of Time* (New York: Princeton Architectural Press, 2010).

[28] Masayuki Sato, "Comparative Ideas of Chronology," *History and Theory* 30 (1991): 275–301.

the Middle Ages. Yet while there has been a lively and sophisticated interest in history and chronography in all three Abrahamic religions, the development of a rigorous technical chronology, based on astronomical dating-techniques and textual criticism, appears to have been an achievement specific to the Christian West. An interesting case in point is the famous chronological manual by the Persian Muslim scholar Abū Rayḥān al-Bīrūnī, composed around the year AD 1000, which covered the various calendar systems and chronographic traditions of the Near East at a depth that far surpassed any of the Christian works produced up to that time. Despite this fact, however, al-Bīrūnī's work apparently lacked the methodological insight that chronological knowledge could be augmented or corrected by technical, astronomical means. Instead, al-Bīrūnī denied that a reliable chronological system could "be obtained by way of ratiocination with philosophical notions, or of inductions based upon the observations of our senses," leaving solely the comparison of written traditions as a source of knowledge.[29]

Returning to the late antique context, one can note that at least some of the early Christian interest in chronography and time reckoning was related to the fact that Christianity itself was a relatively new religion with strongly historical overtones. The central events of the history of salvation around which Christian dogma was constructed—Christ's birth, death, and resurrection—were not relegated to some mystical pre-history, but instead were considered to be concrete events that had taken place within an existing framework of profane history and which could thus be synchronized with the regnal and political chronology of their day. When, at the end of the fourth century, Augustine of Hippo pondered the value of historical education in his famous work *On Christian Teaching*, he emphasized this point by writing:

> Whatever the subject called history reveals about the sequence of past events is of the greatest assistance in interpreting the holy books, even if learnt outside the church as part of primary education. Many problems

[29] Al-Bīrūnī, *The Chronology of Ancient Nations*, trans. C. Edward Sachau (London: Allen, 1879), 3. On Jewish chronography, see Edgar Frank, *Talmudic and Rabbinical Chronology* (New York: Feldheim, 1956); Yosef Hayim Yerushalmi, *Zakhor* (Seattle: University of Washington Press, 1982); Chaim Milikowsky, "*Seder 'Olam* and Jewish Chronography in the Hellenistic and Roman Periods," *Proceedings of the American Academy for Jewish Research* 52 (1985): 115–39; Mitchell First, *Jewish History in Conflict* (Northvale, NJ: Aronson, 1997).

> are often investigated by us using Olympiads and the names of consuls. Ignorance of the consulships in which the Lord was born and died has led many to the erroneous idea that the Lord suffered at the age of 46, because it was said by the Jews that their temple (which represented the Lord's body) was built in forty-six years [John 2:19–20]. We have it on the authority of the gospel [Luke 3:23] that he was baptized at the age of about 30; the number of years that he lived after that could be inferred from the pattern of his activities, but is in fact more clearly and reliably established, beyond any shadow of doubt, by a comparison of secular history with the gospel.[30]

This comparison of sacred history and secular sources was all the more important given the fact that even devout believers such as Augustine had to admit that their own Scriptures were remarkably terse when it came to elucidating the chronological dimensions of Jesus's life. The study of profane chronology and its application to the sacred text could thus serve as a welcome tool of exegetical clarification and—more significantly—reassurance of the historicity of the events described in the Gospels. Augustine's concern over the correct dates of Christ's life thus provides an important link between late antique Christian scholarship and the work of chronologers from more recent periods. For after nearly twenty centuries of research, it still holds true that, as the Egyptologist Leo Depuydt put it in 2002, "no single matter of chronology has been examined so thoroughly as the date of death of Jesus of Nazareth. It is chronology's premier topic."[31]

The present book is founded on the conviction that this remarkable continuity between late antique and modern scholarship regarding the life of Jesus can be fruitfully exploited to shed new light on the history of chronology as a scientific discipline. At its core, the present study

---

[30] Augustine, *De doctrina christiana* (2.28), CCSL 32:62: "Quicquid igitur de ordine temporum transactorum indicat ea quae appellatur historia, plurimum nos adiuvat ad libros sanctos intellegendos, etiamsi praeter ecclesiam puerili eruditione discatur. Nam et per olympiadas et per consulum nomina multa saepe quaeruntur a nobis et ignorantia consulatus, quo natus est dominus et quo passus est, nonnullos coegit errare, ut putarent quadraginta sex annorum aetate passum esse dominum, quia per tot annos aedificatum templum esse dictum est a Iudaeis, quod imaginem dominici corporis habebat. Et annorum quidem fere triginta baptizatum esse retinemus auctoritate evangelica, sed postea quot annos in hac vita egerit, quamquam textu ipso actionum eius animadverti possit, tamen ne aliunde caligo dubitationis oriatur, de historia gentium collata cum evangelio liquidius certiusque colligitur." Translation according to Augustine, *On Christian Teaching*, trans. R. P. H. Green (Oxford: Oxford University Press, 1997), 55.

[31] Leo Depuydt, "The Date of Death of Jesus of Nazareth," *Journal of the American Oriental Society* 122 (2002): 466.

deals with the different attempts to find the precise date of the crucifixion of Jesus Christ, as they were produced in Christian scholarship up to the end of the sixteenth century, when the work of a number of scholars, among them Joseph Justus Scaliger, transformed historical chronology into a fully fledged discipline. It is my contention that a look at these various attempts reveals crucial developments in the methods and concepts of historical dating, pointing to a pre-modern tradition of astronomical and mathematical dating techniques that began during late antiquity, when the lunisolar cycles developed for finding the date of Easter were first used retrospectively to assign calendar dates to events in the past. An exploration of this method can help to shed new light on the conceptual and disciplinary background on which early modern scholars in Scaliger's day could still rely. The first historian to pay proper attention to the chronological methods that arose out of the traditions of Christian world chronography and Easter reckoning (the computus) was Venance Grumel, whose handbook *La chronologie* remains an indispensible guide on the reckoning of time in Byzantium and the Christian East. It is Grumel's particular achievement to have recognized—more clearly than others—that the majority of world eras and chronological systems developed by Christians during late antiquity were based on and constrained by the use of Easter cycles. Their natural sequence of varying combinations between weekdays, lunar ages, and calendar dates imposed decisive limitations on the chronological choices of Christian authors, for instance when it came to deciding on the age of the world and the years of Christ's incarnation and death.[32]

Computistical chronography—the method of applying Easter reckoning to chronological problems—remains an understudied phenomenon in late antique and medieval scientific literature, to which I shall turn in the first four chapters of this book. After a concise introduction to the astronomical and scriptural background of the problems of Easter and the crucifixion date in Chapter One, I shall start Chapter Two by taking a closer look at the roots of this specifically Christian chronographic tradition and its later influence on the chronological

---

[32] Venance Grumel, *La chronologie* (Paris: Presses universitaires de France, 1958). See also Grumel, "Les premières ères mondiales," *Revue des Études Byzantines* 10 (1952): 93–108. Grumel's insights were in many ways developed further by Alden A. Mosshammer, *The Easter Computus and the Origins of the Christian Era* (Oxford: Oxford University Press, 2008).

systems used by churches in the Eastern half of the Roman Empire. The main story begins with the 112-year Easter table ascribed to Hippolytus (ca. 222), which contains the first clear examples for the application of lunisolar cycles to historical dating. As an analysis of the chronological methods of Hippolytus and his successors reveals, the use of Easter cycles had a strong influence on the choice of dates for the creation of the world and—more importantly—for the Passion and other dates in the life of Jesus Christ in ecclesiastical tradition. This influence is particularly palpable in the case of the chronological systems attached to the so-called Alexandrian and Byzantine world eras, which, despite their many differences, were both based on the 19-year cycle of the Alexandrian church and became firmly entrenched in Eastern chronography over the course of the fifth, sixth, and seventh centuries.

A close study of the principles of computistical chronography reveals a law-like relation, according to which changes in the computational basis—i.e. the introduction of new Easter cycles—affected the date of Christ's Passion and related dates accepted by an individual church or community. Chapter Three explores the vicissitudes of the Passion date in the light of these computistical developments, which led the Western churches from the 84-year *Supputatio Romana* to the Alexandrian 19-year cycle, which was to dominate the Latin Middle Ages. What made the Latin tradition stand out in comparison to the chronological systems of the Christian East, is the lack of any fully coherent system, owing to a number of specific chronological problems encountered by Western computists during the early medieval centuries. The persistence of certain chronological views on the Passion along with the introduction of the Dionysiac incarnation era (which is at the root of the BC/AD-system still in use) made it impossible to find a satisfactory set of dates for the incarnation and crucifixion of Jesus, thereby creating a lasting state of confusion that was first made explicit in the Venerable Bede's famed handbook *De temporum ratione* (725), the undisputed milestone of Western computistics. This confusion over the historical Passion date was mirrored by similar problems regarding the date of the world's creation. Illuminative witnesses to this confusion can be found in the Carolingian 'Seven-Book-Computus' of 809/12 and the chronicle of Claudius of Turin (814), which have both been neglected by previous scholarship. In the final part of Chapter Three, I shall provide a thorough analysis of both sources, showing that early medieval thought about the chronology of salvation was in a state of temporary 'crisis', which was to have a lasting effect on the development of computistical chronography in the West.

Although the significance and extent of this crisis has not been properly appreciated in the existing secondary literature, some of the medieval reactions to the problems at hand are relatively well-known. These reactions are found in the works of the 'critical computists' (to use a label first introduced by Joachim Wiesenbach), a group of eleventh- and twelfth-century monastic scholars who were united in their willingness to criticize ecclesiastical tradition, represented by Dionysius Exiguus and his Christian era of the incarnation, and their marked reliance on the lunisolar cycle as a natural tool of chronological investigation. As a result of this approach, the 'critical computists' suggested extensive revisions of the accepted dates of Christ's birth and death and sometimes of the whole framework of ancient Roman chronology. Their work has recently received comprehensive treatment at the hands of the Belgian scholar Peter Verbist, whose monograph *Duelling with the Past* puts a strong emphasis on the critical and scientific rationality discernible in the arguments of these proto-scholastic authors.[33] In Chapter Four, I shall mainly confine myself to summarizing the state of research on the 'critical computists', while adding some new observations on the works of Heriger of Lobbes (died 1007) and Heimo of Bamberg (died 1139).

For all their importance as representatives of a distinctive phase in the history of technical chronology, the project of the 'critical computists' was ultimately doomed to failure, owing to their over-reliance on the data provided by the conventional Easter computus. In reality, the 19-year lunisolar cycle used by the church was not astronomically accurate enough to yield reliable data for ancient history, a problem that was aggravated by the fact that the Easter cycle bore no sufficient historical relation to the methods of calendation used by the Jews at the time of Jesus. According to Verbist, these problems were not perceived by Western computists until the thirteenth century, when doubts of this kind were first expressed by the chronicler Alberic of Trois-Fontaines (died after 1252).[34] As I shall show in Chapter Five, this

[33] Peter Verbist, *Duelling with the Past* (Turnhout: Brepols, 2010). Before Verbist, the standard work on the subject was Joachim Wiesenbach, *Sigebert von Gembloux: Liber decennalis* (Weimar: Böhlau, 1986).

[34] Verbist, *Duelling*, 237. See also Wiesenbach, *Sigebert*, 111.

view is mistaken. In reality, Christian chronology had already entered a new phase during the twelfth century as a result of the exposure to new astronomical sources that had been translated from Arabic and imported to Western Europe from the Iberian Peninsula. At roughly the same time, Christian scholars became acquainted with the conjunction-based calendar of the Jews, which had been developed during the early Middle Ages, confronting them with a new type of lunisolar calendation that was almost wholly independent of the Christian Easter computus. Drawing on the much-neglected *Compotus emendatus* of Reinher of Paderborn (1170/71) and the still-unedited *Compotus Constabularii* (1175), I shall demonstrate how Christian computists began to seize on these new pieces of information and their improved astronomical skills to revisit the question of the Passion date. As a result of these new influences, they were able to suggest fresh solutions while simultaneously displaying a heightened awareness for the chronological and exegetical problems posed by contradictory Gospel accounts.

These new currents of thought within Western computistics reached their peak with the works of the English Franciscan Roger Bacon, who will be the protagonist of Chapter Six. Bacon's thoughts on historical chronology have hitherto received little attention, despite the fact that they take up considerable space in his famed *Opus majus* and further texts from his pen. As I shall show, the English Franciscan was the first author to use advanced astronomical tools to establish 3 April, AD 33, as the likeliest date of the crucifixion—a view still widely shared by scholars today. Even more significantly, Bacon saw his application of astronomy and Jewish calendation to the Passion chronology as part of a larger research programme that championed astronomy and astrology as supreme tools for the clarification and correction of biblical chronology. In detailing this programme, Bacon became the first author explicitly to point out the possibility of using dated eclipses of the sun and moon to establish an absolute chronology for past events. As the works of the French astronomer Jean des Murs and other late medieval and Renaissance authors show, Bacon's opinions on historical chronology and the Passion date did not go down unnoticed and became an important source of inspiration for subsequent works in the field.

The final two chapters are dedicated to a number of chronological studies of the Passion date undertaken during the fifteenth and sixteenth century, many of which are closely linked to the history leading up to the emergence of scientific chronology. In Chapter Seven, I focus on the ongoing controversies over the date of the crucifixion that repeatedly prompted Catholic scholars to write down their opinions during the period in question. Whereas chronological disputations and dating attempts up to Bacon's time had mostly taken place within texts that belong to the Western computistical tradition, the source basis in this subsequent period consists of a variety of genres, from exegetical works to written polemics and tracts on calendar reform as well as early chronological handbooks. Scholars during this period retained their medieval predecessors' interest in the Jewish calendar, especially its postponement rules (*deḥiyyot*) and their relevance for an interpretation of the events surrounding the crucifixion of Jesus. In addition, however, they brought to bear an increasing number of profane sources and information on Greek and Roman chronology that pointed to a conflict between the emerging framework of history and the date of Christ's birth according to the era of Dionysius Exiguus. Looking at the works of chronologers such as Gerhard Mercator, Paul Crusius, and Heinrich Bünting, I shall use the final chapter to shed light on some of the steps that enabled early modern scholars to arrive at our present state of knowledge concerning the timeline of ancient history. An important role in this gradual reconstruction was played by ancient records of eclipses of the sun and the moon, whose dates could be reliably calculated with the known astronomical tools. As I hope to show, the quest for the dates of Jesus's life was at the frontline of these attempts to import astronomy into historical dating in more than one respect. Not only were investigations of the years of Christ's birth and death among the earliest chronological arguments to make use of eclipse-dating, but the development of this method itself is firmly entangled with the history of dating the Passion, as it is going to be told in this book.

While the primary purpose of my study is to highlight the specific contributions made by pre-modern Christian scholarship to the emergence of scientific chronology, the account presented here can also be put to a more mundane use. Students of sources in which the dates of Christ's birth and Passion are discussed can use it as a resource to gain a better understanding of the context in which a specific dating-attempt was made. By comparing the arguments found in a particular source to the wider history of the problem, they can obtain a yardstick to determine in how far the claims made by a particular author were original or unusual relative to the background of their time. In a similar vein, it can be of use to modern chronologers, who like to compare their own contributions to the chronology of the New Testament with past endeavours. Since the search for the dates of Christ's life is an ongoing project in some quarters, an understanding of past attempts will be greatly facilitated by a concise look at the present state of research and the chronological and textual problems involved. To this, I turn in the following chapter, which will serve as a general introduction to the subject and thus help set the scene for the historical investigations to follow.

CHAPTER ONE

FROM ASTRONOMY TO THE CRUCIFIXION AND BACK

Owing to its size and brightness in the night sky and the easily observable regularity of its waxes and wanes, the moon was predestined to become one of mankind's earliest markers of time. The basis for lunar calendation is the cycle of the phases of the moon, whose mean duration is today measured as 29.530589d (mean synodic month). With the spread of agriculture, most civilizations found it desirable to also ground their calendar in the natural cycle of seasons, which are presently found to recur after one tropical solar year of 365.242190d.[1] In Exodus 23:15, for instance, the ancient Israelites are instructed to observe the festival of Passover "in the month of *aviv*, for in that month you came out of Egypt" (see also Exodus 13:14; 34:18; Deuteronomy 16:1). Since *aviv* is an agricultural term, referring to the month of "(fresh) ears" or barley harvest, the calendar year in which such a festival was celebrated had to be either solar or lunisolar.[2] The difference between a lunar year of twelve lunations (12 × 29.530589d = 354.367068d) and the tropical solar year is 10.875122d, meaning that an additional month will have to be inserted (intercalated) after three years, sometimes after two years, if the individual lunar months are to remain aligned to the seasons of the year.

Lunar and lunisolar calendars can take two basic forms, empirical or calculated. In the former case, the beginning and length of the lunar months is decided according to certain observational criteria, whereas calculated calendars attempt to map out the days of the lunar year in

[1] For general information, see Friedrich Karl Ginzel, *Handbuch der mathematischen und technischen Chronologie*, 3 vols. (Leipzig: Hinrichs, 1906–14); Alan E. Samuel, *Greek and Roman Chronology* (Munich: C. H. Beck, 1972); Elias J. Bickerman, *Chronology of the Ancient World*, rev. ed. (London: Thames & Hudson, 1980); Bonnie J. Blackburn and Leofranc Holford-Strevens, *The Oxford Companion to the Year* (Oxford: Oxford University Press, 1999); Leofranc Holford-Strevens, *The History of Time* (Oxford: Oxford University Press, 2005). All values given are accurate for 1 January 2000. The duration of the day and the mean synodic month undergo small secular changes, which can be ignored for present purposes.

[2] On the background, see Jan A. Wagenaar, *Origin and Transformation of the Ancient Israelite Festival Calendar* (Wiesbaden: Harrassowitz, 2005), 25–32.

advance, by introducing a fixed order of 'full' (30-day) and 'hollow' (29-day) months. To construct an arithmetically ordered lunisolar calendar, in which the difference between the solar and the lunar year is compensated by means of regular intercalation, is anything but a trivial task. Since the two main parameters involved are mathematically incommensurable (365.242190 ÷ 29.530589 = 12.368266...), lunisolar cycles can only ever be approximations of varying success. For cycles of less than 100 years, the most exacting scheme that can be achieved is to distribute seven intercalary months (embolisms) over the span of 19 years. This 19-year cycle was known in Babylonia since the beginning of the fifth century BC, while in Greece it was reportedly introduced by the Athenian astronomers Meton and Euctemon in 432 BC. In Babylonia, the year began at the first new moon after the spring equinox and comprised twelve lunar months, which usually began at the evening of the first visibility of the lunar crescent. Although Babylonian astronomy attained a high degree of sophistication that made it possible to compute the day of first visibility beforehand, the calendar retained an empirical element for most of its history. This holds especially true for the pre-Persian period, when the months were still fixed ad hoc, by way of observation. From early cuneiform texts, we learn that the king proclaimed the beginning of a new month after receiving testimony from messengers that the new moon had been sighted. In a similar vein, it was the task of astronomers to decide whether the first month Nisanu should be postponed by means of intercalation, lest it anticipated the spring equinox.[3]

Following the conquest of Judea and the period of Exile, the Babylonian calendar began to exert a strong influence on the Jewish calendar, as can be seen from the Aramaic-Babylonian origin of the Hebrew month names, which are first attested in the post-exilic books of Zechariah (1:7; 7:1) and Esther (2:16; 3:7; 3:13; 8:12; 9:1). It is striking to observe how early Rabbinic sources still reflect some of the same empirical procedures found in ancient Babylonian sources. From the Mishnah (tractate *Rosh Hashanah*) we can infer that in the century preceding the destruction of the Second Temple (AD 70) the beginning of the month depended on the sighting of the new moon crescent. Witnesses were formally interrogated by a rabbinic court to see

---

[3] Ben Zion Wacholder and David B. Weisberg, "Visibility of the New Moon in Cuneiform and Rabbinic Sources," *Hebrew Union College Annual* 42 (1971): 227–42.

if the moon had been sighted on the 30th evening of the month. If such a sighting was reliably attested, the moon was 'sanctified' and the 30th day was reckoned as the first day of a new month. Otherwise it was treated as the 30th day of the old month and the new month began one day later. Like the beginning of the month, intercalation was decided empirically, based on the observation of certain signs of spring. If these were not sighted by the time Adar (the twelfth month of the Jewish calendar, as enumerated from Nisan) drew to a close, it could be decided to insert a second Adar, thereby extending the year by another 30 days. This procedure is nicely exemplified in a letter by Rabban Gamaliel II (fl. AD 80–120) to the Jews of the Dispersion, cited in Tosefta Sanhedrin 2:6, in which he justifies the addition of 30 days to the year by pointing out that the "pigeons are still tender, the lambs are thin, and the spring-tide has not yet come."[4]

It would seem that a rabbinic calendar of this type was already operating in Jerusalem earlier in the first century AD, at the time when Jesus of Nazareth was crucified on orders from the Roman authorities. Contrary to what our Common Era, which purports to count the years from Jesus's conception or nativity, suggests, we know remarkably little about the date and circumstances of his birth. The infancy stories found in the Gospels of Matthew and Luke are of doubtful historicity and cannot serve to provide his biography with any secure chronological footing.[5] The Lucan version begins with the annunciation of the birth of John the Baptist (1:5–23) to the priest Zechariah, who served at the temple in Jerusalem as a member of the priestly division of Abijah. His wife Elizabeth conceives shortly afterwards (1:24) and in the sixth month of her pregnancy, the Annunciation of Jesus's birth to Mary takes place (1:26–38). Since the 24 divisions of Jewish temple priests presumably served their shifts according to a fixed cycle, scholars from Joseph Scaliger onwards have repeatedly

[4] Translation according to Jacob Neusner, *The Tosefta*, 2 vols. (Peabody, MA: Hendrickson, 2002), 1148. See in general Benedict Zuckermann, *Materialien zur Entwicklung der altjüdischen Zeitrechnung im Talmud* (Breslau: Preuss & Junger, 1882), 7–57; Eduard Mahler, *Handbuch der jüdischen Chronologie* (Leipzig: Fock, 1916), 336–439; William Moses Feldman, *Rabbinical Mathematics and Astronomy*, 3rd ed. (New York: Hermon Press, 1978), 178–84; Sacha Stern, *Calendar and Community* (Oxford: Oxford University Press, 2001), 157–64; Roger T. Beckwith, *Calendar and Chronology, Jewish and Christian* (Leiden: Brill, 1996), 278–89.

[5] See in general Raymond E. Brown, *The Birth of the Messiah*, 2nd ed. (New York: Doubleday, 1993).

sought to exploit this information in order to find an approximate date of Jesus's birth.[6] Their approach, however, is no less speculative than the countless attempts to date the nativity by reference to the 'star of Bethlehem' and the journey of the Magi. Astronomical interpretations of the phenomenon, solely attested in the Gospel of Matthew (2:1–23), have included a comet in 12 BC, a conjunction of Jupiter and Saturn in Pisces in 7 BC, conjunctions of Jupiter and Venus in 3/2 BC as well as a supernova in 5/4 BC.[7] Due to the legendary nature of the account, it is doubtful whether the star of Bethlehem can provide any useful indication of the year of Jesus's birth. That said, a reasonable *terminus ante quem* of 4 BC may be established, provided it can be trusted that Herod the Great, who died in that year, was still alive at the time of Christ's birth (Matthew 2:19; Luke 1:5).[8]

Speculations concerning Jesus's birth year, problematic as they may be, are not unimportant when it comes to interpreting the data found in the third chapter of Luke's Gospel, which is the most chronologically informative part of all the primary sources on Jesus's life. According to Luke 3:23, Jesus "was about thirty years of age" when he received baptism at the hands of John. In the same chapter, this event is closely linked to the 15th year of Tiberius, to which Luke dates the activities of John the Baptist (3:1).[9] The emperor Tiberius is known to have begun his reign a few weeks after the death of Augustus, on 17 September, AD 14. His 15th year thus began in the autumn of AD 28 and ended in AD 29, if his regnal years were counted from the actual day of accession. Since, however, the way of counting the years of a particular ruler could differ depending on the calendar used, the range of possible dates indicated by the '15th year of Tiberius'

[6] See my "From Sukkot to Saturnalia: The Attack on Christmas in Sixteenth-Century Chronological Scholarship," *Journal of the History of Ideas* 72 (2011): 503–22. See also Beckwith, *Calendar and Chronology*, 71–92; Jack Finegan, *Handbook of Biblical Chronology*, rev. ed. (Peabody, MA: Hendrickson, 1998), 275–79.

[7] For more information with ample references to older literature, see Brown, *The Birth*, 170–73, 610–13; Finegan, *Handbook*, 306–20; Kocku von Stuckrad, *Das Ringen um die Astrologie* (Berlin: de Gruyter, 2000), 555–86. See also R. M. Jenkins, "The Star of Bethlehem and the Comet of AD 66," *Journal of the British Astronomical Association* 114 (2004): 336–43, who makes a convincing case against all attempts to identify the star with an astronomical event at Herod's time.

[8] Brown, *Birth*, 166–67, 607–8; Finegan, *Handbook*, 291–301.

[9] This, at least, is how late antique and medieval commentators usually saw it. Since we cannot tell, from Luke's statements, for how long John had been baptizing when he met Jesus, the chronological link is much less tight.

technically extends from October AD 27 to December AD 29. If Jesus was born before 4 BC, as is often supposed, this would mean that Luke's statement "about thirty years" was only very approximate. Furthermore, while Luke's mention of the 15th year of Tiberius provides a useable *terminus post quem* for the death of Jesus, the exact duration of his public ministry is difficult to derive from the Gospels. John records at least four visits to Jerusalem during this time (2:13; 5:1; 7:10; 12:12), three of which seem to have taken place at the time of Passover (2:13; 5:1, 11:55). This suggests a public ministry of two years at the very least. By contrast, the only stay in Jerusalem mentioned in the synoptic Gospels (Matthew 26:1; Mark 14:1; Luke 22:1) is the one at whose end the crucifixion took place, creating the impression that Jesus preached for little more than a year. However, since the synoptic account of the public ministry lacks chronological structure, it is likely that it can be re-interpreted in favour of John's 'long' chronology.[10]

We are much better informed when it comes to the chronological circumstances of the Passion, which is the focal point of all four Gospel narratives. All evangelists agree in placing Jesus's death on a Friday afternoon (Matthew 27:62; 28:1; Mark 15:42; Luke 23:54; John 19:31) and in having the Last Supper take place on the night before. In the synoptic Gospels, it is clearly implied that Jesus celebrated a Passover meal with his disciples on the evening before his death (Matthew 26:17–19; Mark 14:12–16; Luke 22:7–15). During the first century of the Christian era, Passover commenced on the 14th day of Nisan, the first spring month in the Jewish calendar. In the afternoon of this day, the sacrificial lambs were sacrificed at the temple of Jerusalem and carried away to be roasted and eaten as part of a family meal in the evening. This evening was also the beginning of the feast of unleavened bread, which extended from the 15th day of the first month until the evening of the 21st. The first day of this feast, the 15th of Nisan, was a high day, on which no work was permitted. If the word of the synoptic Gospels can be trusted, this would have also been the day on which Jesus was crucified and buried. However, the synoptic consensus is challenged by the Gospel of John, whose Passion narrative contains several hints to the effect that the events described took place

[10] For a comprehensive treatment of the problem, see George Ogg, *The Chronology of the Public Ministry of Jesus* (Cambridge: Cambridge University Press, 1940). See also Finegan, *Handbook*, 329–53.

on the "preparation day for Passover" (19:14), which is the day *before* 15 Nisan.[11] This becomes particularly clear from John's claim that, on the morning of the trial of Jesus, the Jews refrained from entering the Roman praetorium "lest they should be defiled; but that they might eat the passover" (18:28), which entails that the ritual eating of lamb lay still ahead. According to this scheme of events, the Last Supper was eaten one day earlier than the regular Passover meal and Jesus was crucified on the same afternoon on which the sacrificial lambs were slain in the Temple. The evangelist underscores this simultaneity by associating the soldiers' failure to break Jesus's legs (19:32–36) with the Mosaic injunction that no bone of the Passover lamb shall be broken (Exodus 12:46).

Modern attempts to explain the discrepancy by harmonizing both chronologies or by proving one of them false have been far too numerous to recount them here at any length.[12] It is worth observing, however, that the Johannine version is in many ways the more realistic of the two. Since 15 Nisan was a day of sabbatical rest (Exodus 12:16), it is quite unlikely that it would have been host to all the activities recorded in the Passion narratives: a capital trial is conducted, Jesus is sentenced to death and executed, crowds are gathering to witness the events, Simon of Cyrene comes in from field work (Mark 15:21), Joseph of Arimathea buys a linen cloth (Mark 15:46), Jesus is hastily buried in the same evening, etc. These are only some of the reasons to suppose that the synoptic version, which makes the Last Supper into a Passover Seder, is a secondary tradition, which may have crept into the Gospel narratives under the influence of an annual meal of

---

[11] Some pertinent passages are John 13:1: "Now before the feast of the passover, when Jesus knew that his hour was come that he should depart out of this world unto the Father, having loved his own which were in the world, he loved them unto the end." John 19:14: "And it was the preparation of the passover, and about the sixth hour: and he saith unto the Jews, Behold your King." John 19:31: "The Jews therefore, because it was the preparation, that the bodies should not remain upon the cross on the sabbath day, (for that sabbath day was an high day,) besought Pilate that their legs might be broken, and that they might be taken away." All biblical citations used in this book follow the King James Version.

[12] See the helpful overviews in Ogg, *Chronology*, 208–42; Raymond E. Brown, *The Death of the Messiah*, 2 vols. (New York: Doubleday, 1994), 1350–78; Finegan, *Handbook*, 353–69. The latest entry in this discussion is Colin J. Humphreys, *The Mystery of the Last Supper* (Cambridge: Cambridge University Press, 2011).

remembrance, celebrated among the earliest Christian communities on Passover eve (14/15 Nisan).[13]

Notwithstanding the problems posed by the Passion chronology, the presence of implicit lunar calendar data (14 or 15 Nisan) along with a clearly attested weekday (Friday) in the four Gospels makes it—in theory at least—feasible to attempt an astronomical investigation of the date of the crucifixion. Since the calendar at that time was based on the sighting of the new moon crescent, a retrospective determination of the evening of first visibility can yield dates that were likely equivalents to 1 Nisan in the period in question. The relevant range of years for such an investigation is conveniently provided by Pontius Pilate's term as Prefect of Judaea (AD 26–36/37). That Jesus was crucified under Pilate is also briefly mentioned by Tacitus (*Annals* 15.44) and by Josephus, in his *Jewish Antiquities* (18.64), provided that this part of the *Testimonium Flavianum* can be trusted to have belonged to the original, uninterpolated, part of the text. Out of all the instances of 14 and 15 Nisan calculated for this period, only those falling on a Thursday (in the case of 14 Nisan) or Friday (in the case of both) can be relevant for the crucifixion. Additional constraints are provided by the year of Jesus's baptism and the duration of his public ministry (approximately three years), which make dates earlier than AD 28 and later than AD 35 into unlikely candidates. Some further uncertainty is caused by the fact that, given the ad hoc criteria of intercalation in use at the time, it is virtually impossible to state with any conviction that a particular spring lunation near the equinox was considered by the Jews to be Nisan rather than the embolismic Adar II. Moreover, astronomical calculations can only show that the moon was *visible* on a particular day, but never whether there was an actual *sighting* (which could be prevented by adverse atmospheric conditions), let alone if the

---

[13] This hypothesis is convincingly argued in Strobel, *Ursprung und Geschichte des frühchristlichen Osterkalenders* (Berlin: Akademie Verlag, 1977), 17–69. See also Eduard Schwartz, "Osterbetrachtungen," *Zeitschrift für die neutestamentliche Wissenschaft* 7 (1906): 22–26; Brown, *Death*, 1369–73; Fergus Millar, "Reflections on the Trials of Jesus," in *A Tribute to Geza Vermes*, ed. Philip R. Davies and Richard T. White (Sheffield: JSOT Press, 1990), 335–81. Although the present discussion expresses preference for the Johannine account, the reverse scenario, according to which John re-dated the crucifixion to have it coincide with the Passover sacrifice on 14 Nisan, should not be dismissed out of hand.

Jewish authorities at the time really deemed it fit to 'sanctify' the new moon for this exact day.[14]

In spite of all these vagaries, the repeated attempts to calculate the Passion date during the twentieth century have left us with a surprisingly small number of possible candidates. It is now generally agreed that the two most likely dates are 7 April, AD 30, and 3 April, AD 33, which were both instances of 14 Nisan according to modern calculation.[15] A further possible scenario was recently proposed by Leo Depuydt, who argues that the beginnings of the month in first-century Jerusalem could sometimes fall earlier than the evening of first visibility. In this scenario, 18 March (or, possibly, 15 April), AD 29, would be a further conceivable candidate for the crucifixion.[16] Out of these dates, 3 April, AD 33, appears to be the likeliest for reasons both astronomical and historical. The year AD 33 is at sufficient distance from the 15th year of Tiberius to accommodate the Johannine 'long' chronology of the public ministry, whereas the time span implied by the other two dates may be deemed too short to account for all the events related in the Gospels. In addition, 3 April, AD 33, was the date of a partial lunar eclipse that may have been visible over Jerusalem during the evening of the crucifixion. It is conceivable that the memory of this phenomenon influenced certain passages in the Gospels and the Acts of the Apostles, which record a "darkness" during the crucifixion (Matthew 27:45; Mark 15:33; Luke 23:44–45) and a "moon turning to blood" (Acts 2:20).[17]

---

[14] See the cautionary remarks in Beckwith, *Calendar and Chronology*, 276–96. All attempts at calculation will be void if the weekday reported by the evangelists (Friday) cannot be trusted as historical. Such doubts are voiced in Schwartz, "Osterbetrachtungen," 27–33.

[15] See, e.g., John K. Fotheringham, "The Evidence of Astronomy and Technical Chronology for the Date of the Crucifixion," *Journal of Theological Studies* 35 (1934): 146–62; Strobel, *Ursprung*, 70–100; Bradley E. Schaefer, "Lunar Visibility and the Crucifixion," *Quarterly Journal of the Royal Astronomical Society* 31 (1990): 53–67; Ogg, *Chronology*, 277; Brown, *Death*, 1373–76; Finegan, *Handbook*, 359–65.

[16] Leo Depuydt, "The Date of Death." Another, less convincing, alternative (30 March, AD 36) was suggested by Nikos Kokkinos, "Crucifixion in A.D. 36: The Keystone for Dating the Birth of Jesus," in *Chronos, Kairos, Christos*, ed. Jerry Vardaman and Edwin M. Yamauchi (Winona Lake, IN: Eisenbrauns, 1989), 133–63.

[17] Colin J. Humphreys and W. Graeme Waddington, "Dating the Crucifixion," *Nature* 306 (1983): 743–46; Humphreys and Waddington, "Astronomy and the Date of the Crucifixion," in Vardaman and Yamauchi, *Chronos, Kairos, Christos*, 165–81. Schaefer, "Lunar Visibility," 59–65, disputes that the lunar eclipse of 3 April, AD 33, was actually visible over Jerusalem. See also Clive Ruggles, "The Moon and the Crucifixion," *Nature* 345 (1990): 669–70, but cf. Humphreys and Waddington, "Crucifixion

It is worth underlining that all the dates mentioned would likely have corresponded to the 14th day of Nisan, thereby lending support to the chronological account found in the Gospel of John. Implicit in the latter is a powerful typological image, according to which Christ, as the sacrificial Lamb of God, was slain on the same day as the Passover lamb. This notion is already present in the Pauline epistles (1 Corinthians 5:7), where Jesus is directly identified with the Passover lamb. It also helps explaining why, during the first three centuries of Christian literature, the day of Christ's crucifixion was commonly associated with 14 Nisan, the day of preparation for Passover. By contrast, the earliest Christian 'Easter' celebrations, which emerged within a Jewish milieu, seem to have taken the form of a Passover meal on the evening between 14 and 15 Nisan (a practice later decried as 'Quartodecimanism'). From the second century onwards, however, Christian communities in Rome and elsewhere began to put greater emphasis on the weekday of the resurrection, making it binding practice to celebrate Easter on the Sunday that followed the Jewish Passover.[18]

While this step helped to emancipate the Christian Pasch from its Jewish predecessor, the fact remained that 14 Nisan was a full moon date in the Jewish calendar. For Christians who had converted from paganism, this created something of a problem. Ever since the Julian reform of 46/45 BC, the Roman calendar, which was used by most subjects of the empire in everyday affairs, had been purely solar, with an average year length of 365.25d. It was thus only natural if the Julian calendar became the framework for the emerging Christian liturgical year, with the birth of Jesus ending up on 25 December, the traditional Roman date of the winter solstice. Yet the question remained how Easter, as a date tied to the moon, could be accommodated within this framework. In this context, it is striking to observe that some Christian communities reacted to the problem by redefining 14 Nisan as the 14th day of the first month in their local solar calendar, thereby divorcing the remembrance of the crucifixion from the

---

Date," *Nature* 348 (1990): 684. For the moon "turned into blood," see also the account of the crucifixion-eclipse in the *Anaphora Pilati*, translated by James Keith Elliott, *The Apocryphal New Testament* (Oxford: Clarendon Press, 1993), 212.

[18] On Easter and the Passion day in the early Church, see Thomas J. Talley, *The Origins of the Liturgical Year*, 2nd ed. (Collegeville, MN: The Liturgical Press, 1991), 2–27; Karl Gerlach, *The Antenicene Pascha* (Louvain: Peeters, 1998). For the wider context, see Paul F. Bradshaw and Lawrence A. Hoffman, eds., *Passover and Easter*, 2 vols. (Notre Dame, IN: University of Notre Dame Press, 1999).

full moon. The fifth-century church historian Sozomen reports that the Montanist sect of Asia Minor used to set Easter on April 6 or the Sunday thereafter, following a solar calendar whose first month of spring began on 24 March, while from Epiphanius of Salamis we learn that certain groups from the same region always celebrated the Pasch on 25 March.[19]

For all others, who wanted to retain the lunar nature of 14/15 Nisan, it was necessary to keep track of the Jewish Passover. As long as the lunar months were determined empirically, it was impossible to foretell with any certainty the date of Passover, forcing Christians to await the respective decisions of the Jews. This ongoing dependency of Christian liturgy upon Jewish calendation must have been a continuing source of unease for the young church, especially in times and places where Christians were eager to emancipate themselves from the Synagogue. Moreover, since the Jewish calendar seems to have remained empirical for some time after the destruction of the Second Temple, observations of the lunar crescent at different longitudes and latitudes (and under different atmospheric conditions) as well as diverging decisions concerning intercalation could give rise to local variants, in which the beginning of Nisan would fall on different days in Julian calendar. Given such circumstances it was difficult, if not impossible, for the church to retain its desired goal of unity by celebrating Easter on the same date throughout the world.[20]

It is thus no surprise if Christians, from the early third century onwards, were no longer prepared to "walk in blindness and stupidity behind the Jews," as it was put by the anonymous North African author of the *De pascha computus* (243), the earliest preserved treatise on Easter reckoning.[21] This outspoken will to emancipation gave rise

[19] Sozomen, *Historia ecclesiastica* (7.18.12–14), GCS 50:329; Epiphanius, *Panarion* (50.1.6–8), GCS 31:245–46. See Schwartz, "Osterbetrachtungen," 12–13; Strobel, *Ursprung*, 368–74; Thomas J. Talley, "Further Light on the Quartodeciman Pascha and the Date of the Annunciation," *Studia Liturgica* 33 (2003): 151–58; Talley, *Origins*, 7–9; Roger T. Beckwith, *Calendar, Chronology, and Worship* (Leiden: Brill, 2005), 99–102.

[20] This goal was first explicitly stated in the canons of the Council of Arles (AD 314). See *Concilium Arelatense* (1), CCSL 148:9. See further Timothy C. G. Thornton, "Problematical Passovers: Difficulties for Diaspora Jews and Early Christians in Determining Passover Dates During the First Three Centuries A.D.," *Studia Patristica* 20 (1989): 402–8. On Jewish calendar diversity, see Stern, *Calendar*, 47–154.

[21] *De pascha computus* (1), CSEL 3.3:248–49: "Deo inspirati volumus amantibus et adpetentibus studia divina ostendere numquam posse Christianos a via veritatis errare

to the development of lunisolar cycles, adapted for the Julian calendar and specially designed for the calculation of the moveable feast days. While several varieties of such cycles were proposed during late antiquity, they were all based on the same insight that, after a given number of years, the Easter full moon was bound to return to the same Julian calendar date as in some previous year. The interval between two such years comprises both a whole number of Julian years (365.25d) and a whole number of lunar months (29.5306d). The ratio between both periods has to come as close as possible to the decimal fraction contained in the excess of twelve lunar months compared with a Julian year, which is 0.3685 (365.25 ÷ 29.5306 = 12.3685). The latter can be approximated using the following common fractions: 3/8 (0.3750); 4/11 (0.3636); 7/19 (0.3684); 31/84 (0.3690). As can be seen from these values, the most exacting approach will be to intercalate seven lunar months in every 19 years, i.e. to equate 19 Julian years or 6939.75d with 12 × 19 + 7 = 235 lunations. Moreover, since 235 lunar months are one day in excess (12 lunar years of 354d + 7 embolismic months of 30d + 4.75 Julian leap days = 6940.75d) over 19 solar years, one lunar day will have to be dropped at the end of each 19-year cycle. To medieval computists, this mechanism was known as the 'leap of the moon' or *saltus lunae*. While the 19-year cycle or enneacaidecaëteris was by far the best possible approximation to astronomical reality within reasonable length and eventually emerged as the sole standard of Easter reckoning in East and West, alternative cycles, based on lunisolar ratios of 3/8 (octaëteris) and 31/84 (*Supputatio Romana*), were also in use during late antiquity and sometimes competed with the 19-year cycle, which was championed by the Alexandrian church since the late third century.[22]

et tanquam ignorantes quae sit dies Paschae, post Iudaeos caecos et hebetes ambulare." See Eviatar Zerubavel, "Easter and Passover: On Calendars and Group Identity," *American Sociological Review* 47 (1982): 284–89; Marcel Simon, *Verus Israel* (London: Littman Library, 1996), 310–22. On the wider context, see Israel Jacob Yuval, "Easter and Passover as Early Jewish-Christian Dialogue," and "Passover in the Middle Ages," in Bradshaw and Hoffman, *Passover*, 1:98–160; Yuval, *Two Nations in Your Womb* (Berkeley: University of California Press, 2006), chap. 2 and 5.

[22] On the early history of Easter computation, see Charles W. Jones, *Bedae Opera de Temporibus* (Cambridge, MA: The Medieval Academy of America, 1943), 3–122; Olaf Pedersen, "The Ecclesiastical Calendar and the Life of the Church," in *Gregorian Reform of the Calendar*, ed. George V. Coyne, Michael A. Hoskin, and Olaf Pedersen (Vatican City: Specola Vaticana, 1983), 17–74; Faith Wallis, *Bede: The Reckoning of Time* (Liverpool: Liverpool University Press, 1999), xxxiv–lxiii; Georges Declercq,

Given the many different flavors of lunisolar calendation, it is easy to overlook that the chief purpose of these devices was to calculate a Jewish date—the 14th of Nisan, which in Christian usage became referred to as the 'Easter full moon'. At first glance, it would seem that Easter cycles did not fulfill this objective very successfully. Since the Jewish calendar of the Mishnah followed empirical criteria, the Christian Easter full moon was bound to deviate from 14 Nisan, as it was observed by the Jews, in many years. This obvious discrepancy has prompted scholars to suppose that some Easter cycles were themselves based on local Jewish calendars, which are otherwise unattested in our sources. It is indeed striking to observe that the author of the aforementioned *De pascha computus* seemed convinced that the primitive 8-year cycle advocated by him had already been used by the ancient Israelites of the Old Testament. An allusion to this idea can be also found in the *Chronographiae* of Julius Africanus (ca. 221), who ascribed the octaëteris to "the Greeks and the Jews."[23] It is doubtful, however, if first- to third-century Jews ever used any such device. The same must be said concerning the idea, which was proposed by Otto Neugebauer on the basis of late Ethiopic sources, that the 19-year cycle of the Alexandrian church had been taken over from the local Jewish community.[24] Instead of acknowledging any such dependencies, Christian writers from the fourth century onwards are often found polemicizing against the calendar of contemporary Jewry. A prominent early example for this stance is a lost tractate *On Easter* by Peter, the patriarch of Alexandria from ca. 300–310, which is fragmentarily preserved in the seventh-century *Chronicon Paschale*. Peter accused the Jews of regularly celebrating their Paschone month too early, which in his mind was

Anno Domini (Turnhout: Brepols, 2000); Declercq, "Dionysius Exiguus and the Introduction of the Christian Era," *Sacris Erudiri* 41 (2002): 165–246; Mosshammer, *Easter Computus*, 59–316; Leofranc Holford-Strevens, "Paschal Lunar Calendars up to Bede," *Peritia* 20 (2008): 165–208; Warntjes, *Munich Computus*, xxx–li.

[23] *De pascha computus* (6–7), CSEL 3.3:252–55; Julius Africanus, *Chronographiae* (F 93), GCS-NS 15:282; *Prologus Paschae*, ed. Krusch, *Studien* (I), 231. See Jones, *Bedae Opera*, 14; Stern, *Calendar*, 23, 63–64.

[24] See Otto Neugebauer, *Ethiopic Astronomy and Computus* (Vienna: Verl. der Österr. Akad. der Wiss., 1979), 8–9, 27–28, 129–49. For further such theories, see Franz Rühl, "Der Ursprung der jüdischen Weltära," *Deutsche Zeitschrift für Geschichtswissenschaft*, n.s., 2 (1897–98): 201–2; Eduard Schwartz, *Christliche und jüdische Ostertafeln* (Berlin: Weidmann, 1905); Strobel, *Ursprung*, 301–24, 430–40, 446–49; Raymond Mercier, "The Dates in Syriac Martyr Acts," *Analecta Bollandiana* 117 (1999): 47–66. The latter's views are refuted by Sacha Stern, "Near Eastern Lunar Calendars in the Syriac Martyr Acts," *Le Muséon* 117 (2004): 447–72.

a violation of the rules once established by Moses. As he saw it, the Jewish calendar had changed for the worse after the destruction of the Second Temple, whereas his church still preserved the legitimate reckoning, which had been observed at the time of Jesus.[25]

The discrepancies, which Peter and other authors are found complaining about, seem to have mainly arisen from the 'rule of the equinox', which was an element characteristic of Alexandrian (and later Western) Easter reckoning. According to this rule, the Easter full moon cannot fall earlier than 21 March, the supposed date of the vernal equinox.[26] If the Jews disregarded this rule in Peter's time, they were bound to celebrate Easter a month earlier on many occasions. Sacha Stern has recently confirmed this picture by showing that before the destruction of the Second Temple, the Jewish Passovers tended to fall in April rather than March, presumably in order to give pilgrims enough time to reach Jerusalem. By contrast, the available evidence for the fourth century points towards Passovers' being celebrated in March and hence often ahead of the equinox. As a result, those Christians who insisted on following the Jews instead of observing the Alexandrian rules would in many cases have celebrated on a different date than the mainstream church.[27] Their practice was condemned at the Council of Nicaea (325), as can be inferred from a letter sent by Constantine the Great to Syrian and Palestinian bishops, but apparently lingered on in parts of the East, as can be seen from the fact that John Chrysostom and an anonymous Anatolian homilist still felt the need to inveigh against dissenters in their own communities as late as 387.[28] Christian

---

[25] *Chronicon Paschale*, PG 92:69–76. On this fragment, see Gerlach, *Antenicene Pascha*, 295–99.

[26] The first clear reference to this rule can be found in a paschal treatise by Anatolius of Laodicea, cited in Eusebius, *Historia ecclesiastica* (7.32.14–19), GCS 9.2:722–26. On Anatolius's statements, see Schwartz, *Ostertafeln*, 138–50; Gerlach, *Antenicene Pascha*, 292–95; Stern, *Calendar*, 50–55; Mosshammer, *Easter Computus*, 135–36.

[27] Stern, *Calendar*, 55–98. See also Schwartz, "Osterbetrachtungen," 8–10; Venance Grumel, "Le problème de la date pascale aux IIIe et IVe siècles," *Revue des Études Byzantines* 18 (1960): 163–78.

[28] John Chrysostom, *Adversus Judaeos* (3), PG 48:861–72; Fernand Floëri and Pierre Nautin, eds., *Homélies pascales*, vol. 3, *Une homélie anatolienne sur la date de Paques en l'an 387* (Paris: Éditions du Cerf, 1957), 120–25. For Constantine's letter, see Eusebius, *Vita Constantini* (3.18), GCS 7:87. See further Marcel Metzger, ed., *Les constitutions apostoliques*, 3 vols. (Paris: Éditions du Cerf, 1985–87), 2:266–68 (5.17.1–3); 3:276 (8.47); Eusebius, *De solemnitate Paschali* (9–12), PG 24:704–5; Epiphanius, *Panarion* (70.9–13), GCS 37:241–47; Athanasius, *De synodis* (5), PG 26:688; Athanasius, *Epistola ad Afros Episcopos* (2), PG 26:1032; Athanasius, *Lettres festales et pastorales*

writers in the Latin West, by contrast, had little exposure to these kinds of polemics, which mostly stem from the Greek East. A general lack of available sources on the Jewish calendar other than the Old Testament soon led to a situation where early medieval scholars ceased to differentiate between the biblical calendar of the Hebrews and the calendar of the Jews that they may or may not have encountered in their own time. Instead, they went on to by and large identify Jewish lunar months with the lunations of the Alexandrian 19-year cycle, sometimes following an assumption that this cycle had already been revealed to Moses at Mt. Sinai.[29]

The construction of new Easter cycles virtually ceased after the mid-fifth century, although debates over the acceptance of particular cycles continued well into the early medieval period, especially on the British Isles, where the Celtic churches long upheld specific computational traditions that separated them from Rome. This apparent halt in technical development, however, did not lead to the kind of intellectual stagnation one might expect. Hundreds of surviving manuscripts from the seventh to the tenth century testify to the importance of the computus—the reckoning of Easter—to scientific learning and monastic education in this period, making it clear "that almost every person educated in a monastic school was taught at the least the basics of this science."[30] The monastic scholars who copied late antique Easter cycles and applied them to the dating of ecclesiastical feasts soon realized that their sources left many essential questions unanswered. Authorities such as Dionysius Exiguus, whose 95-year Easter table became the basis for most of subsequent medieval computistics, had provided

---

en Copte (24), CSCO 151:11; Athanasius ap. *Chronicon Paschale*, PG 92:76; Socrates, *Historia ecclesiastica* (1.9.12–14; 5.22.19–20), GCS-NF 1:30, 298–99; Theodoret, *Historia ecclesiastica* (1.9.12), GCS 44:41; *Epistola Proterii*, ed. Krusch, *Studien* (I), 271, 275–76; Photius, *Bibliotheca* (115), ed. René Henry, 9 vols. (Paris: Société d'Édition Les Belles Lettres, 1959–91), 2:68. On the general context, see Gerlach, *The Antenicene Pascha*, 257–317.

[29] See my "Between Crucifixion and Calendar Reform: Medieval Christian Perception of the Jewish Lunisolar Calendar," in *Living the Lunar Calendar*, ed. Jonathan Ben-Dov, Wayne Horowitz, and John M. Steele (Oxford: Oxbow Books, forthcoming), and below, chapters three (section 3) and four.

[30] Warntjes, *Munich Computus*, xxxiii. See also Evelyn Edson, "World Maps and Easter Tables: Medieval Maps in Context," *Imago Mundi* 48 (1996): 25–42; John J. Contreni, "Counting, Calendars, and Cosmology: Numeracy in the Early Middle Ages," in *Word, Image, Number*, ed. John J. Contreni and Santa Casciani (Florence: SISMEL, 2002), 43–83, and the essays collected in Wesley M. Stevens, *Cycles of Time and Scientific Learning in Medieval Europe* (Aldershot: Variorum, 1995).

the necessary information for finding the date of Easter, but they generally remained silent on points such as the exact position of the embolismic months in the Julian calendar, the date of the 'leap of the moon' or the exact sequence of 'full' and 'hollow' months in the lunar year. Interest in even the most recondite aspects of Easter computation led to the development of a vibrant early medieval textbook tradition, in which different theories were compared and expounded. Moreover, the need to elucidate every facet of the lunisolar cycle led computistical texts to attract entire clusters of background knowledge from basic mathematics and astronomy to medicine and natural philosophy, making the computus synonymous with natural science during the early Middle Ages. The main contributions to the development of these computistical textbooks were made by Irish monastic scholars in the seventh and early eighth century, who sometimes worked against a contemporary backdrop of heated controversies over the correct computation of Easter. This early medieval tradition of computistical textbooks reached its acme with *De temporum ratione*, written by the Northumbrian monk Bede the Venerable in 725, who became the foremost authority for Western computistics and Christian chronology until the Gregorian reform of the calendar in 1582.

CHAPTER TWO

THE ORIGINS OF COMPUTISTICAL CHRONOGRAPHY

1. *The Beginnings*

For the first two centuries of the Christian era, there is only meager evidence that any attempts were made to fix the dates of the life of Jesus Christ. Judging from the dearth of useable chronological information in the source texts, i.e. the four canonical Gospels, one may presume that this is partly due to a relative disinterest of early Christians in questions of this kind. Given the widespread expectation of an imminent second coming of Christ, there was indeed little motivation for believers to pay any prolonged attention to historical and chronological matters. As a result, the same indifference towards time that kept St. Paul from dating his letters may have also been responsible for the apparent failure to preserve the historical date of the crucifixion for posterity.[1] Under these circumstances, it is not surprising if the actual historical date of the crucifixion of Jesus was soon forgotten, only to later give way to a multitude of different local traditions, none of which can be considered reliable. Our earliest witness in this regard is Clement of Alexandria, whose *Stromata*, written shortly after the assassination of Emperor Commodus (31 December 192), contain some interesting remarks on the chronology of Christ's life. Concerning the date of his death, Clement notes:

> And treating of His Passion in very precise detail, some say that it took place in the 16th year of Tiberius, on the 25th of Phamenoth; and others the 25th of Pharmouthi; and others say that the Saviour suffered on the 19th of Pharmouthi.[2]

[1] See Pedersen, "The Ecclesiastical Calendar," 22; Gerald J. Whitrow, *Time in History* (Oxford: Oxford University Press, 1988), 65.

[2] Clement of Alexandria, *Stromata* (1.146.3), GCS 15:90: τό τε πάθος αὐτοῦ ἀκριβολογούμενοι φέρουσιν οἳ μέν τινες τῷ ἑκκαιδεκάτῳ ἔτει Τιβερίου Καίσαρος Φαμενὼθ κε′, οἳ δὲ Φαρμουθὶ κε′· ἄλλοι δὲ Φαρμουθὶ ιθ′ πεπονθέναι τὸν σωτῆρα λέγουσιν. For useful discussions of Clement's chronological statements, see Roland Bainton, "Basilidian Chronology and New Testament Interpretation," *Journal of Biblical*

Presuming a regular count of Tiberius's regnal years from his accession in the fall of AD 14, a crucifixion in his 16th year would imply that it took place in the spring of AD 30. Unfortunately, an interpretation of the corresponding Egyptian calendar dates (25 Phamenoth, 19 Pharmouthi, and 25 Pharmouthi) is complicated by our lack of knowledge if Clement (or his sources) used the old Egyptian 'wandering year' or the reformed Alexandrian calendar as a frame of reference. In the reformed calendar introduced after Augustus's conquest of Egypt (30 BC), a leap day was added at the end of every fourth year. This was done in order to adapt the Egyptian calendar, whose traditional year length was 365d, to the Julian year of 365.25d, which had been introduced in Rome in 45 BC. As a result, the correlation of dates between both calendars became static, with the Alexandrian year always beginning on 1 Thoth, which was equivalent to 29 or 30 August, depending on whether or not it was preceded by a leap day. In the case of Clement's dates for the Julian common year AD 30, we get:

25 Phamenoth = 21 March
19 Pharmouthi = 14 April
25 Pharmouthi = 20 April

Out of these dates, only 14 April was a Friday in AD 30; 20 April would have been a Friday in AD 31, whereas 21 March only fell on this weekday in AD 27 and 32. A different and arguably more attractive result can be attained, if the dates cited by Clement are interpreted as being based on the old Egyptian solar year of 365 days. Some Alexandrian astronomers, such as Claudius Ptolemy, kept this old calendar going even after the Augustan reform, because its lack of any leap days made it easier to count the exact number of days between observations. As a result of this refusal to intercalate, the old Egyptian dates steadily drifted through the Alexandrian/Julian calendar (hence the term 'wandering' or 'revolving year') at a rate of one day in four years, coming full circle after 1460 years.[3] One such calendrical period can be supposed to have started in 25 BC, when both the reformed Alexandrian

Literature 42 (1923): 81–134; George Ogg, "A Note on *Stromateis* 1.144.1–146.4," *Journal of Theological Studies* 46 (1945): 59–63; Strobel, *Ursprung*, 151–59.

[3] On the background, see Leo Depuydt, "Calendars and Years in Ancient Egypt: The Soundness of Egyptian and West Asian Chronology in 1500–500 BC and the Consistency of the Egyptian 365-Day Wandering Year," in *Calendars and Years*, ed. John M. Steele (Oxford: Oxbow Books, 2007), 35–81.

and the old Egyptian calendar had their 1 Thoth fall on 29 August. 54 years later, by the year AD 30, the Egyptian dates had receded by 54 ÷ 4 = 13.5 days, yielding the following results:

25 Phamenoth = 8 March
19 Pharmouthi = 1 April
25 Pharmouthi = 7 April

Out of these dates, 8 March was a Friday in AD 28, 1 April in AD 31, and 7 April in AD 30. Moreover, the latter date was also a full moon and is today regarded as one of the astronomically and historically possible dates of the crucifixion. This coincidence has repeatedly invited scholars to speculate on the provenance of Clement's dates. For the Protestant theologian Erwin Preuschen, who may have been the first to analyse the crucifixion dates found in the *Stromata* on the basis of the Egyptian 'wandering year', the dating to 25 Pharmouthi represented an authentic historical record of the day of Christ's death. August Strobel, who wrote what is perhaps the most extensive study of early calendrical traditions connected to Christ's Passion, was likewise convinced that the Passion had actually taken place on 7 April, AD 30, and that some memory of this date had been therefore preserved in the earliest Christian Easter cycles.[4] Clearly, however, the mere fact that 7 April, AD 30, can be detected as a possible second-century dating tradition, recorded in Clement's *Stromata*, fails to have any bearing on the actual historicity of the latter date. Just as modern scholars have used astronomical calculations to establish this particular date as a possible candidate for the crucifixion, the possibility should not be excluded that second-century Christians, who were themselves uncertain of the day of the Lord's Passion, employed similar technical means in their attempts to arrive at the truth.

An interesting, but little-noticed, suggestion as to how such a calculation might have been performed was made in 1980 by the Austrian astronomer Konradin Ferrari-d'Occhieppo, who is otherwise famous for his work on the 'Star of Bethlehem'. He pointed to a demotic papyrus from the Fayum, now preserved in the Carlsberg collection in Copenhagen (P. Carlsberg 9), which was written in or after AD 144 and contains a 25-year lunar cycle, based on the Egyptian 'wandering

[4] Erwin Preuschen, "Todesjahr und Todestag Jesu," *Zeitschrift für die neutestamentliche Wissenschaft* 5 (1904): 1–17; Strobel, *Ursprung*, 450–56.

year'. According to the indications found in the papyrus, the 16th year of Tiberius (AD 29/30) was the eleventh year of this cycle, which had a 'new moon' on the 10th day of the eighth month of the Egyptian calendar, also known as Pharmouthi. This makes 25 Pharmouthi the 16th day of the lunation, a date which indeed corresponds to 14 or 15 Nisan if one takes into account that Egyptian lunar months were counted from the morning of the moon's last visibility before conjunction, whereas the Jewish lunar month began with the evening of its first visibility. In Ferrari-d'Occhieppo's view, it is hence conceivable that second-century Christian users of such a 25-year cycle might have employed it to calculate the Passion date, setting it on Friday, 25 Pharmouthi = 7 April in the 16th year of Tiberius (= AD 30), the date which is also found in Clement's *Stromata*.[5]

Our search for early Christian applications of technical chronology reaches much firmer grounds once we enter the third century. Luckily for us, some evidence for chronological activity in this period is engraved into stone. In 1551, a heavily damaged marble statue of a person sitting on a throne was discovered outside Rome, near the Via Tiburtina. Early drawings of the statue, which is now located at the entrance of the Vatican Library, indicate that it originally depicted a female person. Under the direction of the antiquary Pirro Ligorio (1500–1583), however, the statue was eventually restored as the stereotypical image of a male bishop and soon became known as the 'statue of St. Hippolytus'.[6] In making this identification, Ligorio and his contemporaries were partly inspired by the 112-year Easter table that had been engraved in two parts on the flanks of the throne, the left side showing the Julian calendar dates and weekdays of the Easter or

[5] Konradin Ferrari-d'Occhieppo, "Die Osterberechnung als Kalenderproblem von der Antike bis Regiomontanus," in *Regiomontanus-Studien*, ed. Günther Hamann (Vienna: Verl. der Österr. Akad. der Wiss., 1980), 96–98. On the papyrus and the calendar contained therein, see Otto Neugebauer and Richard A. Parker, *Egyptian Astronomical Texts*, vol. 3, *Decans, Planets, Constellations, and Zodiacs* (Providence: Brown University Press, 1969), 220–25; Richard A. Parker, *The Calendars of Ancient Egypt* (Chicago: University of Chicago Press, 1950), 9–29; Alan E. Samuel, *Ptolemaic Chronology* (Munich: C. H. Beck, 1962), 54–61.

[6] For the archeological evidence, see Margherita Guarducci, "La statua di 'sant' Ippolyto' in Vaticano," *Rendiconti della Pontificia Accademia Romana di Archeologia* 47 (1974–75): 163–90; Guarducci, *Epigrafia greca*, 4 vols. (Rome: Istituto Poligrafico dello Stato, 1967–78), 4:535–45. See further Jean Michel Hanssens, *La liturgie d'Hippolyte* (Rome: Pontificium Institutum Orientalium Studiorum, 1959), 217–31.

Paschal full moons, the right side the corresponding Easter Sundays. Both lists of data were based on the *octaëteris*, a primitive 8-year cycle, which, for the purposes of said table, was doubled to yield a 16-year cycle. The 112-year structure arose from combining this 16-year cycle with the sequence of the seven weekdays, which are represented on the left-hand table by seven columns of Greek numeral letters indicating the day of the week.[7]

The earliest mentions of a 16-year Easter cycle can be found in the *Ecclesiastical History* of Eusebius of Caesarea, written shortly after 300, and in the *De viris illustribus* of Jerome of Stridon (ca. 392), who partly drew on the information provided by Eusebius. According to the latter, a 16-year 'list' or κανών could be found in a treatise *On Easter* (περὶ τοῦ πάσχα), written by a prolific Christian author named Hippolytus, who worked during the reign of Alexander Severus (222–35). In addition to the 'canon', the treatise is also said to have contained a 'chronography' (χρόνων ἀναγραφὴν) down to the first year of the latter emperor's reign.[8] Most strikingly, the first year of Alexander Severus (222/23) is also the starting point of the inscribed Easter table, as identified by the accompanying text. Moreover, the backside of the statue's plinth displays a partly illegible listing of book titles, some of which match the corpus of works that Eusebius and Jerome ascribe to the aforementioned Hippolytus. In particular, the list makes mention of an ἀπόδειξις χρόνων τοῦ πάσχα καὶ τὰ ἐν τῷ πίνακι, which obviously refers to the inscribed tables, but perhaps also connects these tables to the work *On the Easter* mentioned by Eusebius.

In the centuries following the discovery of the statue, further works bearing the authorship of Hippolytus were identified and numerous hypotheses on the author and his provenance were proposed. In 1853, Ignaz von Döllinger suggested that Hippolytus was a Roman Presbyter,

---

[7] The inscription was first edited by Martin Smetius, *Inscriptionum antiquarum quae passim per Europam, liber* (Leiden, 1588), 37v–38r. Further transcriptions can be found in PG 10:875–84; Henri Leclercq, "Hippolyte (statue et cimetière de saint)," in *Dictionnaire d'archéologie chrétienne et de liturgie*, 15 vols. (Paris: Letouzey et Ané, 1903–51), 6:2426–34; Guarducci, *Epigrafia*, 4:539–43.

[8] Eusebius, *Historia ecclesiastica* (6.22), GCS 9.2:568; Jerome, *De viris illustribus* (61), ed. Aldo Ceresa-Gastaldo (Florence: Nardini, 1988), 162–64. The references to Hippolytus in Eusebius and Jerome are analysed at length in John A. Cerrato, *Hippolytus between East and West* (Oxford: Oxford University Press, 2002), 26–68.

who led the opposition against the Roman bishop Callistus, making him the first 'antipope' of ecclesiastical history, before he was reconciled to the mainstream church under Pontianus and eventually exiled to Sardinia in 235. Döllinger's reconstruction remains influential to this day, even though subsequent research has seriously undermined his conclusions. In the 1940s, Pierre Nautin advanced the opinion that the late antique *Corpus Hippolyticum* was the work of at least two different authors, one of whom came from the Greek East and was responsible for the exegetical works attributed to Hippolytus, whereas the other worked in Rome and produced the inscribed paschal 'canon'. More recently, Alan Brent has argued that the works recorded on the statue can be situated within the milieu of a Hippolytan 'school' of several authors, working in Rome during the early decades of the third century.[9]

The historical and text critical issues surrounding the Hippolytan corpus are far too complex to be treated here in any detail. For present purposes, it shall suffice to simply refer to the two tables, inscribed in the statue's throne, as the 'canon of Hippolytus'. Aside from the evidence already mentioned, this approach seems legitimate in the light of the testimony of the Syriac bishop and chronicler Elias of Nisibis (975–ca. 1049), which has too often been neglected by modern scholars. In his *Chronography* (1019), Elias describes a 112-year Easter table, ascribed to "bishop Hippolytus," which is close to identical with the one found on the left flank of the throne.[10] That this Hippolytan Easter table was composed soon after the year 222, the starting point of the 112-year cycle, is borne out not only by the third-century provenance of the inscription, but also by the astronomical data. The Easter full moon in the first year of the cycle is marked as Saturday, 13 April,

[9] Johann Joseph Ignaz von Döllinger, *Hippolytus und Kallistus* (Regensburg: Manz, 1853); Pierre Nautin, *Hippolyte et Josipe* (Paris: Éditions du Cerf, 1947); Allen Brent, *Hippolytus and the Roman Church in the Third Century* (Leiden: Brill, 1995). On the history and state of research on the *Corpus Hippolyticum*, see also Cerrato, *Hippolytus*, 69–123; John F. Baldovin, "Hippolytus and the *Apostolic Tradition*: Recent Research and Commentary," *Theological Studies* 64 (2003): 522–29; Ronald E. Heine, "Hippolytus, Ps.-Hippolytus and the Early Canons," in *The Cambridge History of Early Christian Literature*, ed. Frances Young, Lewis Ayres, and Andrew Louth (Cambridge: Cambridge University Press, 2004), 142–51; Mosshammer, *Easter Computus*, 118–21.

[10] Elias of Nisibis, *Opus chronologicum*, CSCO 63**:120–21. See Marcel Richard, "Notes sur le comput des cent-douze ans," *Revue des Études Byzantines* 24 (1966): 261–66.

which is perfectly congruent with modern calculations of the opposition of sun and moon for that year.[11] This perfect match at the beginning of the cycle can only have been achieved by observation, for the underlying 8-year cycle is far too inaccurate to keep the tabulated full moons close to their astronomical equivalents for any length of time. The Hippolytan version of the octaëteris equates eight lunar and eight Julian years by inserting three intercalary months of 30 days, thereby producing: 8 × 354 + 3 × 30 = 99 × 30 = 2922d = 8 × 365.25d. Yet 99 lunar months equal 99 × 29.53806 = 2923.53d, which means that the full moons calculated on the basis of the octaëteris will be out of step by approximately 1.5 days after only eight years. After a full run of 112 years, this discrepancy will have accrued to a staggering 21 days.

The error rate of the plain octaëteris is so excessive that some scholars have felt compelled to assume that third-century Christians must have applied periodic adjustments to prevent their cycles from completely falling out of touch with the moon.[12] As we shall see, however, those who composed the 'canon of Hippolytus' (and those who inscribed it on the base of a marble statue), seem to have done so in the conviction that it formed a perfect cycle making all Easter full moons recur on both the same calendar dates and the same weekdays after every 112 years. In fact, they even used this principle to calculate the dates of Easter or Paschal full moons which lay centuries in the past. This becomes manifest from a number of additions in the margins of the table of 112 Easter full moons, which mention different historical Passovers from Old Testament history (the Exodus, 'In the Desert', Joshua, Hezekiah, Josiah, Esdras), along with the birth (*génesis*) and Passion (*páthos*) of Christ. The presence of these glosses has puzzled early commentators on the Hippolytan 'canon'. Joseph Scaliger, who published a pioneering study of the inscribed Easter table in 1595,

---

[11] See George Salmon, "Some Notes on the Chronology of Hippolytus," *Hermathena* 1 (1873): 88, who notes that "the table gives accurately the astronomical full moons for the years 217–223, inclusive." See now also Ulrich Voigt, "Über das Jahr 1 des Alexander Severus auf der Passatafel des Hippolytus," *Scholion* 5 (2008): 99–121. For ancient lunar phases and eclipse times, see also the data provided on the NASA website, http://eclipse.gsfc.nasa.gov/phase/phases0201.html.

[12] Grumel, *Chronologie*, 17; August Strobel, *Texte zur Geschichte des frühchristlichen Osterkalenders* (Münster: Aschendorff, 1984), 61–63; Strobel, *Ursprung*, 125n3; Marcel Richard, "Le comput pascal par octaétéris," *Le Muséon* 87 (1974): 307–39; Richard, "Notes," 266–72. See now also Ulrich Voigt, "Geminos von Rhodos und die Oktaëteris," *Beiträge zur Astronomiegeschichte* 10 (2010): 7–34.

surmised that the entries might have designated liturgical 'readings' (*lectiones*) that were supposed to be held on the weekdays and calendar dates in question.[13] Only at the beginning of the eighteenth century did the astronomer Francesco Bianchini (1662–1729) advance the hypothesis that these glosses were historical in that they meant to indicate the actual date and year of the events they specified.[14]

Bianchini's conjecture is corroborated by the testimony of the *Liber generationis*, an early fourth-century Latin redaction of a Greek world chronicle from ca. 235.[15] Towards the end, the *Liber generationis* contains a concise listing of intervals of years between the creation of Adam and the thirteenth year of Alexander Severus (234/35), which is based on the very same biblical Passover dates that are found recorded on the Hippolytan statue. Beginning with the Irish mathematician George Salmon in 1873, a number of scholars have collated the intervals of the *Liber generationis* with those implied on the inscribed Easter table. As demonstrated in the table below, their work uncovered a nearly perfect match between both sources, once multiples of 112 were added to the intervals on the statue.[16]

[13] Joseph Justus Scaliger, *Hippolyti Episcopi Canon Paschalis* (Leiden, 1595), 10: "Haec autem omnia, quae inclusa sunt in cellulis areae Laterculi, sunt lectiones ex utroque Testamento, quarum usus erat eo tempore in Ecclesia. Sed quaedam earum statae sunt ex orbe hebdomadum, et eidem diei hebdomadis haerent, aliquando fortasse eidem diei mensis: quaedam vagae pro ratione Termini."

[14] Francesco Bianchini, *De Kalendario et cyclo Caesarisac De paschali canone S. Hippolyti martyris dissertationes duae* (Rome, 1703), 105–14.

[15] *Liber generationis*, in *Chronica minora saec. IV. V. VI. VII*, ed. Theodor Mommsen, 3 vols. (Berlin: Weidmann, 1892–98), 1:130–31. See also Theodor Mommsen, *Über den Chronographen aus dem Jahre 354* (Leipzig: Weidmann, 1850), 585–98; Heinrich Gelzer, *Sextus Julius Africanus und die byzantinische Chronographie*, 2 vols. (Leipzig: Teubner, 1880–98), 2:1–23.

[16] The following table is based on the cogent reconstructions by Marcel Richard, "Comput et chronographie chez saint Hippolyte," pt. 1, *Mélanges de science religieuse* 7 (1950): 250–57, and Hanssens, *Liturgie*, 259–70, 280–81. Earlier attempts were made by George Salmon, "Some Notes," 93–99; Salmon, "The Commentary of Hippolytus on Daniel," *Hermathena* 8 (1893): 169–74; Bruno Krusch, "Die Chronicae des sogenannten Fredegar," pt. 2, *Neues Archiv der Gesellschaft für ältere deutsche Geschichtskunde* 7 (1882): 456–63; Gelzer, *Sextus Julius Africanus*, 2:3.

Historical Passover dates in the works ascribed to Hippolytus

Liber generationis	Statue	
	[regular]	κατὰ Δανιήλ
Adam	VII/5: 29 Mar (Thu) [5503 BC]	—
+3813	+ 5 (34 × 112 + 5 = 3813)	
Exodus (Exod. 12:1–13, 16)	VII/10: 2 Apr (Sun) [1690 BC]	I/15: 5 Apr (Di) [1557 BC]
+41	+41	+40
Joshua (Josh. 5:10)	III/3: 21 Mar (Fri) [1649 BC]	IV/7: 5 Apr (Wed) [1517 BC]
+864	+ 80 (7 × 112 + 80 = 864)	+ 79 (7 × 112 + 79 = 863)
Hezekiah (2 Chron. 30:21)	I/3: 21 Mar (Sun) [785 BC]	II/6: 18 Mar (Sat) [654 BC]
+113	+ 1 (112 + 1 = 113)	+ 0 (= 112)
Josiah (2 Kings 23:21)	I/4: 9 Apr (Sat) [672 BC]	II/6: 18 Mar (Sat) [542 BC]
+108	+107	+107
Esdras (Ezra 6:19)	VII/15: 5 April (Wed) [565 BC]	II/1: 13 Apr (Fri) [435 BC]
+563	+ 3 (5 × 112 + 3 = 563)	+ 97 (3 × 112 + 97 = 433)
Christ's birth	I/2: 2 Apr (Wed) [2 BC]	
+30	+30	
Christ's Passion	II/16: 25 Mar (Fri) [AD 29]	
+205	+ 93 (112+ 93 = 205)	
13 Alexander	I/13: 29 Mar (Sat) [AD 234]	

The emerging picture was further complemented by the findings of Adolf Bauer, who, in 1905, published fragments from of a Greek world chronicle that he showed to have been the original version of the *Liber generationis*. The preserved table of contents of the Greek text listed an item entitled Ἀπόδειξις περὶ τοῦ Πάσχα at the exact spot where the *Liber generationis* offered its list of intervals between historical

Passovers. As we have seen, a very similar title can be found on the back of the statue. In addition, the inscription also lists a χρονικῶν. For these and other reasons, Bauer went on to identify the work as the '*Chronicle* of Hippolytus', which is now commonly regarded as the earliest example of Christian world chronography alongside the work of Julius Africanus.[17]

In sum, the available evidence strongly suggests that the *Chronicle/ Liber generationis* and the inscribed Easter table are witnesses to one and the same chronological system, which was composed and disseminated, in Rome and possibly elsewhere, during the third decade of the third century. With respect to the references provided by Eusebius, it is also worth speculating whether the 'chronography' (χρόνων ἀναγραφὴν) that he associated with the Hippolytan work *On Easter* might have been a version of the list of historical Passovers found in the chronicle.[18] It further seems possible that this list was composed with the express intention of using it to calculate the calendar dates and weekdays of these individual Passovers. The 112-year Easter table preserved on the statue and in the work of Elias of Nisibis provided all the necessary data for such an undertaking. All one had to do was to determine the interval of the Passover in question from the present year in the cycle, subtract all multiples of 112 from this interval, and then count backwards the remaining number of years in the table. The dates of the birth and death of Jesus can serve to exemplify this approach: according to the Hippolytan 'canon', the birth of Christ took place in the second year of the 112-year cycle, which corresponds to AD 223. In this year, the Easter full moon is noted for Wednesday, 2 April. Yet the intended Passover of Christ's birth must have occurred 224 years earlier, in 2 BC. This is consistent with the intervals listed in the *Liber generationis*, where the distance between Passover of Christ's

---

[17] Adolf Bauer, *Die Chronik des Hippolytos im Matritensis Graecus 121* (Leipzig: Hinrichs, 1905); Hippolytus, *Chronicle* (8, 689–700), GCS 46:6, 116–18. On the chronicle of Hippolytus and its relation to the inscribed 'canon', see most recently Osvalda Andrei, "Dalle *Chronographiai* di Giulio Africano alla *Synagoge* di 'Ippolito': Un dibattito sulla scrittura cristiana del tempo," in Wallraff, *Julius Africanus*, 113–45.

[18] That Eusebius's testimony refers to the same paschal canon as found on the statue is convincingly argued by Richard, "Notes," 257–60. See also Hanssens, *Liturgie*, 254–58; Miroslav Marcovich, "Note on Hippolytus' *Refutatio*," *Journal of Theological Studies*, n.s., 15 (1964): 71–72; Brent, *Hippolytus*, 307–10; Andrei, "Dalle *Chronographiai*," 116–18.

birth and the Passover in the thirteenth year of Alexander Severus (AD 234) is specified as 235 years.

Most of the other intervals found on this list were more or less easily deducible from the chronological material found in the Septuagint. Potential problems were only encountered in the case of the interval between Esdras and the birth of Christ, as the Old Testament offered no continuous chronology for this period. In the *Liber generationis*, this interval is specified as 563 years, but the 'canon' also provides an alternative set of Old Testament data, designated κατὰ Δανιήλ, "according to [the book of] Daniel." Both sets ('Daniel' and 'regular') are essentially the same, except for the interval between Esdras and Jesus, which is a mere 434 years in the case of the chronology 'according to Daniel'. The latter number is obviously related to the 'weeks of years'-prophecy, found in the book of Daniel (9:23–27), which states that 70 × 7 = 490 years would intervene between the Babylonian captivity and the coming of the Messiah. According to the chronology κατὰ Δανιήλ preserved on the statue, 62 × 7 = 434 of these years had elapsed between the end of the exile at time of Esdras and the birth of Christ (Daniel 9:25). Naturally, this is only one of several possible readings of the prophecy, but it is precisely the one found in a Greek *Commentary on Daniel* mentioned by Jerome as belonging to the works of Hippolytus, the same author who also composed the 16-year 'canon'.[19]

Since this commentary was presumably written during the early years of the Severian persecution (202/4), the chronology 'according to Daniel' may be considered the older of the two systems preserved in the margins of the table. The question remains why it was eventually complemented or rather—to judge from the *Liber generationis* and the fact that it is seemingly presented as the 'standard' chronology on the inscription—superseded by an alternative chronology, in which the interval had been raised to 563 years. A cogent explanation for this change was offered by Marcel Richard, who points to the important influence of another chronological idea on third-century Christian chronography. Proponents of this idea limited the duration of the

---

[19] Hippolytus, *Commentarii in Danielem* (4.30–31), GCS-NF 7:264–69; Jerome, *De viris illustribus* (61), ed. Ceresa-Gastaldo, 164. On Hippolytus's computation of the year-week, see also Beckwith, *Calendar and Chronology*, 272–74.

world to 6000 years (in parallel to the six days of creation), but at the same time assumed that the birth of Christ had taken place precisely in the middle of the sixth millennium, leaving only a few more centuries until the end of earthly history. In the aforementioned *Commentary on Daniel*, this thesis is supported by an appeal to the numerology of the ark of the covenant, whose measures of 2 ½ + 1 ½ + 1 ½ = 5 ½ cubits (Exodus 25:10) are read as a typological hint at the five and a half millennia of pre-Messianic history. In a similar vein, the author interprets the time of the crucifixion "about the sixth hour" (John 19:14) as a pronouncement of the same chronological fact.[20] In the *Liber generationis* or '*Chronicle* of Hippolytus', the difference between Adam and Esdras is 4939 years, a number derived from the various chronological intervals found in the Septuagint. With all other intervals remaining the same in both the 'regular' and the 'Daniel' chronology, it was clear that in order to bring the total duration between Adam and Christ to 5500 years, the scripturally unspecified interval between Esdras and Christ had to be raised to over 560 years.[21]

In the case of the Hippolytan chronicle, the interval was chosen to be 563 years, thereby bringing the distance between Adam and Christ to 5501, which meant that the birth of Christ had taken place in the 5502nd year of the world. Why did the author of the chronicle opt for this interval over the more symmetrical 5500 years, found in the *Commentary on Daniel*? Once again, Marcel Richard was able to provide a convincing explanation: if Christ was born in the second year of the 112-year cycle, an interval of 5501 (as opposed to 5500) years would put the creation in the 101st year of this cycle, with an Easter full moon on Thursday, 29 March. Such a combination of data in the year of creation would be in conformity with some widespread assumptions in patristic literature, according to which the world was created at the time of the vernal equinox and the moon was full at the time of its creation. In the present case, if the moon first appeared on the evening of the fourth day (Genesis 1:14), the full moon could be justifiably assigned to the fifth day, i.e. to Thursday. If the latter fell on 29 March, this also meant that the preceding Sunday, on which

[20] Hippolytus, *Commentarii in Danielem* (4.23–24), GCS-NF 7:244–48.
[21] Richard, "Comput," pt. 1, 247–49.

creation began, had been a 25 March, the traditional Roman calendar date for the vernal equinox.[22]

In 1958, eight years after Richard's groundbreaking article, Venance Grumel published his monumental handbook on Byzantine chronology, the first chapters of which contained a comprehensive survey of late antique Christian world eras and their underlying principles of construction. Although Grumel proposed a slightly different (and less convincing) interpretation of the year of creation implicit in Hippolytus's chronology, his other results amply confirmed the basic insights on which Richard's reconstruction rested. According to Grumel, Christian computists of the late antique and early Byzantine period worked hard to construct chronological systems in which core assumptions about the chronological dimensions of the history of salvation would match the combinations of data provided by the Easter cycles they respectively used.[23] One of the central constraints at work in these systems was the 5500-year interval between the creation and the birth of Christ. At the same time, various existing theories about the calendrical parameters of the divine hexaëmeron made it necessary to choose a year of creation in which the appropriate combination of lunar and solar data would be reflected in the Easter cycle. Even more importantly, this year had to be linked, via the aforementioned 5500-year interval, to an appropriate year for Christ's Passion, in which the Easter full moon (*luna* 14 or 15) would fall on both a Friday and on a suitable date in the Julian calendar.

Needless to say, the attempt to accommodate all these data into a single, coherent system could result in a cumbersome chronological tightrope walk. In the case of Hippolytus—the presumed author behind both the chronicle and the 'canon'—these various constraints apparently necessitated a slight deviation from the 5500-year scheme, extending the interval between creation and the birth of Christ by one year. According to the inscribed 'canon', Jesus was born on Wednesday, 2 April, 2 BC, while his crucifixion took place on Friday, 25 March, AD 29. If both dates are taken as equivalent to 14 Nisan (the 'Jewish' equivalent to the Easter full moon), this corresponds to an interval of precisely 30 years. The birth date here is interesting in two respects.

---

[22] Ibid., 239–47. See also Salmon, "Some Notes," 91–92; Hanssens, *Liturgie*, 279; George Ogg, "Hippolytus and the Introduction of the Christian Era," *Vigiliae Christianae* 16 (1962): 5–7.

[23] Grumel, *Chronologie*, 6–17.

Firstly, the 'canon' is a clear witness to an early Christian tradition, in which the nativity of Jesus was not yet assigned to a winter date, that is to either 25 December (Christmas) or 6 January (Epiphany). Instead, Hippolytus seems to have followed certain Jewish traditions, according to which important figures, in particular Isaac the son of Abraham, were born on Passover and also died on that day.[24] Secondly, he assigned the nativity to the year 2 BC, which is identical with the 42nd year of the reign of Augustus, if the latter's reign is counted from 43 BC. The 42nd year of Augustus is also mentioned as the year of Christ's birth in numerous other patristic sources, including the aforementioned Hippolytan *Commentary on Daniel*.[25]

Since the year of the Passion had to allow for an Easter full moon on a Friday, whereas there were no similar constraints on the year of the nativity, there is every reason to believe that Hippolytus first selected his date of the crucifixion and then counted back 30 years to arrive at the date of Christ's birth. This interval was ostensibly derived from the Gospel of Luke (3:23), where it is stated that Jesus was "about thirty years old" at the time of his baptism. Slightly earlier in the text (3:1), the appearance of John the Baptist is linked to the 15th year of the reign of Tiberius, which corresponds to AD 28/29, if the latter's reign is counted in ordinary fashion from his accession in the fall of AD 14. In Hippolytus's scheme of events, according to which the crucifixion took place in the spring of AD 29, that is still in the 15th year of Tiberius, this would imply that Christ's baptism and death were separated by hardly more than a few months. While this is unlikely given the 'long' chronology of Jesus's public ministry, as it is preserved in the Gospel of John, Hippolytus's scheme is quite consistent with that of other ante-Nicene exegetes, including Clement of Alexandria, who assumed that Jesus publicly preached for merely a year and lived to the age of 30.[26]

---

[24] See Strobel, *Ursprung*, 128–33; Talley, *Origins*, 81–83.

[25] Hippolytus, *Commentarium in Danielem* (4.9, 23), GCS-NF 7:214, 244. See also Eusebius, *Historia ecclesiastica* (1.5.2), GCS 9.1:44; Eusebius, *Chronicle*, GCS 47:169; Epiphanius, *Panarion* (51.22.3), GCS 31:284; Epiphanius, *Ancoratus* (60.3), GCS 25:71; Epiphanius, *De mensuris et ponderibus* (13), PG 43:260; Orosius, *Historia adversum paganos* (7.2), CSEL 5:437. Other sources tie the nativity to the consulate of Augustus (XIII) and Silanus, which likewise corresponds to 2 BC. See Epiphanius, *Panarion* (51.22.3), GCS 31:284; *Consularia Italica*, in *Chronica minora*, 1:278; Idatius, *Descriptio Consulum*, PL 51:900. See also Finegan, *Handbook*, 288–91.

[26] Clement of Alexandria, *Stromata* (1.145.3–4), GCS 15:90. On the background, see Ogg, *Chronology*, 62–76, 103–13, 130–39; Strobel, *Ursprung*, 101–3. See also

Another respect in which the dates of the Hippolytan 'canon' fit well into the context of second- and third-century Christian exegesis concerns the lunar date. Since Hippolytus puts the crucifixion on the day of the Passover full moon and hence on 14 Nisan, this means that he followed the Passion chronology of John, which implies that Jesus was crucified on the afternoon of 14 Nisan. As noted in the previous chapter, this preference for the Johannine version was very common among early Christian writers. The seventh-century *Chronicon Paschale*, for instance, preserves a number of citations from ante-Nicene authors, who all speak in favour of a crucifixion on 14 Nisan. Besides Apollinaris of Hierapolis and Clement of Alexandria, these include excerpts from two works by "Hippolytus, bishop of Portus near Rome," one of them entitled Περὶ πάσχα. It may well be that the latter text is identical to the *On the Easter* mentioned by Eusebius.[27]

The question remains whether or not the crucial decision to date the crucifixion on 25 March, AD 29, was in a similar way dependent on pre-existing traditions or exegetical conventions. This question is all the more important given the impressive number of subsequent late antique Latin sources that link either 25 March or AD 29 (usually expressed as the consulate of the two Gemini, C. Fufius and L. Rubellius) or both to the Passion of Christ. The Hippolytan 'canon' thus brings us face to face with the origins of what can be referred to as the 'Latin tradition' for the date of the crucifixion, which dominated chronological statements on the life of Jesus in Western sources up until at least the fifth century.[28] Naturally, this tradition is closely linked to a 'short' chronology of the life of Jesus, seeing how—according to most counts—the spring of AD 29 coincides with the spring of the 15th year of Tiberius, which is mentioned by Luke as the regnal year

Hippolytus, *Commentarium in Danielem* (4.23), GCS-NF 7:244, where the Passion is dated to the 33rd year of Christ's life. This may be a later interpolation, influenced by the 'long' chronology, which became popular after the fourth century. See Marcel Richard, "Comput et chronographie chez saint Hippolyte," pt. 2, *Mélanges de science religieuse* 8 (1951): 19–42.

[27] See *Chronicon Paschale*, PG 92:80–81.

[28] On the popularity of 25 March and the Gemini consulate see Vincenzo Loi, "Il 25 Marzo data pasquale e la cronologia giovannea della passione in età patristica," *Ephemerides Liturgicae* 85 (1971): 48–69. See also Ogg, *The Chronology*, 105–11; Richard, "Comput (II)," 32–39; Damiano Lazzarato, *Chronologia Christi seu discordantium fontium concordantia ad juris normam* (Naples: M. d'Auria, 1952), 349–423. Strobel, *Ursprung*, 370–72, suggests that 25 March used to be a fixed "solar quartodeciman" Easter date in Spain, Gaul, northern Italy, North Africa, and Syria.

in which John the Baptist began his preaching and in which Jesus was presumably baptized. Aside from the Hippolytan 'canon', another early source which contains the exact same dating of the crucifixion is the eighth chapter of the treatise *Adversus Iudaeos*, generally attributed to Tertullian:

> This Passion of Christ was fulfilled within the time of the seventy weeks of years under Tiberius Caesar, in the consulate of Rubellius Geminus and Fufius Geminus, in the month of March, at the time of the Passover, on the eighth day before the Kalends of April, on the first day of unleavened bread, on which they slew the lamb in the evening, just as had been prescribed by Moses.[29]

According to conventional wisdom, *Adversus Iudaeos* belongs to the earlier works of the Tertullianian corpus and was written at the beginning of the third century. It is hence no surprise that it has sometimes been treated as the origin of the Latin tradition of the crucifixion date.[30] This view, however, faces certain problems. Scholars have questioned both the early date and the Tertullianian authorship of the eighth chapter on stylistic and doctrinal grounds.[31] It is also worth observing that in two of his other more securely attributed works, Tertullian seems to follow the synoptic chronology of the Passion in dating the crucifixion to the first day of unleavened bread—an opinion also found in another third-century North African source, the *De pascha computus* of 243 (on which see below). By contrast, the passage in *Adversus Iudaeos* expressly refers to the day of the crucifixion as the day on which the lamb was slain in the evening, implying the same Johannine sequence of events that is also manifest in the Hippolytan 'canon' and thereby supporting the possibility that this passage was composed by a later, non-Tertullian hand.[32]

---

[29] Tertullian, *Adversus Iudaeos* (8.18), CCSL 2:1363: "Quae passio Christi intra tempora LXX ebdomadarum perfecta est sub Tiberio Caesare, consulibus Rubellio Gemino et Fufio Gemino mense martio temporibus paschae, die octava Kalendarum Aprilium, die prima azymorum, qua agnum occiderunt ad vesperam sicuti a Moyse fuerat praeceptum."

[30] See most recently Mosshammer, *Easter Computus*, 29, 48, 326, 406.

[31] Hermann Tränkle, *Q. S. F. Tertulliani Adversus Iudaeos* (Wiesbaden: Steiner, 1964), xi–xxiii. See further Paul Capelle, *Le texte du Psautier Latin en Afrique* (Rome: Pustet, 1913), 6; Hugo Koch, "Tertullianisches," *Theologische Studien und Kritiken* 101 (1929): 462–69; Eligius Dekkers, *Clavis Patrum Latinorum*, 3rd ed. (Steenbrugge: Brepols, 1995), 9n33.

[32] Tertullian, *Adversus Marcionem* (4.40.1–3), CCSL 1:655–56; Tertullian, *De baptismo*

Rejecting the anteriority of the Tertullianian testimony to the Latin tradition has the great advantage of enabling us to shed light on the possible rationale behind its Passion date. Such an explanation becomes feasible once we consider the Hippolytan 'canon' not only as a witness to this particular tradition, but as its actual source of origin. From the foregoing, it should have become clear that the author of the 'canon' had all the tools at his disposal to extract this date for himself from the data posed to him by the Gospels. From the latter, he could glean that Jesus had been crucified on Friday, 14 Nisan, the Jewish date corresponding—in theory at least—to the Easter full moons displayed on his table. All he needed to do was to search for a year in which Friday and the Easter full moon coincided and which was close enough to the spring of the 15th year of Tiberius, the year mentioned in the Gospel of Luke. These conditions were satisfied by the 32nd year of the 112-year cycle, which happened to correspond to AD 29 (the consulate of the Gemini), but not by any of the following years in this cycle.

The attractiveness of the resultant date—25 March, AD 29—was doubtlessly heightened by the fact that 25 March was considered in Roman tradition to be the date of the vernal equinox. The latter bore a number of important implications for the interpretation of the events surrounding Christ's crucifixion. To begin with, the equinox was the astronomical beginning of spring, heralding the start of a period during which the duration of daylight would again exceed the duration of night. Accordingly, the equinox could be seen to mirror the victory of light over darkness or life over death, which was in turn symbolized by Christ's resurrection. Moreover, the date of 25 March linked the crucifixion and resurrection of Christ to the week of creation, which—as mentioned before—was widely held to have occurred at the time of the vernal equinox, when new life sprang forth in nature. The crucifixion/resurrection could similarly be viewed as a *re*-creation of mankind, which was thereby redeemed from its fall on the sixth day.[33]

All of these motives doubtlessly provided good grounds to assign the crucifixion to 25 March and thus help explain how this date could have come to play a dominant role in the chronological traditions on the life of Jesus during subsequent centuries and throughout the

(19.1), CCSL 1:293. See also Loi, "Il 25 Marzo," 51–53; Salmon, "Some Notes," 103–4; Ogg, *Chronology*, 147–49; Grumel, *Chronologie*, 27.

[33] On the general background, see Anscar J. Chupungco, *The Cosmic Elements of Christian Passover* (Rome: Editrice Anselmiana, 1977).

Middle Ages. At the same time, it seems reasonable to suppose that it could not have gained the same momentum had it not been supported by an Easter cycle, by which it could be demonstrated to have actually coincided with a Friday and a full moon in the year of the Passion. As we have seen, the Hippolytan 'canon' provided just this kind of technical chronological support. It would be a mistake, however, to suspect that it was specially constructed in order to produce this result. The fact that the first year of the cycle shows an astronomically correct full moon date, as well as the fact that the Easter full moons of the subsequent years are simply a result of the internal structure of the octaëteris, speaks against this hypothesis. In the end, it remains a most compelling scenario to assume that the author of the 'canon' used his Easter cycle to search for an adequate date of the Passion, which made him arrive at 25 March AD 29.[34]

This understanding of the Latin tradition as a computational product of the Hippolytan 'canon' is supported by a look at another third-century work, the aforementioned *De pascha computus* of 243, written by an anonymous author in Roman North Africa and later falsely attributed to St. Cyprian.[35] Since the Hippolytan Easter table, in its present form, is not accompanied by any longer text explaining its purpose and construction (although Eusebius's remarks strongly suggest that it was originally embedded in such a work), the *De pascha computus*, which is the earliest preserved prose work on Easter computation, can be used to shed further light on the intellectual background that may have spawned the inclusion of chronographic marginalia into the 'canon'.[36] In fact, the 112-year Easter cycle described by the North African author bears such a close resemblance to the tables cast into

[34] See also Declercq, *Anno Domini*, 18–19; Salmon, "Some Notes," 102–3; Salmon, "The Commentary," 175: "We can therefore regard the date March 25 as inseparably connected with the 16 years' cycle of Hippolytus, and may confidently assert that any ancient writer who names March 25 as the day of the Crucifixion is later than Hippolytus, and got the idea from him." It may be worth noting that the Hippolytan Passion chronology (Friday, 25 March, *luna* 14) is also expressly asserted in a short treatise on Easter reckoning, written by the African bishop Quintus Julius Hilarianus in AD 397, which advocates an 8-year cycle very similar to the one used by Hippolytus. See Hilarianus's *Expositum de die paschae et mensis* (15), PL 13:1114.

[35] Edited in CSEL 3.3:248–71. English translation and commentary: George Ogg, *The Pseudo-Cyprianic De Pascha Computus* (London: S. P. C. K., 1955). German translation and commentary: Strobel, *Texte*, 43–67. A new edition of this text is being prepared by Alden Mosshammer.

[36] Schwartz, *Ostertafeln*, 37, referred to it as a "vortrefflichen Commentar zu den lakonischen Notaten Hippolyts in der Ostertafel."

stone that it appears highly likely that he was aware of the Hippolytan precedent. The fact that the Easter full moon dates in *De pascha computus* are consistently lowered by three days compared to the 'canon' suggests that the computist of 243 intended to correct the error that had accrued owing to the defects of the octaëteris since the beginning of the Hippolytan table in 222.

In parallel to the latter, the computist of 243 reserved much space for delineating the intervals between important biblical Passovers while using the 112-year cycle to calculate their calendar dates and weekdays. Although there is some variation between the intervals given by the North African computist and those chosen by Hippolytus, the events listed in *De pascha computus* remain basically the very same as those featured in the margins of the Hippolytan table and in the *Liber generationis*. Most of the calculations are based on the date of the very first Passover in history, which preceded the Israelite Exodus from Egypt. According to the computist, Passover in this year had been celebrated on Monday, 12 April, in the last year of the 112-year cycle. From this point he reckons 1579 years to the Passion of Christ in the eleventh year of his cycle (1579 = 14 × 112 + 11), and 1548 years to his nativity. The year of the latter in the 112-year cycle corresponded to 4 BC, meaning that the Exodus took place in 1552 BC.[37]

While assumptions concerning the creation date chosen by Hippolytus cannot go beyond informed guesswork, the author of *De pascha computus* makes some explicit statements about when and how the world began. The similarities to the views proposed for Hippolytus are again striking: according to the computist of 243, the world was created on 25 March, the date of the vernal equinox, while the moon began its course as a full moon on the fourth day, i.e. on Wednesday, 28 March. The latter date was of crucial importance to the computist, since he also regarded it as the date of Christ's birth. As he saw it, this parallel of calendar dates between the creation of sun and moon and the birth of the Messiah was no mere coincidence, but actually reflected the fact that Jesus was the "sun of righteousness" (Malachi 4:2).[38] The computist's intention to forge a typological link

[37] *De pascha computus* (7–9; 13–18), CSEL 3.3:253–56, 260–66.

[38] Ibid. (3–6; 19), CSEL 3.3:250–53, 266. There is a 76-year interval between the 36th year of the 112-year cycle, which contains the data that match the week of creation, and the 112th year, in which the Exodus took place. In the mind of our computist, this might have gone together with an interval of 34 × 112 + 76 = 3884 years

between creation and nativity via a full moon on Wednesday, 28 March, was strong enough to make him choose 4 BC as the year of Christ's birth, despite the fact that this is a deviation from both Hippolytus and most other patristic writers, who favoured a birth in 3 or 2 BC. If he had followed Clement of Alexandria and Hippolytus in assigning to Jesus a life-span of 30 completed years, the resultant crucifixion date should have fallen in AD 27. In this year, *luna* 14 coincided with Friday, 21 March, which was satisfactory from the point of view of a Johannine Passion chronology. Yet the computist opted for the following year (AD 28), in which *luna* 14 fell on Thursday, 8 April, implying that he favoured the synoptic chronology instead.[39] As Vicenzo Loi has plausibly suggested, this preference for the synoptic account, which is also found in Tertullian's *Against Marcion*, might reflect an exegetical tradition specific to North Africa.[40] In addition, the computist's decision to choose 9 April, AD 28, may in part have been influenced by the insight that AD 27 was unreasonably early, seeing how the 15th year of Tiberius (AD 28/29) was the *terminus post quem* provided by the Gospel of Luke. Extending the lifespan of Jesus from 30 to 31 years was a reasonable move, because it also left more room for the events of his public ministry. In accordance with this chronological decision, our computist marked the year of the crucifixion as the 16th rather than the 15th year of Tiberius.[41]

The crucial detail with respect to our assessment of the Hippolytan 'canon', however, is the North African computist's Julian calendar date for the crucifixion. With 9 April, he chose an altogether insignificant date, both from a typological point of view (i.e. compared to the date of creation or the vernal equinox) and judged from the fact that it spawned no known followers in subsequent Christian literature.

---

between both events. Together with the 1548 years between the Exodus and the birth of Christ, this makes for an interval of 5432 years. See George Ogg, "The Year of the Exodus in the Pseudo-Cyprianic *De Pascha Computus*," *The Expository Times* 60 (1948–49): 226. Richard, "Comput," pt. 1, 246, instead proposes that the computist put the creation and the birth of Christ in the same year of the cycle, assuming an interval of 5488 years and thus coming close to the 5500-year scheme used by Hippolytus. Grumel, *Chronologie*, 17, doubts that there was any specific creation date presupposed by *De pascha computus*.

[39] *De pascha computus* (9), CSEL 3.3:256.

[40] Loi, "Il 25 Marzo," 52–53.

[41] *De pascha computus* (20–22), CSEL 3.3:267–68. The 16th year of Tiberius is also asserted by Julius Africanus, *Chronographiae* (F 93), GCS-NF 15:282, and by Quintus Julius Hilarianus, *Expositum de die paschae et mensis* (15), PL 13:1114.

Indeed, it is safe to say that this date was a mere by product of the arithmetic structure of the octaëteris and was thus simply forced upon the computist, once he had decided on the year of crucifixion. By recognizing the Passion date of the *De pascha computus* as the result of pure computistical investigation, we can see more clearly how the Latin tradition of 25 March AD 29 could have originated from the octaëteris of the Hippolytan 'canon'.[42]

It is worth noting that these third-century efforts to date biblical Passovers by use of lunisolar cycles have not been appreciated by all scholars who wrote about the computistical work associated with the statue of Hippolytus. For Eduard Schwartz, who remains one of the most important authorities on the history of Easter calculations, the mentioned chronological approach was simply "foolish" (*thöricht*), while Friedrich Karl Ginzel, author of what is still the most comprehensive handbook on historical chronology, called Hippolytus's work a "hodgepodge" (*Machwerk*) of little value.[43] What both men were appalled by was the crudeness of the 8-year lunisolar cycle, which underlies the Hippolytan 112-year Easter table and which caused the displayed full moon dates to fall behind the astronomical moon by three full days in every 16 years. With such an error rate, the very idea of using this cycle to calculate full moon dates that go back several decades, let alone centuries or millennia, must appear delusive to any modern observer.

But let us grant for a moment that the 8-year cycle is an exact representation of astronomical reality (and let us also bear in mind that later Christian writers would use far more sophisticated cycles)—should it then not be possible to appreciate the rational impetus behind Hippolytus's approach? Unlike the chronographer, who adds up intervals between important events, the moon itself is infallible in its course about the earth and the resulting cycle of its waxings and wanings. Similarly, the correlations between dates in the Julian calendar and the day of the week never allow for any deviation from the rigid 28-year solar cycle which they describe. Lunisolar calendar cycles thus provide an excellent opportunity to generate new chronographic data, otherwise unknown, or to check upon already existing traditions, i.e.

---

[42] The thesis advanced by Strobel, *Ursprung*, 169, and Strobel, *Texte*, 66, according to which the computist intentionally reinstated a version of the original Passion date (7 April, AD 30), fails to convince.

[43] Schwartz, *Ostertafeln*, 29; Ginzel, *Handbuch*, 3:237.

to see whether they conform to the infallible course of the heavenly luminaries. Nowhere in Christian chronography was this idea more palpable and more important than in the case of Christ's crucifixion, for which the four canonical Gospels provided valuable data concerning its weekday and lunar age. To third-century Christian computists, this meant that there was clearly a limited number of days in history to which the crucifixion could be reasonably assigned. From their vantage point, the Easter cycle could thus be perceived as an ideal companion to the methods of the Christian chronographer, who could use it to improve and fine-tune his chronology of salvation.

We are thus face to face with a specific heuristical *modus operandi*, which I should henceforth like to refer to as 'computistical chronography' (as in the title of this chapter). In contrast to the disapproval expressed by Schwartz and Ginzel, I should like to claim that the Hippolytan method of finding past dates bears a close resemblance to the techniques of modern-day historians, who—to give just one example—can use the astronomical data on the statue reliably to translate the 1st year of the reign of Alexander Severus into our present system of dating, which enables us to estimate how long ago the inscription we look at was incised. The similarity of both approaches is no coincidence, but, as I shall endeavour to show, is owed to a genetic relation between them. As will become clear from the following chapters, the Hippolytan 'canon' is merely the starting point of a vibrant Christian scholarly tradition, which stretches throughout late antiquity and the earlier part of the Middle Ages. Over the course of this and the next two chapters, I shall trace the further development and subsequent ramifications of computistical chronography, up to its ultimate demise in twelfth-century Western Europe. As we shall see, this demise was only the beginning of fruitful new developments in the history of technical chronology.

2. *Julius Africanus, Annianus, and the Alexandrian 19-year Cycle*

At roughly the same time as Hippolytus forged his chronological system, an Eastern writer named Julius Africanus produced the founding text of Christian world chronography, the so-called *Chronographiae* (ca. 221). Modern scholars, from Joseph Scaliger to Heinrich Gelzer, have worked hard to reconstruct the contents of this highly important work from the few remaining fragments, most of which are preserved in the chronicles of Eusebius of Caesarea (ca. 325) and Georgius Syncellus (ca. 810). Similar to Hippolytus, Julius Africanus held that precisely five and a half millennia had separated the creation of Adam from the incarnation and birth of Jesus Christ, meaning that he dated these events to *annus mundi* or AM 5501. From the extant fragments, we can also conclude that Africanus believed the crucifixion to have taken place in the spring of the second year of the 202nd Olympiad (or Ol. 202.2), in what he designated as the 16th year of Tiberius. The Olympiad date strongly points to the spring of AD 31 (seeing how, according to the regular count, Ol. 202.2 began in the summer of AD 30), although this should have already been the 17th year of Tiberius's, if the latter's reign was counted, in regular fashion, from the autumn of AD 14.[44]

As Venance Grumel has observed, the year AD 31 has 25 March fall on a Sunday, which may well have been Africanus's intended date for the resurrection. From the extant remains of his *Chronographiae*, one can also infer that Africanus treated the day of the resurrection of Christ as the beginning of a new year of the world, as he seems to have put the Passion in AM 5531, whereas the resurrection, two days later, is already dated AM 5532. This indicates that Africanus, just like Hippolytus and the computist of 243, considered the world to have been created on 25 March and he may well have associated the same date with Christ's incarnation. These speculations are important, because they may help explain the origin of a dating tradition which has been just as ubiquitous in the Eastern churches as the dating of the crucifixion to 25 March has been in the West: the Passion on 23 March, implying a resurrection on 25 March.[45]

Grumel attempted to lend further credence to the hypothesis, according to which Africanus's intended Passion date was 23 March, AD 31, by showing how he could have arrived at this date by computistical means. It is indeed striking to observe that Africanus's world chronicle

[44] Gelzer, *Sextus Julius Africanus*, 1:19–52; Mosshammer, *Chronicle*, 146–57; Mosshammer, *Easter Computus*, 328–29, 385–421.

[45] On this date, see Grumel, *Chronologie*, 23, 28, 30, 91, 112, 121–24; Loi, "Il 25 Marzo," 60–62. Among the Byzantine sources attesting to a Resurrection on 25 March are: *Chronicon Paschale*, PG 92:537; Georgios Hamartolos, *Chronicon Breve* (112), PG 110:372; Georgios Kedrenos, *Historiarum compendium*, PG 121:372, 381; *Synaxarium Constantinopolitanum*, AASS 67:555–58; Matthaeus Blastares, *Syntagma alphabeticum*, PG 145:73; Nicephorus Gregoras, *Byzantina Historia* (8.13), ed. Ludwig Schopen, 3 vols. (Bonn: Weber, 1829–55), 1:371.

contained a lengthy exposition of the 'weeks of years'-prophecy in the book of Daniel, in which he covered both the octaëteris and the enneacaidecëteris as modes of lunisolar reckoning. Although he does not explicitly connect these cycles to Easter reckoning, the passage makes Africanus the earliest Christian author to ever mention the 19-year lunisolar cycle.[46] As noted by Grumel, calculations on the basis of the 'classical' Alexandrian 19-year cycle, as it has been in use since at least the fifth century, would indeed have lent some support to 23 March, AD 31, as the historical date of the crucifixion. In this year, the Alexandrian Easter full moon falls on Saturday, 24 March, making the preceding Friday correspond to *luna* 13. The fact that this is not in perfect conformity with the Johannine chronology may merely be taken to indicate that Africanus used a slightly different version of the 19-year cycle. However, it is also worth pointing out that Africanus's chronicle contained the peculiar statement that "the Hebrews celebrate the Passover on *luna* 14, and what happened to the Savior occurred on the day before the Passover."[47] While the statement that the crucifixion took place on the day before Passover reflects the Johannine chronology, the claim that the Jewish Passover corresponds to *luna* 14 would seem to direct the Passion date towards *luna* 13, which would have been precisely the lunar age on 23 March, AD 31, in the Alexandrian 19-year cycle, had the latter already existed in the early third century.[48]

While the specifics of Africanus's Passion chronology will probably remain a matter of speculation, we are on much firmer grounds when it comes to the reception and modification of his work at the hands of later Eastern writers. The resurrection date on 25 March played a crucial role in the chronological system underlying the lost chronicle of the Alexandrian monk Annianus (fl. 400), which later provided a template for the work of the ninth-century Byzantine chronicler Georgius Syncellus.[49] According to Annianus/Syncellus, the world was

[46] Julius Africanus, *Chronographiae* (F 93), GCS-NF 15:282.

[47] Ibid. (F 93), GCS-NF 15:276: Ἑβραῖοι γὰρ ἄγουσι τὸ πάσχα κατὰ σελήνην ιδʹ, πρὸ δὲ μιᾶς τοῦ πάσχα τὰ περὶ τὸν σωτῆρα συμβαίνει. Translation ibid. On the history of the 19-year cycle in Alexandria, see now Mosshammer, *Easter Computus*, 135–203.

[48] Grumel, *Chronologie*, 22–24; Mosshammer, *Easter Computus*, 418–20.

[49] See Georgius Syncellus, *Ecloga chronographica*, ed. Alden A. Mosshammer (Leipzig: Teubner, 1984). For the sake of convenience, I shall hereafter refer to the translation by William Adler and Paul Tuffin, *The Chronography of George Synkellos* (Oxford: Oxford University Press, 2002), but provide the page-numbers according to

created on Sunday, 29 Phamenoth/25 March, while Jesus was incarnated in the womb 5500 years later, on Monday, 25 March AM 5501, and rose from the dead on Sunday, 25 March AM 5534. Just as has been suspected for the chronicle of Africanus, Annianus's system thus uses 25 March as an epoch for its count of the years of the world. A further notable point of correspondence between both chronographies consists in the fact that they date the incarnation and birth of Christ to AM 5501. However, while Africanus had assigned to Jesus a lifespan of only 31 years, Annianus added two more years to the tally, dating the resurrection to the beginning of AM 5534 in the 19th year of the reign of Tiberius. Syncellus expressly faulted Africanus for his 'error' of two years, claiming that it was

> abundantly clear that starting about the beginning of his 30th year...he was baptized and began to teach and treat every disease and every infirmity over a period of three years. Therefore, from the time of his divine conception, beginning in AM 5501, on the first day of the first-created month of Nisan, 25 March, until his life-bringing Resurrection, which occurred on the same day, 25 March, at the beginning of AM 5534, the duration of time is thirty-three years and one day. And there are forty days from that day up to his divine and exalted bodily assumption into heaven. So from Adam up to this same day, there are 5533 years and 40 days.[50]

This criticism of Africanus, which may already have been present in Annianus's chronicle, was a palpable reaction to the chronicle of Eusebius, who had given special attention to the question of the duration of Jesus's public ministry. While the synoptic Gospels provided no clear chronological structure for the time between baptism and crucifixion, the repeated occurrences of Jewish feasts (2:13; 5:1; 6:4; 11:55) in the Johannine account suggested that this period stretched over roughly three years. Eusebius supplemented this exegetical point with a typological one, as he set the duration of the ministry to three and a half years, which was exactly one half of the last prophetic 'week of years' mentioned in the book of Daniel. So powerful was Eusebius's

---

Mosshammer's edition, as they are marked by Adler and Tuffin. For Syncellus's use of Annianus, see ibid., lxiii–lxix, lxxi, lxxiv–lxxv, 1–2, 34–37, 376–77, 381, 388–90, 395, 397. The two most important accounts of Annianus's (and Panodorus's) chronological system are Grumel, *Chronologie*, 83–97, and Mosshammer, *Easter Computus*, 198–203, 357–84. I have followed the latter wherever opinions diverge.

[50] Adler and Tuffin, *Chronography*, 394–95. See also Julius Africanus, *Chronographiae* (T 93b), GCS-NF 15:286.

defence of a 'long' chronology of the life of Jesus, that later chroniclers and scholars almost unanimously adhered to this view, first in the East, but from the sixth century onwards also in the Latin-speaking West.[51] Yet Annianus's chronicle was far more than a mere re-writing of the chronicle of Africanus, combined with some Eusebian insights. As Syncellus's dating of the incarnation to the consulate of Sulpicius Camerinus and C. Poppaeus (AD 9) makes clear, his chronicle proposed a radical shift of the life of Jesus Christ within the framework of ancient history. With AM 5501 = AD 9 as the year of the nativity, the resulting dates for the creation and the resurrection became 25 March, 5492 BC, and 25 March, AD 42, both of which were indeed Sundays. The rationale behind this step becomes apparent by looking at the corresponding lunar age, which on 25 March, AD 42, was *luna* 17, making the Friday of the Passion coincide with the synoptic *luna* 15.

This preference for the synoptic account fits well into the post-Nicene Alexandrian context, within which Annianus composed his chronicle. As mentioned in the previous chapter, early Christian scholars had generally favoured the Johannine chronology, which is already privileged in 1 Corinthians (5:7), where the death of Jesus is likened to the slaying of the sacrificial Passover lamb on the afternoon of 14 Nisan. For all the typological and symbolic attractiveness inherent to John's chronology, the synoptic version of events, testified to by three out of four evangelists, could not simply be ignored, which partly explains why the fourth century saw a notable swing of the exegetical pendulum towards a crucifixion on 15 Nisan. An important testimony is provided by the treatise *On the celebration of Easter* (Περὶ τῆς τοῦ Πάσχα ἑορτῆς) by Eusebius of Caesarea, who tried to harmonize the discrepancy between John and the synoptic Gospels by reverting to anti-Jewish polemic. According to the Johannine account, the Jews avoided the Roman *praetorium* on the day of the crucifixion, because they did not want to become unclean before "eating the *pascha*" (18:28). Since Jesus and his disciples had apparently held a Passover meal on the preceding evening, this implied that either the Jews or Jesus had sinned against the Mosaic precepts by not celebrating Passover on the right day. Unsurprisingly, Eusebius laid the blame on the doorstep

[51] Eusebius, *Chronicle*, GCS 47:174; Eusebius, *Historia ecclesiastica* (1.10), GCS 9.1:72–77; Eusebius, *Demonstratio Evangelica* (8.2.107), GCS 23:389. See also Ogg, *Chronology*, 98–103, 119–28; Richard, "Comput," pt. 2, 28–32; Strobel, *Ursprung*, 104–9; Mosshammer, *Easter Computus*, 337.

of the Jews, who, he claimed, had deliberately postponed the feast in order to be able to capture and kill Jesus.[52] That Eusebius's view was far from being completely idiosyncratic among fourth-century scholars can be seen from the Gospel commentaries of John Chrysostom (ca. 345–407), who upheld the idea that the Jewish temple priests changed the Passover date for their own sinister purposes. In fact, Chrysostom made this claim in the teeth of his own commentary on the Gospel of John, where he surmised that Jesus may have been the one who anticipated the meal, because he had foreseen his imminent death and desired to eat the Passover with his disciples one last time.[53]

The anti-Jewish readings proposed by Eusebius and Chrysostom made it possible to date the crucifixion to 15 Nisan while accepting the claim, implicit in John's Gospel, that the paschal lambs were slain on the afternoon of Christ's death. By the end of the fourth century, the synoptic date (15 Nisan) was officially endorsed by the Alexandrian church, whose Easter limits likewise prescribed that Easter Sunday could not fall earlier than *luna* 15. Theophilus, the patriarch of Alexandria (385–412), justified this rule in a letter to Emperor Theodosius (ca. 395), in which he pointed out that the Last Supper had taken place on *luna* 14, that the Passion had occurred on *luna* 15, and that Jesus had risen from the dead on *luna* 17. The same sentiment was repeated in the fifth century by the patriarch Proterius and was to become the dominant medieval tradition in the Latin West, once the Venerable Bede endorsed the same chronology in his highly influential *De temporum ratione*.[54] It is this Alexandrian preference for the synoptic chronology, which evidently formed the background for Annianus's re-calculation of the Passion date. Since 23 March, AD 31, the date chosen by Julius Africanus, corresponded to *luna* 13 rather than *luna* 15 in the Alexandrian 19-year cycle and since this year was deemed

[52] Eusebius, *De solemnitate Paschali* (9–12), PG 24:704–5. See Strobel, *Ursprung*, 24–27; Gerlach, *The Antenicene Pascha*, 272–75.

[53] John Chrysostom, *Homiliae in Matthaeum* (84.2), PG 58:754; Chrysostom, *Homiliae in Joannem* (83.3), PG 59:452.

[54] Theophilus of Alexandria, *De observatione sancte pasche*, ed. Krusch, *Studien* (I), 225; Proterius of Alexandria, ibid., 272; Bede, *De temporum ratione* (47; 61), CCSL 123B:432, 452. See also Wiesenbach, *Sigebert*, 69–72; Declercq, *Anno Domini*, 15–17; Olive M. Cullen, "A Question of Time or a Question of Theology: A Study of the Easter Controversy in the Insular Church" (PhD diss., St. Patrick's College, Maynooth, 2007), 170–78.

too early from the point of view of an Eusebian 'long' chronology of the life of Jesus, Annianus had to look for the closest year in the Easter cycle to contain the desired data. As it turned out, the next year in the vicinity, in which *luna* 15 coincided with Friday and 23 March, was AD 42, which was equivalent to AM 5534 in Annianus's chronological system. This conclusion automatically led to a shift of the incarnation/nativity to AD 9 and of the creation of the world (which had to have taken place 5500 years earlier) to 5492 BC. The resulting world era was eventually synchronized with the beginning of the Alexandrian civil calendar on 1 Thoth, thereby producing what is referred to in modern scholarship as the 'Alexandrian world era', starting on 29 August, 5493 BC.

As we have seen, Annianus thus built his entire chronology around the 19-year lunisolar cycle, which the Alexandrian church used to calculate the date of Easter. The merger of chronography and computistics went so far that the first year of Annianus's world era (AM 1) was simultaneously the beginning of a 19-year lunisolar cyce and—by consequence—a 532-year Easter cycle. This meant that the location of a given year of the world in the Easter cycle could be easily determined by dividing it by 19 or 532. While the early stages of the development of the Alexandrian 19-year cycle remain somewhat in the dark, Annianus's chronicle (as preserved by Syncellus) gives us the first definite glimpse of the 'classical' form of this cycle, which was to dominate medieval Easter reckoning. The first year of this version of the cycle corresponds to the first year of the regnal era of Diocletian (AD 284/5), which is separated by an interval 304 × 19 = 5776 years from the beginning of the world in Annianus's scheme. This is very probably more than a coincidence and it is reasonable to suppose, as Alden Mosshammer has done, that Annianus re-calibrated a previous version of the enneacaidecaëteris by tying its beginning to his own world era.[55]

It is has sometimes been assumed that the Alexandrian world era had previously already existed in another chronicle, written by the monk Panodorus, with whom Annianus was roughly contemporary.[56] Our only remaining bits of information on Panodorus's chronological system are Syncellus's critical comments on the latter, which indicate

[55] Mosshammer, *Easter Computus*, 190–203.

[56] Previous scholarship on this question is conveniently summarized ibid., 365–67.

that he made a sophisticated attempt to synthesize biblical and profane chronology, incorporating regnal data from the *Astronomical Canon*, otherwise known from the works of Claudius Ptolemy. According to Syncellus, Panodorus, "in his pursuit of agreement with the pagan scholars regarding the motion of the celestial spheres," reckoned 5493 instead of 5500 years between the creation and Christ.[57] Seeing how the Alexandrian world era can be plausibly explained as having been derived from a Passion date in AD 42 = AM 5533/34, it seems more probable that it was Panodorus who corrected Annianus's system, rather than vice versa. If Panodorus was as well-acquainted with pagan scholarship as Syncellus claims, he would have likely taken offence at the inconsistency of Annianus's system with the chronology of the Roman emperors, thereby providing him with a motive for intervention.[58]

From Syncellus's comments, it further appears that Panodorus moved the Passion from AM 5534 to 5525, which would mean that he chose the year AD 33.[59] Yet in this year neither 23 March nor 25 March fell on a Friday, nor was there a full moon close to either of these dates. This calendrical problem (added to the inelegance of a world era of 5493 years compared with Annianus's 5500) may explain why Panodorus's system failed to have any larger impact on subsequent Christian chronographers. Instead, he was largely eclipsed by Annianus, whose chronology was soon adapted into Eastern church tradition. By the 550s, we find a date according to the incarnation of Christ in the biography of St. Saba, written by Cyril of Scythopolis, which matches Annianus's incarnation date in AD 9. A century later, in about 640, Maximus the Confessor referred to Annianius's system as the "reckoning and the tradition of the Church," thereby indicating its widespread acceptance.[60] The influence of this system in the Byzantine East would last well until the ninth century, when it was still used by Georgius Syncellus and his continuator Theophanes Confessor. While it fell out of use in Byzantium after that time, it has remained the standard of Christian chronology in both the Coptic and

[57] Adler and Tuffin, *Chronography*, 377–78. See also ibid., 34–37, 42, 89, 396–97.

[58] See Mosshammer, *Easter Computus*, 368–71.

[59] Adler and Tuffin, *Chronography*, 36. Syncellus claims that Panodorus chose 20 March = 20 Phamenoth as the date of the Passion, but this date is clearly corrupt.

[60] Cyril of Scythopolis, *Vita Sabae*, ed. Eduard Schwartz (Leipzig: Hinrichs, 1939), 183; Maximus Confessor, *Computus ecclesiasticus* (17), PG 19:1233.

Ethiopic Churches until the present day.[61] The presence of Annianus in Ethiopia is not surprising, seeing how the country received Christianity via Egypt, which is reflected in the fact that the Ethiopic liturgical calendar is a replica of the Alexandrian calendar and makes use of the same Easter cycle. A striking case in point is the Ethiopic 'era of Grace' (*mehrat*), which begins in AD 360/1 and thus coincides with the start of a new 532-year cycle, the twelfth since the beginning of the Alexandrian world era and the same one in which Annianus composed his work.[62]

Ethiopic sources may also help us form a better idea of what the chronicle of Annianus could have looked like. According to Syncellus, Annianus composed an Easter table for 11 × 532 years, "along with accurate scholia."[63] This note makes Annianus the earliest known author to employ the 532-year 'great' Easter cycle, after which all Easter Sundays return to the same dates in the Julian calendar. The scarcity of witnesses for this cycle in late antique sources may be less a sign of ignorance, but rather a reflection of the fact that its excessive length, which encompassed numerous generations of men, made it impractical for conventional usage, which is why lists of 95 or 100 years were preferred. A cycle of 532 years, however, was an excellent tool for structuring a world chronicle, which by definition encompassed the history of several millennia. Indeed, it appears that Annianus's Easter table was more than just an addendum to an otherwise conventional work of chronography. From Syncellus's words, it cannot be ruled out that the Easter table represented the actual core of Annianus's work, whose margins contained 'scholia' or historical notes, not unlike the Hippolytan 'canon'. It is the latter principle, which is often found in Ethiopic sources, where 532-year Easter tables contain a special column for historical events, called *tārik* ('history').[64]

[61] Neugebauer, *Ethiopic Astronomy*, 8, 23, 56–66. On Annianus's influence on Syriac chronography, see Daniel Serruys, "Les Canons d'Eusèbe, d'Annianos et d'Andronicus d'après Élie de Nisibe," *Byzantinische Zeitschrift* 22 (1913): 16–28; Kevin van Bladel, *The Arabic Hermes* (Oxford: Oxford University Press, 2009), 139–47.

[62] Otto Neugebauer, "Abū-Shāker and the Ethiopic Hasāb," *Journal of Near Eastern Studies* 42 (1983): 58; Neugebauer, *Ethiopic Astronomy*, 116–20.

[63] Adler and Tuffin, *Chronography*, 382. Ibid., 35: "That this day of the Resurrection was also the first-formed day he has demonstrated in the Paschal tables of 532 years that he compiled together with some learned observations."

[64] Neugebauer, *Ethiopic Astronomy*, 67; Neugebauer, *Chronography in Ethiopic Sources* (Vienna: Verl. der Österr. Akad. der Wiss., 1989).

Interestingly, Ethiopic tables often follow the same basic chronological structure of world history as Annianus's and Syncellus's chronicles. One important difference concerns the beginning of the Ethiopic world era, which starts on 1 Meskerem/Thoth (= 29 August), 5943 BC, rather than from 25 March, 5492 BC. This shift was obviously owed to a synchronization of the Annian world era with the beginning of the Ethiopic/Alexandrian calendar year.[65] It is this same shift which explains why Ethiopic sources date the birth of Jesus to AD 8 as opposed to AD 9, thereby extending his lifespan by one year. The discrepancy is easily accounted for once we consider that the actual year for the incarnation/nativity found in the Annian chronicle was not 'AD 9' but the year 5501 of the Alexandrian world era. Thanks to the earlier start of the Ethiopic world era, Christmas of AM 5501 would now already fall in the preceding year, i.e. AD 8, whereas the original Annian year AM 5501 only began in the following spring. Aside from this discrepancy, the Annian and the Ethiopic chronologies for the life of Jesus are one and the same.[66]

3. *The Chronicon Paschale and the Byzantine World Era*

For all its lasting impact on Christian Northeast Africa, it is not difficult to see why the Annian or Alexandrian system eventually fell out of favour in Constantinople and other places, where the classical heritage was more vigorously cultivated. For scholars mindful of the chronology of the Roman emperors, it must have been difficult to accept that the 19th year of Tiberius, in which Jesus was crucified according to Annianus and Syncellus, should have corresponded to the year we now call AD 42, by which time Claudius had long succeeded to the principate. It is not clear if Annianus was actually ignorant of this conflict of chronologies, or if he willingly rejected profane sources that gave countervailing evidence. In any case, the core data on which Annianus based his work were the ecclesiastical tradition that Jesus rose from the dead on 25 March, the synoptic Gospel tradition that this was on a Sunday and on *luna* 17, and the 532-year Easter cycle, which

[65] Neugebauer, *Ethiopic Astronomy*, 77–78, 115–20.

[66] See Marius Chaîne, *La Chronologie des temps chrétiens de l'Égypte et de l'Éthiopie* (Paris: Geuthner, 1925), 11; Otto Neugebauer, *Abu Shaker's "Chronography"* (Vienna: Verl. der Österr. Akad. der Wiss., 1988), 120; Neugebauer, *Ethiopic Astronomy*, 171.

told him in which year these parameters would coincide. It is very likely that, in Annianus's eyes, these considerations enjoyed greater epistemic authority than the claims of profane historical records. As his case shows, and as we shall see again in the case of the 'critical computists' of the High Middle Ages, computistical chronography can quickly turn into a Procrustean bed for recorded history.

In any case, the chronological strain created by Annianus's system could hardly have been ignored by subsequent users of his chronicle and by the seventh century, Byzantine chronography was ready for a return to the more reasonable dates once proposed by Julius Africanus. Our most important witness for this development is the so-called *Chronicon Paschale*, whose present version was redacted in 629/30.[67] As its name implies, paschal computations were of special importance to its author, who remains anonymous. Just as had been the case with Annianus and his Alexandrian world era, the *Chronicon Paschale* let the first year of the world coincide with the beginning of a 532-year cycle. Its year count commenced on Wednesday, 21 March, 5509 BC, which was meant to correspond to the fourth day of creation, on which the heavenly luminaries had started their perennial course. The decision to place the creation of the world 17 years earlier than Annianus, and thereby create a new version of the 532-year cycle, was not the result of computistical whim, but evidently arose from a reconsideration of the dates of Christ's crucifixion and resurrection. The author of the *Chronicon Paschale* decided to date these events to the year AM 5540, which means that he acquiesced with Julius Africanus in putting them on 23 and 25 March, AD 31.[68]

He even managed to bring these dates in line with the Johannine chronology by means of a brilliant technical argument: knowing that the Easter full moon (*luna* 14) of the Alexandrian 19-year cycle fell on Saturday 24 March in AD 31, he proposed that the 'leap of the moon' and the lunar bissextile days, which were intercalated in certain years of the 19-year cycle, should be divided into fractions or *lepta* and equally distributed over the 235 lunar months of the cycle. This made a great deal of sense, seeing that the length of the individual lunar months was meant to be equal and only the conventions of the calendar, where the

---

[67] On the following, see Joëlle Beaucamp et al., "Temps et histoire I: Le prologue de la Chronique Pascale," *Travaux et mémoires* 7 (1979): 223–301; Mosshammer, *Easter Computus*, 286–311.

[68] *Chronicon Paschale*, PG 92:529–41.

day was the basic unit, made it necessary to discern 'hollow' (29-day) from 'full' (30-day) months. By assigning an equal number of *lepta* to each month and by measuring their increment, our computist was able to show that by the eleventh year of the cycle, in which the Passion had taken place, the additional fractions had accrued to a full day, meaning that the real lunar age on 23 March was already *luna* 14.[69] Like Hippolytus and the computist of 243 (and presumably also Annianus), the author of the *Chronicon Paschale* applied his computistical method to far more dates than just the death and resurrection of Jesus. As he announced in the preface to his chronicle, his work included calculated Julian calendar dates and weekdays for the Passover preceding the Exodus, for the day on which the angel announced the conception of John the Baptist to the priest Zechariah, for the incarnation of Christ, and for his baptism in the river Jordan.[70]

If it had been his main intention to safeguard the traditional 5500-year scheme, it might have sufficed for him to begin his world era in 5503 BC, eleven years earlier than Annianus. However, his chosen starting point in 5509 BC had the great advantage of having 21 March fall on a Wednesday. Since 21 March was considered the date of the vernal equinox in the Alexandrian tradition of Easter computation, this meant that the sun had been created at the cardinal point of the year, when night and day were of equal length. The author of the *Chronicon Paschale* responded to this felicitous coincidence by making AM 1 = 5509 BC the beginning of a 19-year lunisolar cycle (and also of a 532-year Easter cycle), which began two years later compared to the one used by Annianus and the Alexandrians. In addition to all of this, 1 September, 5509 BC, was also the beginning of a new cycle of indictions, meaning that the 532-year Easter cycle and the 15-year indictional cycle could be combined to form a 7980-year cycle, starting from that very date, which is today commonly known as the epoch of the Byzantine world era. This world era had a great impact on the Byzantine world as an official means of dating, its first documented use being found in a document of the Trullan Synod, which convened in 691/92. Since the *Chronicon Paschale* contains an earlier version of this world era, beginning in spring 5509 BC, this has sometimes been

---

[69] See Arthur Mentz, *Beiträge zur Osterfestberechnung bei den Byzantinern* (Königsberg: Leupold, 1906), 50–59; Mosshammer, *Easter Computus*, 302.

[70] *Chronicon Paschale*, PG 92:92, 224–25, 480–84, 513.

dubbed the 'proto-Byzantine' era, yet it seems clear that the Byzantine era proper can be directly traced back to that of the *Chronicon*.[71] Aside from an early 'experiment' in the work of a certain "George, monk and presbyter" (638/39), who combined the new world era with the incarnation and Passion dates of Annianus (putting them in AM 5517 = AD 9 and AM 5550 = AD 42 respectively), the Byzantine era has consistently been associated with a Passion date on 23 March, AD 31, for most of its history. In some parts of Europe, such as Russia, it remained in official use until the early modern period.[72]

---

[71] Mosshammer, *Easter Computus*, 311–16.

[72] Franz Diekamp, "Der Mönch und Presbyter Georgios," *Byzantinische Zeitschrift* 9 (1900): 24–32, 44–50; Grumel, *Chronologie*, 116–17; Mentz, *Beiträge*, 59–61. For further information, see Daniel Serruys, "Des quelques ères usitées chez les chroniqueurs byzantins," *Revue de Philologie*, n.s., 31 (1907): 151–89; Ginzel, *Handbuch*, 3:288–98; Anthony Bryer, "Chronology and Dating," in *The Oxford Handbook of Byzantine Studies*, ed. Elizabeth Jeffreys, John Haldon, and Robin Cormack (Oxford: Oxford University Press, 2008), 31–37.

CHAPTER THREE

THE CRISIS OF COMPUTISTICAL CHRONOGRAPHY IN THE EARLY MIDDLE AGES

1. *The Passion Date in the Latin West and the Era of Dionysius Exiguus*

As the previous chapter has shown, Eastern Christian attempts to calibrate the chronology of salvation by using computistical methods were mostly based on a 19-year cycle that probably originated in Alexandria, the earliest version of which was created by Anatolius of Laodicea in the 270s.[1] Compared to the octaëteris-based 112-year cycles that have come down to us from the early decades of the third century, the 19-year cycle or enneacaidecaëteris was a considerable astronomical improvement. Where the octaëteris, with its crude equation of 99 lunations and 2922 days, fell behind the actual lunations at a rate of approximately 1.53 days in every eight years, the latter managed to reduce this error to just one day in roughly every three centuries. Things looked somewhat different in the Latin-speaking West, where the 112-year cycle was first succeeded by a 84-year cycle, based on an equation of 1039 months with 84 Julian years or 30681d, which—owing to its endorsement by the Roman Church—is also known as the *Supputatio Romana.* Although this lunisolar cycle was significantly less accurate than that of the Alexandrians (its error rate being approximately 1.28 days per cycle), its 84-year form had the distinct advantage of being a multiple of the 28-year solar cycle, which marks the return of the weekdays on the same dates in the Julian calendar. As a result, the *Supputatio Romana* was a true 'Easter cycle', in so far as it encompassed not only a full cycle of the calendrical Easter full moons, but also that of the corresponding Easter Sundays.[2] By

[1] For an edition and study of the Latin *De ratione paschali,* which is probably based on Anatolius's original treatise, see Daniel McCarthy and Aidan Breen, *The Ante-Nicene Christian Pasch* (Dublin: Four Courts Press, 2003).

[2] Schwartz, *Ostertafeln,* 40–58, 89–104; Ginzel, *Handbuch,* 3:238–45; Holford-Strevens, "Paschal Lunar Calendars," 173–87; Mosshammer, *Easter Computus,* 204–39.

contrast, the Alexandrian enneacaidecaëteris had to be expanded into an unwieldy 532-year cycle, which made the production and dissemination of tables covering a full Easter cycle much less convenient.

As in the case of the 19-year cycle, it is worth asking in what way the switch from the octaëteris to the 84-year cycle affected the dating of Christ's Passion among the churches of the West. As we have seen, the Hippolytan date of the crucifixion had been Friday, 25 March, AD 29, with an Easter full moon (*luna* 14) on that same day. By contrast, the *Supputatio Romana* indicated a full moon on Tuesday, 22 March, for all years which cyclically corresponded to AD 29.[3] The implication was that the lunar age on 25 March, the putative date of the crucifixion, had been *luna* 17, which is significantly outside the range permitted by the Gospel texts. Only in the 57th year of the 84-year cycle, which cyclically corresponds to years AD 18, 101, 184, etc. could one find a combination of *luna* 15, 25 March, and Friday, which would have conformed to the synoptic chronology. Such a date is indeed assigned to the Passion of Jesus in the *Cologne Prologue*, a presumably late fourth-century preface to an 84-year cycle, named after a famous ninth-century computistical manuscript in the Cologne cathedral library, in which it is uniquely preserved. The *Prologue* dates the crucifixion to 25 March and *luna* 15, notwithstanding the fact that the 84-year cycle originally attached to it did not allow for such a combination in AD 29 or any of the surrounding years.[4]

By the fifth century, the intrinsic deficiencies of the 84-year cycle, as well as the chronological problem just outlined, began to inspire pious attempts to modify the traditional *Supputatio Romana*. Our most important witness to this development is a North African computist, writing in 455, whose work is known as the *Computus Carthaginiensis*. One characteristic feature of this text is its use of two distinct versions of the 84-year cycle, referred to as the *circulus primus* and *secundus* respectively. The starting year of the *circulus secundus* is AD 439, the 58th year of the conventional Roman *Supputatio*, with an Easter full moon on Wednesday, 12 April. According to the computist

[3] Here and elsewhere I follow the Roman 84-year cycle as reconstructed by Mosshammer, *Easter Computus*, 209–13.

[4] MS Cologne Erzbischöfliche Diözesan- und Dombibliothek, 83^{II}, 193v–97r. See the edition (as *Prologus paschae*) in Krusch, *Studien* (I), 227–35 (especially pp. 228, 230, 232). See also ibid., 32–37. The Cologne Prologue contains chronographical elements suggesting that there may have been a connection between the 84-year cycle and the world era used by its author. See Grumel, *Chronologie*, 18–22.

of 455, this year is cyclically related to the year of the Exodus, as they are separated by exactly 25 × 84 = 2100 years. The deliberate shift of the 84-year cycle's starting point by 57 years, which characterizes the *circulus secundus*, did not leave unaffected the ages of the moon on individual dates of the Julian calendar. Since the 84-year cycle has a fixed sequence of embolisms and 'leaps of the moon' (*saltus lunae*), which were now displaced compared to the traditional *Supputatio Romana*, most of the Easter full moons of the *circulus secundus* are found to occur one day later than in the old cycle. The same modified sequence of epacts also formed the basis of the *circulus primus*, which is identical to the *circulus secundus* except that it begins with AD 449, designated as the tenth year of the Vandal king Geiseric. As the computist himself indicates, this year was chosen because it is a cyclical equivalent to AD 29 (29 + 5 × 84 = 449), the commonly accepted year of the Passion. Thanks to the mentioned changes in the lunar epacts, both the *circulus primus* and *secundus* have Friday, 25 March, AD 29, coincide with *luna* 16 instead of *luna* 17, as had been the case in the old *Supputatio Romana*. This result was somewhat closer to the Gospel data, although still not fully satisfactory.[5]

In addition to his own computistical innovations, the North African author also refers to the work of a certain Augustalis, whose cycle is said to have begun in AD 213, which is the 84th year of the *Supputatio Romana*. In this year, *luna* 15 falls on Friday, 26 March, which is another close call compared to the Hippolytan Passion date. Augustalis designated his starting-year year as the 186th year since the Passion, which implies that Jesus was crucified in AD 28. However, as the computist of 455 correctly points out, AD 213 cannot count as a cyclical equivalent to AD 28, since both dates are not separated by a multiple of 84 years.[6] A far more coherent attempt to reform the 84-year cycle was undertaken around the same time by the author of another Easter

[5] *Computus Carthaginiensis*, ed. Krusch, *Studien* (I), 279–97. See also ibid., 164–77; Schwartz, *Ostertafeln*, 67–69; Ginzel, *Handbuch*, 3:243–45; Strobel, *Ursprung*, 137, 270–74; Immo Warntjes, "The Munich Computus and the 84 (14)-year Easter Reckoning," *Proceedings of the Royal Irish Academy* 107C (2007): 69–71; Mosshammer, *Easter Computus*, 217–20, 224–28.

[6] *Computus Carthaginiensis*, ed. Krusch, *Studien* (I), 289–90. Krusch (ibid., 5–10) claims that Augustalis's date of Easter Sunday in the first year of his cycle was 28 March, *luna* 16. However, the computist of 455 clearly attests to *luna* 17, which is in harmony with the Roman *Supputatio*. See now also Alden A. Mosshammer, "The Computus of 455 and the Laterculus of Augustalis, with an Appendix on the Fractional Method of Agriustia," in *The Easter Controversy of Late Antiquity and the Early*

table, now uniquely found in a fragmented fifth-century manuscript, preserved at the Stiftsbibliothek in the East German town of Zeitz (hence the name *Zeitzer Ostertafel*). Once again, special significance was attached to the date of the Passion, as becomes apparent from the fact that the Zeitz Easter table begins in AD 29, the consulate of the Gemini, and stretches over four consecutive 84-year cycles. Compared to the original Roman *Supputatio*, the epacts are consistently lowered by two days, which brought the Easter full moons of the Zeitz table closer to the Alexandrian reckoning and to astronomical reality. In addition, 25 March in AD 29 now coincided with *luna* 15, thereby finally conforming to the synoptic Passion chronology.[7]

The production of a reformed 84-year table in the middle of the fifth century, as attested by the Zeitz fragments, has to be viewed in the context of ongoing disputes between Rome and Alexandria over the correct method of calculating the date of Easter. Besides the differing computational frameworks (the 84-year cycle used in Rome and the West vs. the more refined Alexandrian 19-year cycle), the most controversial issue concerned the permitted calendrical limits for Easter Sunday. Ever since the construction of the Hippolytan 'canon' in the early third century, Roman Easter reckoning had been based on the idea that Easter Sunday could not fall earlier than *luna* 16 and not later than *luna* 22, whereas the Alexandrian limits stretched from *luna* 15 to 21. In addition, the Roman Church found it necessary to restrict Easter to the period before 21 April, the date of the Parilia, on which Romulus's foundation of the City was traditionally celebrated. Since these Roman festivities were accompanied by circus games and other merriments, an interference of the Parilia with the highest Christian feast day was to be avoided. By contrast, the Alexandrian Easter could be celebrated as late 25 April, which meant that in some years the

---

Middle Ages, ed. Immo Warntjes and Dáibhí Ó Cróinín (Turnhout: Brepols, forthcoming).

[7] *Liber paschalis codicis Cizensis*, in *Chronica minora*, 1:507–8. See also Theodor Mommsen, *Zeitzer Ostertafel vom Jahre 447* (Berlin: Dümmler, 1863); Krusch, *Studien* (I), 116–23; Krusch, *Neue Bruchstücke der Zeitzer Ostertafel vom Jahre 447* (Berlin: de Gruyter, 1933); Schwartz, *Ostertafeln*, 71–72; Eef Overgaauw and Frank-Joachim Stewing, *Die Zeitzer Ostertafel aus dem Jahre 447* (Petersberg: Imhof, 2005); Eef Overgaauw, "Auseinandersetzung um den Ostertermin: Die Berliner und Zeitzer Fragmente der Zeitzer Ostertafel (447)," in *Handschriften und frühe Drucke aus der Zeitzer Stiftsbibliothek*, ed. Frank-Joachim Stewing (Petersberg: Michael Imhof, 2009), 14–17; Warntjes, "The Munich Computus," 71.

dates of Easter Sunday, as calculated by both churches, were separated by a whole month.[8]

This was the case in 444, when the computistical differences with Alexandria prompted Pope Leo I to seek epistolary advice from Paschasinus, who was then bishop of Lilybaeum in Sicily. In his reply to Leo, Paschasinus openly advertised the Alexandrian 19-year cycle as the only sound approach to Easter reckoning, referring to it as the "calculation of the Hebrews, that is the calculation according to the [Mosaic] law" (*Hebraeorum, hoc est legalem supputationem*).[9] The pressure exerted by the Alexandrian patriarchs and their supporters eventually compelled the Roman episcopal see to review its traditional practice of Easter reckoning. By 457, Leo's archdeacon and later successor Hilarus had commissioned the acclaimed mathematician Victorius of Aquitaine to inquire into the problem. Victorius reacted by composing a 532-year Easter table, based on the 19-year cycle of the Alexandrians, which covered the years AD 28 to 559. Although Victorius's *Cursus paschalis* was Alexandrian on the surface, it contained a number of elements which prevented a full harmonization between Eastern and Western calculations. Unwilling completely to abandon the traditional Roman limits, his table eschewed Easter dates on 25 April and generally listed double dates when the 'Greek' computation called for Easter Sunday on *luna* 15, whereas the 'Latins' would have postponed the feast to *luna* 22.[10]

An even more serious problem arose from Victorius's noteworthy attempt to connect the Easter computus to the larger chronographic framework of the history of salvation. As his template, he chose the work of his compatriot Prosper of Aquitaine, who had written a continuation to the chronicle of Eusebius, which circulated in the West in a Latin translation, made by Jerome in ca. 382. Besides bringing the latter chronicle up to the year 455, Prosper also introduced some new

[8] On the Easter conflict between Rome and Alexandria, see Schwartz, *Ostertafeln*, 50–58; Jones, *Bedae Opera*, 26–28, 55–61; Blackburn and Holford Strevens, *Oxford Companion*, 792–94, 807–8; Declercq, *Anno Domini*, 54–60, 72–82.

[9] Paschasinus, *Epistola ad papam Leonem*, ed. Krusch, *Studien* (I), 248. For an illuminating discussion of Paschasinus's letter, see Leofranc Holford-Strevens, "Church Politics and the Computus: From Milan to the Ends of the Earth," in Warntjes and Ó Cróinín, *Easter Controversy* (forthcoming).

[10] For details, see Schwartz, *Ostertafeln*, 73–80; Jones, *Bedae Opera*, 61–67; Pedersen, "The Ecclesiastical Calendar," 46–49; Wallis, *Bede*, l–lii; Blackburn and Holford-Strevens, *Oxford Companion*, 793, 808–9; Declercq, *Anno Domini*, 82–95; Declercq, "Dionysius Exiguus," 181–87; Mosshammer, *Easter Computus*, 239–44.

elements. In particular, he took exception to the 'long' three-year duration of the public ministry of Jesus previously defended by Eusebius, which separated Christ's baptism in the 15th year of Tiberius from his crucifixion in the 18th year. In keeping with the canonical Western date for the crucifixion, Prosper made both events fall in the same year, equating the latter with the consulate of the Gemini. In addition, Prosper made this year the starting point of a year-by-year listing of consulates, which he supplemented with a count of years since the Passion, leading up to the consulate of Valentinian III and Anthemius in AD 455, which was numbered 428. This correlation implied that the consulate of the Gemini took place in the year AD 28, a manifest error, which can be attributed to the use of a corrupt consular list.[11]

Victorius eagerly incorporated the chronological framework provided by Prosper's chronicle into his *Cursus paschalis*. In choosing the starting point of his 19-year cycle, he did not follow standard Alexandrian usage, but instead sought to tie it to the year of creation, which according to Prosper's chronicle had taken place 5228 years before the crucifixion in AD 28, i.e. in 5201 BC. In this year, the Easter full moon had fallen on Thursday, 29 March, which implied that the creation had begun on 25 March, the traditional date of the vernal equinox. As we have seen, versions of the same chronology of the hexaëmeron had previously been found attractive by other Christian chronologers such as Hippolytus, the computist of 243, and Annianus. In addition to the creation of the world, Victorius also provided a Julian date for the first Passover before the Exodus (Friday, 25 March, AM 3690) and for Christ's Passion in the year AM 5229 = AD 28. In the latter year, the Easter full moon had occurred on Friday, 26 March, a date reasonably close to the canonical 25 March, AD 29.[12]

---

[11] Prosper, *Chronicon*, in *Chronicaminora*, 1:409–10: "Quidam ferunt anno XVIII Tiberii Jesum Christum passum, et argumentum huius rei ex evangelio adsumunt Iohannis, in quo post XV annum Tiberii Caesaris triennio dominus praedicasse intellegatur. Sed quia usitatior traditio habet dominum nostrum XV anno Tiberii Caesaris duobus Geminis consulibus crucifixum, nos sine praeiudicio alterius opinionis successiones sequentium consulum a supra scriptis consulibus ordiemur, manente adnotatione temporum quae cuiusque imperium habuit." This passage is based on Eusebius, *Chronicle*, GCS 47:173–74. On Prosper's chronicle, see also Mark Humphries, "Chronicle and Chronology: Prosper of Aquitaine, his Methods and the Development of Early Medieval Chronography," *Early Medieval Europe* 5 (1996): 155–75.

[12] Victorius of Aquitaine, *Cursus Paschalis* (8–9), ed. Krusch, *Studien* (II), 23–25. Victorius (ibid., 24) states that 5228 world years had been completed (*peractus*) at the Passion. As his other chronological statements show, this should be understood as

Unfortunately, Victorius's desire to merge computus and chronography into a coherent system led him to modify the 19-year cycle in a way that seriously marred his project of reconciling the computational traditions of Rome and Alexandria. By choosing the year of creation as the natural starting point of the 19-year lunisolar cycle, he shifted the beginning of this cycle in such a way that it corresponded with the seventh year of the Alexandrian cycle. This in itself would not have been a great problem, had Victorius left the 'leap of the moon' or *saltus lunae* in its accustomed place. Yet according to his reasoning, the leap of the moon could only take place in the last year of a 19-year sequence. The resulting displacement of the *saltus* by six years vis-à-vis the original cycle meant that the Victorian lunar ages differed from the Alexandrian ones by one day in no less than 13 out of 19 years. Bearing these problems in mind helps explaining why Victorius's 532-year table, despite its widespread use during the ensuing three centuries, did not remain the standard of Easter reckoning of the Roman Church.[13] Instead, his work was eventually superseded by the Easter table of the Scythian monk Dionysius Exiguus, who, in 525, tabulated Easter data for 95 years, from AD 532 to 626. This latter table was little more than the Latin continuation of a previous 95-year table (AD 437 to 531) that Dionysius ascribed to the Alexandrian patriarch Cyril.[14]

It is worth pausing for a moment to analyse the chronological consequences that resulted from the introduction of the Alexandrian 19-year cycle in the Latin West. Up until the spread of the Dionysiac Easter table and its continuations, Western computists had attempted, in one way or another, to connect their Easter cycles to the Hippolytan crucifixion date of 25 March, AD 29, despite the fact that this sometimes posed difficulties. As we have seen, the Alexandrian 19-year cycle was most reasonably combined with a Passion on 23 March, AD 31,

meaning that AM 5229 = AD 28. It should be noted that Victorius dated the Passion to the consulate of the Gemini, so for him there was no noticeable difference between the traditional year (AD 29) and his own Passion year (AD 28), although Prosper's chronicle technically led him to misdate this consulate.

[13] The Victorian table was formally adopted for Gaul by the Synod of Orléans in 541. See *Concilium Aurelianense* (1), CCSL 148A:132. On its reception in the West, see Charles W. Jones, "The Victorian and Dionysiac Paschal Tables in the West," *Speculum* 9 (1934): 408–21; Jones, *Bedae Opera*, 65–68; Warntjes, *Munich Computus*, xxxviii–xli.

[14] Jones, *Bedae Opera*, 68–75; Pedersen, "The Ecclesiastical Calendar," 49–54; Wiesenbach, *Sigebert*, 47–52; Declercq, *Anno Domini*, 97–112; Declercq, "Dionysius Exiguus," 187–209; Mosshammer, *Easter Computus*, 5–9, 59–106.

the date probably first chosen by Julius Africanus and later favoured by Byzantine chroniclers. In the case of Africanus, this date had been tied to a version of the 'short' chronology of the life of Jesus, which assigned the Passion to the 16th year of Tiberius, when Jesus was 31. Accordingly, Africanus must have put the nativity of Christ in the year we today refer to as 1 BC. As Alden Mosshammer pointed out most recently, there is thus a conspicuous and probably telling affinity between Africanus's nativity date and the *Annus Domini* or Common Era that we still use today.[15] The latter is traceable back to the aforementioned Easter table of Dionysius Exiguus, where one column was reserved for numbering the years "from the incarnation of our Lord." In creating this column, Dionysius abandoned the Alexandrian usage of counting years in Easter tables according to the regnal era of the emperor Diocletian—a custom that displeased the Scythian monk owing to the latter's reputation as a fierce persecutor of Christians.

The precise rationale for Dionysius's equation of the 248th year of the Diocletian era with the year 532 "from the incarnation" has long puzzled modern scholars and will probably never be explained to everybody's satisfaction.[16] Even so, it is clear that Dionysius's new era, once it began to find acceptance in the Latin West, put additional constraints on the range of Passion dates that could be legitimately chosen out of the raw data provided by the Gospels and the 19-year cycle. In particular, it meant that the widespread tradition of assigning the Passion to the consulate of the Gemini in AD 29 (in which the Easter full moon fell on Friday, 15 April, according to the Alexandrian cycle) could no longer be upheld. As we have seen in our analysis of the Hippolytan 'canon', a Passion in AD 29 could only be reasonably combined with a nativity in 2 BC or earlier. After all, the Gospel of Luke had set an important lower limit for the age of Jesus at the time of his death by remarking that he was "about thirty years old" at the time of his baptism. Even the shortest possible estimate of the duration of Christ's public ministry thus had to respect the fact that Jesus was at

[15] Mosshammer, *Easter Computus*, 340, 420–437.

[16] Besides Mosshammer's theory, noteworthy attempts to explain the era's origins have included Gustav Oppert, "Über die Entstehung der Aera Dionysiana und den Ursprung der Null," *Verhandlungen der Berliner Gesellschaft für Anthropologie, Ethnologie und Urgeschichte* 32 (1900): 102–36; Declercq, *Anno Domini*, 112–47; Declercq, "Dionysius Exiguus," 230–42; Daniel McCarthy, "The Emergence of *Anno Domini*," in *Time and Eternity: The Medieval Discourse*, ed. Gerhard Jaritz and Gerson Moreno-Riaño (Turnhout: Brepols, 2003), 31–53. For further opinions on the matter, see the overview in Mosshamer, *Easter Computus*, 339–56.

least in his 30th year (and probably much older) when he died. Yet this lower limit could not be properly squared with a Passion in the spring of AD 29, if the *Annus Domini* was supposed to be a reliable reflection of Christ's nativity date. Depending on whether Jesus was thought to have been born on 25 December in 1 BC or AD 1, a Passion in AD 29 would have led to a lifespan of just over 28 or 27 years respectively.

By default, the next best option for Western computists would have been to opt for the existent Greek tradition and date the Passion to 23 March, AD 31. Aside from the fact, however, that the lunar age on 23 March (*luna* 13) was not in exact conformity with the parameters posed by the Gospels, the resultant lifespan of Jesus was still considerably too small if the Eusebian 'long' chronology of the life of Jesus was accepted. While Prosper of Aquitaine had still objected to the notion that Christ was crucified as late as the 18th year of Tiberius's reign, the 'long' chronology had begun to establish itself by the early sixth century, as witnessed by the chronicle of Cassiodorus (ca. 519), who dated the birth of Christ to 3 BC and his death to AD 31. Clearly, this chronology could not be reconciled with the incarnation era introduced by his friend and contemporary Dionysius Exiguus.[17] Worse even, none of the ensuing years yielded a satisfactory combination of data, since neither *luna* 14 nor *luna* 15 fell on a Friday in these years. During the relevant year range of AD 26–36, the dates and weekdays of the Alexandrian Easter full moons (*luna* 14) were as follows:

AD	Julian date	Weekday
26	18 April	Thursday
27	07 April	Monday
28	27 March	Saturday
29	15 April	Friday
30	04 April	Tuesday
31	24 March	Saturday
32	12 April	Saturday
33	01 April	Wednesday
34	21 March	Sunday
35	09 April	Saturday
36	29 March	Thursday

[17] Cassiodorus, *Chronica*, in *Chronica minora*, 2:135–37. See also the preface to Cassiodorus, *De artibus ac disciplinis liberalium litterarum*, PL 70:1149; *Chronicon Gallicani*, in *Chronica minora*, 1:639.

As a result, Western adherents to the Alexandrian enneacaidecaëteris, as represented by the Dionysiac Easter table, were faced with a serious chronological problem. Owing to the combination of the Dionysiac era of the incarnation with the full moon data of the 19-year cycle, one of the crucial dates in the history of salvation, namely the Passion of Jesus Christ, effectively ceased to be a dateable event. This inability of the Dionysiac Easter table to provide anything close to a satisfactory historical date for Christ's Passion naturally benefitted the cause of all those who sought to maintain their allegiance to the Victorian table. As we have seen, the Victorian *Annus Passionis* had the historical Passion fall on Friday, 26 March, *luna* 14, which neatly corresponded to Johannine chronology. The Dionysiac table, by contrast, showed *luna* 13 for the same date in AD 28, thereby foiling any attempt to plausibly assign the Passion of Jesus to this year.

The competition between both tables was particularly severe on the British Isles, where the calculation of Easter had originally been based on the so-called *latercus*, a variant of the 84-year cycle with lunar limits of *luna* 14–20 that had been brought to the island Celts from Gaul in the fifth century. In a synod held at Mag Léne in ca. 630, the Irish churches decided to convert to the 'Roman' reckoning, which was subsequently understood to be represented by the Victorian table, but the monks of Iona and other northern Irish monasteries refrained from change and instead held to their old *latercus*. During the 680s, Adomnán (679–704), abbot of Iona, was able to convince most of the remaining Irish adherents of the 84-year cycle (except Iona itself) to convert to the Dionysiac table, thereby creating a new temporary schism, because the Victorian table was at that time still being used in parts of the Irish South.[18]

Textual reverberations of the arguments that must have been going back and forth between both parties at the end of the seventh century can still be detected in early medieval computus manuscripts. One noteworthy example is a short treatise *De comparatione epactarum Dionysii et Victorii* uniquely preserved in the aforementioned ninth-century manuscript from Cologne's cathedral library (cod. 83^{II}),

[18] On the background, see Jones, *Bedae Opera*, 78–104; Thomas M. Charles-Edwards, *Early Christian Ireland* (Cambridge: Cambridge University Press, 2000), 391–415; Dáibhí Ó Cróinín, *Early Irish History and Chronology* (Dublin: Four Courts Press, 2003); Caitlin Corning, *The Celtic and Roman Traditions* (New York: Palgrave Macmillan, 2006); Warntjes, "Irische Komputistik," 18–21.

which is also home to the *Cologne Prologue*. The text had been largely ignored by modern scholars, until it was recently edited by Immo Warntjes, who cogently argues for a northern Irish provenance and a date of composition in ca. 689.[19] Its anonymous author, who was obviously an adherent to the Dionysiac reckoning, tried to remove the discrepancy that existed between both tables with regards to the Passion date by a remarkable feat of twisted computistical logic. The background to his argument was provided by the conventional division of the enneacaidecaëteris into an *ogdoas*, which covered the first eight years of the cycle, and the *hendecas*, which comprised years 9 to 19. Seizing on this division, the author made the entirely fictitious claim that the first year of the *hendecas* in the Victorian cycle actually corresponded to the beginning of the *ogdoas* in the Dionysiac cycle. As we have seen, the true correlation between both cycles was 1/19 (Victorius) = 7/19 (Dionysius). By contrast, the explanations found in the treatise *De comparatione epactarum* amounted to a correlation of 9/19 (Victorius) = 7/19 (Dionysius), meaning that the sequence of Easter full moons in the Dionysiac table was effectively displaced by eight years vis-à-vis the Victorian one. Instead of comparing the Victorian *Annus Passionis* with the tenth year of the Dionysiac cycle, where 26 March coincided with *luna* 13, the anonymous Irish author thus set his sights on the second year, in which the Easter full moon occurred on 25 March, conveniently yielding a 'synoptic' lunar age of *luna* 15 for Friday, 26 March.

It seems clear that the *De comparatione epactarum* was written in response to allegations made by Irish adherents of the Victorian reckoning who defended their position by pointing to the fact that the Dionysiac table failed to provide any satisfactory Passion date. That this response was taken seriously in at least some quarters can be seen from the fact that the treatise's peculiar correlation of cycles was discussed by the author of the so-called *Munich Computus* (719), who likewise hailed from Ireland and was an advocate of the Dionysiac reckoning. As the Munich computist also realized, however, there was a sounder way of comparing both tables, in which the *ogdoas* and the *hendecas* of each cycle were laid out side by side. The downside of this

[19] MS Cologne, Erzbischöfliche Diözesan- und Dombibliothek, 83^{II}, 176v–78r, edited in Warntjes, *Munich Computus*, 322–26. See also ibid., ci–cii, clii–clviii; Schwartz, *Ostertafeln*, 94–96.

approach was that the table of Dionysius no longer showed *luna* 17, but *luna* 15 on Easter Sunday in the Victorian Passion year, which went against scriptural and ecclesiastical tradition with regard to the age of the moon at the crucifixion.[20] By altering the synchronism between the Victorian and Dionysiac tables, the author of *De comparatione epactarum Dionysii et Victorii* had in fact a-historically dated the Passion to a year in the Dionysiac cycle which in reality corresponded to AD 20. In addition, both he and the Irish 'Munich computist' of 719 failed to notice that the very early Victorian *Annus Passionis* (AD 28) could not properly be reconciled with the *Annus Domini* (AD 1) count presupposed by Dionysius Exiguus. As we shall see further below in this chapter, such arguments in the teeth of sound calendrical arithmetic were no rare occurrence in early medieval computistical texts, especially when hallowed chronological traditions were at stake.

2. *The Venerable Bede*

An important watershed in the development and dissemination of early medieval computistics came with the composition of the short treatise *De temporibus* (703) and its 'bigger brother', the comprehensive handbook *De temporum ratione* (725), both written by the Anglo-Saxon scholar Bede ('the Venerable') at the monastery of Jarrow in Northumberland.[21] In his works on computus, Bede clearly rejected the Victorian system and instead advocated the Alexandrian 19-year cycle, as represented by the Dionysiac table, as the sole basis for the computation of Easter Sunday.[22] Apart from his shaping influence on Western computistics, Bede's works also had a seminal impact

---

[20] See Warntjes, *Munich Computus*, 194–203. Even a century later, when the Cologne codex 83^{II} was compiled in AD 805, the text of *De comparatione epactarum* was still deemed important enough to be both copied as part of the collection and used as a source for a Carolingian computistical handbook, which is found in the same manuscript. See the *Computus Coloniensis* (5.4; 5.5), edited in Borst, *Schriften*, 2:930–33.

[21] On Bede's computistical works and their transmission, see Jones, *Bedae Opera*, 125–72. For commentaries on his *De temporum ratione*, see Wallis, *Bede*; Roland-Pierre Pillonel-Wyrsch, *Le calcul de la date de Pâques au Moyen Âge* (Fribourg: Academic Press, 2004).

[22] For Bede's criticism of Victorius, see his *De temporum ratione* (42; 51), CCSL 123B: 409–12, 437–41.

on Christian chronography. At the beginning of the eighth century, most computists still relied on the universal chronology found in the chronicle of Eusebius of Caesarea, which had been slightly modified by Prosper of Aquitaine in the middle of the fifth century and was disseminated in the West through the prologue of Victorius's paschal work.[23] In this system, which was based on the number material found in the Septuagint, the birth of Christ took place in approximately AM 5199. By contrast, the world chronicle included as an appendage to Bede's *De temporibus* referenced an alternative chronology, according to which Christ's birth was separated from the creation of Adam by only 3952 years.[24] Unsurprisingly, this radically different estimate of the age of the world, which was based on the much smaller figures found in the Hebrew version of the Old Testament (as translated by Jerome), was bound to be greeted with resistance and Bede was soon horrified to hear that a certain David had accused him of heresy in the presence of Wilfrid, the bishop of Hexham. Despite these drawbacks, Bede remained steadfast in his allegiance to the *Hebraica veritas*, i.e. the Vulgate chronology, which he went on to defend both in a letter to the monk Plegwin (708) and in his magnificent *De temporum ratione*

---

[23] See, e.g., the computistical manual of AD 737, edited in Borst, *Schriften*, 1:375–423, and discussed in Bruno Krusch, "Das älteste fränkische Lehrbuch der dionysianischen Zeitrechnung," in *Mélanges offerts à M. Émile Chatelain* (Paris: Champion, 1910), 232–42. See also the formula for AD 743 discussed in Kerstin Springsfeld, "Eine Beschreibung der Handschrift St. Gallen Stiftsbibliothek, 225," in *Computus and its Cultural Context in the Latin West, AD 300–1200*, ed. Immo Warntjes and Dáibhí Ó Cróinín (Turnhout: Brepols, 2010), 224. On the wider context, see now also James T. Palmer, "Computus after the Paschal Controversy of AD 740," in Warntjes and Ó Cróinín, *Easter Controversy* (forthcoming). The Victorian tradition of aligning the 532-year cycle with Eusebian world chronology was continued in Ireland by the authors of the pseudo-Augustinian *De mirabilibus sacrae scripturae* (AD 654) and a lost 'Victorian' computus of AD 689. For further details, see Warntjes, *Munich Computus*, cxxiv–cxxv, 314–17.

[24] Bede, *De temporibus* (22), CCSL 123C:607: "Sexta aetas continet annos praeteritos DCCVIIII. Octavianus ann. LVI. Huius anno XLII Dominus nascitur, conpletis ab Adam annis IIIDCCCCLII, iuxta alios VCLXVIIII." The mention of 709 years is puzzling in light of the fact that the *annus praesens* of Bede's work was 703. As Masako Ohashi has shown, the additional six years probably reflect a later interpolation by scribes who wished to honour the fact that Victorius's Passion date (AD 28) was six years earlier than that proposed by Bede (AD 34). See Masako Ohashi, "'Sexta aetas continent annos praeteritos DCCVIIII' (Bede, *De temporibus*, 22): A Scribal Error?," in Jaritz and Moreno-Riaño, *Time and Eternity*, 55–61; Ohashi, "The *Annus Domini* and the *Sexta Aetas*: Problems in the Transmission of Bede's *De Temporibus*," in Warntjes and Ó Cróinín, *Computus*, 190–203.

(hereafter: DTR), in which he greatly expanded on the chronological themes of *De temporibus*.[25]

Bede's famous involvement in a controversy surrounding the Vulgate chronology, which rejuvenated the world by more than 1200 years, has led many scholars to naturally assume that the Northumbrian monk was also the originator of this chronology. Recent research by Daniel McCarthy, however, has seriously questioned this supposition. Basing himself on a comparison of the chronological structure of Bede's chronicles with that found in medieval Irish annals, which are likewise grounded in the Vulgate chronology, McCarthy has forcefully argued that both chronologies can be traced back to a common Irish source. According to his conjecture, this source ultimately derives from a lost recension of Jerome's translation of Eusebius's chronicle produced by Rufinus of Aquilea at the beginning of the fifth century.[26] McCarthy's claim that Bede's chronology of the world was dependent on a previously existent Irish tradition receives some support from the *laterculus Malalianus*, an obscure Latin text, partly based on the Byzantine *Chronographia* of John Malalas, which was probably composed by Bishop Theodore of Canterbury, who came from Tarsus, in the late seventh century. Malalas followed a Syriac chronographic tradition, according to which the sixth millennium since the world's creation had already been fulfilled with Christ's Passion, which he dated on 25 March. The author of the *laterculus* openly endorsed this chronology, adding some disparaging remarks about the Irish, who were unwilling to agree in this question. It is quite possible that it is the much shorter Vulgate chronology rather than the Septuagint-based world era of Eusebius-Jerome that was targeted here.[27]

---

[25] Bede, *Epistola ad Plegvinam*, CCSL 123C:617–26; Bede, *De temporum ratione* (praefatio; 67), CCSL 123B:263–65, 535–37. On the background, see von den Brincken, *Studien*, 107–20; Paolo Siniscalco, "Le età del mondo in Beda," *Romanobarbarica* 3 (1978): 297–331; Martin Haeusler, *Das Ende der Geschichte in der mittelalterlichen Weltchronistik* (Cologne: Böhlau, 1980), 26–32.

[26] See most recently Daniel McCarthy, *The Irish Annals* (Dublin: Four Courts Press, 2008); McCarthy, "Bede's Primary Source for the Vulgate Chronology in his Chronicles in *De Temporibus* and *De Temporum Ratione*," in Warntjes and Ó Cróinín, *Computus*, 159–89.

[27] See Jane Stevenson, *The "Laterculus Malalianus" and the School of Archbishop Theodore* (Cambridge: Cambridge University Press, 1995), 120: "Iam ne nos fallant multoloquio suo Scottorum scolaces." Ibid., 124: "In sex millia autem annorum concordant omnes apparuisse Dominum; quamuis Scotti concordare nolunt, qui sapientia se existimant habere, et scientiam perdederunt." Ibid., 154: "Igitur expletum est sextum millarium aetatis huius mundi, aetiam quamuis contradicant qui hoc percipere

That Bede's dating of the birth of Jesus to AM 3952 may not have been the result of his own personal calculations becomes even more likely from a look at some troubling chronological contradictions contained in his work that for the most part have escaped the attention of modern scholars. In his discussion of the date of the creation of the world, found in the sixth chapter of DTR, Bede notes that the moon must have been full at the time of its creation, for the Creator "would never make something in an imperfect state." In contrast to authors such as Victorius of Aquitaine and the computist of 243, he furthermore denied that the Sunday on which God had begun his work had fallen on the day of the vernal equinox. Instead, he tied the latter date to the fourth day, since this was the day on which the sun and the moon had been created (Genesis 1:14). Following the Alexandrian rules, according to which the vernal equinox fell on 21 March, Bede thus arrived at his own chronology of the hexaëmeron, which began on Sunday, 18 March, and ended on Friday, 23 March, the day on which Adam was formed from the dust of the earth. By implication, the corresponding lunar ages were *luna* 11–16, since the moon was thought by Bede to have been created full (i.e. *luna* 14) on the fourth day.[28]

From a computistical point of view, the creation of the world thus should have take place in a year, in which the Easter full moon fell on Wednesday, 21 March. This is the case in years 54, 149, 369, and 491 of the 532-year Easter cycle advocated by Bede, meaning that a world era fully consistent with his own chronological teachings should have begun in one of these four years. Yet if Jesus was born in AM 3952 (= 228 mod 532), this could only mean that the world had been created in years 305, 306, or 307 of the cycle, depending on whether the nativity was assigned to 2 BC (532/532), 1 BC (1/532) or AD 1 (2/532).

---

nolunt." See also ibid., 10, 26–27, 177–78, and Stevenson, "Theodore and the *Laterculus Malalianus*," in *Archbishop Theodore*, ed. Michael Lapidge (Cambridge: Cambridge University Press, 1995), 204–21.

[28] Bede, *De temporum ratione* (6), CCSL 123B:290–92. The idea that the fourth day of creation coincided with the equinox, albeit on 25 March, had previous to Bede already been advanced by the Irish computus of 719, edited in Warntjes, *Munich Computus*, 144–49. The prevalence of Bede's chronology of creation during the Middle Ages is documented in Ferdinand Piper, *Die Kalendarien und Martyrologien der Angelsachsen* (Berlin: Verlag der Königlichen Geheimen Ober-Hofbuchdruckerei, 1862), 3–7, 86–87. See also Piper, "Der erste Tag der Welt," in *Königlich Preussischer Staats-Kalender für das Jahr 1856* (Berlin: Verlag der Königlichen Geheimen Ober-Hofbuchdruckerei, 1856), 6–35; Jones, *Bedae Opera*, 338.

Only in year 306 did 21 March fall on a Wednesday, but in none of the three years (nor, for that matter, in any other year in the vicinity) did it also coincide with *luna* 14. As we shall see, this discrepancy was a source of confusion for subsequent computists, who puzzled over the creation date no less than they did over the date of Christ's Passion.

Out of the three possible dates just mentioned, the one in the middle, which puts creation in year 306, is certainly the most attractive. Not only does the weekday on 21 March match Bede's claims about the chronology of creation, but year 306 is also the second year of the 19-year lunisolar cycle and the first year after a Julian leap year. The same parameters apply to the first year of the Christian era, which corresponds to year 2/532. This means that many of the same calendar algorithms or *argumenta* that early medieval computists used in relation to years of the Dionysiac era could also be applied to years AM without any need for elaborate reworking.[29] For instance, the position in the 19-year cycle could be determined in both cases by diving the present year by 19 and adding 1 to the remainder. It is thus not surprising if, as we shall see further below, subsequent Carolingian computists interpreted Bede's world era in such a way that the world was created in 3952 BC = 306/532, making the birth of Christ fall in 1 BC.

Whether Bede himself also subscribed to this view is a difficult question to answer, given the ambiguous nature of his own remarks in DTR. It has occasionally been suggested that he identified the year of the nativity as 2 BC, which was the most common date in late antique patristic writing.[30] This, however, seems to be a misunderstanding, based on Bede's *Greater Chronicle* (chapter 66 of DTR), where the year of the nativity is synchronized with the year 752 since the foundation of Rome (as found in Orosius) and the 42nd year of the reign of Augustus Caesar (as found, for instance, in Eusebius's chronicle).[31] While these years jointly point to a birth in 2 BC (given that both eras

[29] On the early medieval *argumenta* and the history of their transmission, see now Immo Warntjes, "The *Argumenta* of Dionysius Exiguus and their Early Recensions," in Warntjes and Ó Cróinín, *Computus*, 40–111.

[30] Georg Wetzel, *Die Chronicen des Baeda Venerabilis* (Halle: Plötz, 1878), 50–51, 59; Blackburn and Holford-Strevens, *Oxford Companion*, 780.

[31] Bede, *De temporum ratione* (66), CCSL 123B:495: "Anno Caesaris Augusti XLII, a morte vero Cleopatrae et Antoni, quando et Aegyptus in provinciam versa est, anno XXVII, olympiadis centesimae nonagesimae tertiae anno tertio, ab urbe autem condita anno DCCLII, id est eo anno quo conpressis cunctarum per orbem terrae gentium motibus firmissimam verissimamque pacem, ordinatione Dei, Caesar conposuit, Iesus Christus filius Dei sextam mundi aetatem suo consecravit adventu." See Orosius,

are counted in a standard way), there is nothing that would indicate that Bede possessed the kind of knowledge of ancient chronology necessary to draw this conclusion. Lacking clear and independent information on the proper beginning of either the Roman foundation era or Augustus's regnal years, Bede will most likely have tried to determine their beginning by seizing on the one era, whose starting point he could claim to know with certainty: the Dionysiac era of Christ's incarnation.

As Bede explained in chapter 47 of DTR, Dionysius Exiguus had designated the second year of his Easter table, which contains the Easter Sunday of AD 1, as the first year *ab incarnatione Domini*.[32] If Jesus was incarnated in the womb in this very year, he should have been born at the end of AD 1. Unfortunately, Bede's subsequent remarks make it everything but clear if he himself accepted this view. As he went on to relate in an oft-cited passage from the same chapter, two of Bede's brethren had made an interesting chronological observation during a visit to Rome in the year 701 of the Dionysiac era. According to their report, the Easter candles in Santa Maria Maggiore carried an inscription, which designated the present year as the year 668 from the Passion of Christ. For Bede, this 33-year difference between both year counts was clear proof that the Roman Church approved of dating the crucifixion to AD 34.[33] He lent further support to this view by citing evidence for the duration of Jesus's life, arriving at the conclusion that he was aged 33½ years at the time of his death. This figure was essentially a combination of the 30 years, mentioned by Luke as Jesus's age at the time of his baptism (3:23), and the 3 years which Eusebius had assigned to the public ministry. As mentioned in the previous chapter, Eusebius's estimate was partly based on the repeated mentions of Passovers in the Gospel of John, which according to the bishop of Caesarea indicated a time span of about three years. The more precise figure of 3½ years could be found in Eusebius's *Demonstratio evangelica*,

Historia adversum paganos (6.22; 7.2), CSEL 5:426, 436–37; Eusebius, *Chronicle*, GCS 47:169.

[32] Bede, *De temporum ratione* (47), CCSL 123B:427–29: "Qui in primo suo circulo quingentesimum tricesimum secundum dominicae incarnationis annum in capite ponendo, manifeste docuit secundum sui circuli annum ipsum esse, quo eiusdem sacrosanctae incarnationis mysterium coepit.... Denique Dionysius ipse nobis quodammodo tacite quae dicimus in paschalibus quae scripsit argumentis ostendit...significans illo incarnato unum circuli decemnovenalis annum iam fuisse completum."

[33] Ibid., CCSL 123B:431.

where the duration of the public ministry was linked to the prophetic 'weeks of years' from the book of Daniel. In chapter 47 of DTR, Bede took over Eusebius's basic insight, but slightly expanded on him in his prophetic derivation of the same figure by alluding to the book of Revelations alongside Daniel. Indeed, both works contain repeated references to "three-and-a-half times" (Revelation 11:3; 12:6, 14; 13:5; Daniel 7:25; 12:7), which Bede obviously interpreted as a typological indication of the duration of Christ's public ministry.[34]

The one problem, however, which seems to have escaped Bede, was that a lifespan of 33½ years, when subtracted from a crucifixion in spring AD 34, could not possibly be squared with a birth on 25 December, AD 1, but instead pointed towards the autumn of 1 BC. Conversely, a birth at the end of AD 1 would have implied a Passion in AD 35, if the sum of 33½ years was maintained. Subsequent Carolingian computists generally preferred the former set of dates (1 BC and AD 34), thereby respecting what Bede himself had portrayed as the opinion of the Roman Church regarding the Passion year. This view could be easily reconciled with the Dionysiac era by assuming that the latter was meant to begin with the birth of Christ on 25 December 1 BC, which became a widely used epoch of the Christian era during the Middle Ages. Yet none of this was of any help in solving another chronological problem, which Bede influentially exposed in chapter 47 of DTR. According to his own teachings, the crucifixion of Jesus should have had taken place on 25 March, the 15th day of the lunar month. The latter date was based on the synoptic Gospels, whose authority was so great that—in the words of Bede—"no Catholic" was allowed to doubt its veracity.[35] However, even a casual glance at the Easter table revealed that in AD 34, the Easter full moon had fallen on 21 March, making 25 March equivalent to *luna* 18, and that the concurrent was IV, meaning that 25 March was a Thursday and not a Friday.

[34] Ibid., CCSL 123B:430–31: "Habet enim, ni fallor, ecclesiae fides dominum in carne paulo plus quam XXXIII annos usque ad suae tempora passionis vixisse, quia videlicet XXX annorum fuerit baptizatus, sicut evangelista Lucas testator, et tres semis annos post baptisma praedicaverit, sicut in evangelio suo Iohannes non solum commemorato redeuntis paschae tempore perdocet, sed et idem in apocalypsi sua. Danihel quoque in suis visionibus prophetice designat." See Eusebius, *Demonstratio Evangelica* (8.2.107), GCS 23:389.

[35] Bede, *De temporum ratione* (47), CCSL 123B:432: "Nam quod dominus XV luna feria sexta crucem ascenderit, et una sabbatorum, id est die dominica, resurrexit a mortuis, nulli licet dubitare catholico."

Bede obviously felt that something had to be done about this glaring contradiction, for he went on to cite Theophilus of Caesarea, "an old teacher of the church, who lived closed to the time of the Apostles" and who had written "a synodal letter along with the other bishops of Palaestina against those who celebrate on the 14th day of the moon together with the Jews."[36] Contrary to what Bede himself may have believed, the text from which he went on to cite was not an authentic transcript of the acts of the late second-century Council of Caesarea, but a pseudepigraphic computistical treatise, which had probably first originated in North Africa in the fifth century. The anonymous author of this text tried to defend the Alexandrian Easter limits, according to which Easter Sunday could not fall earlier than 22 March, by evoking the chronology of the Passion, which in his scheme of things began with the Last Supper on 22 March and ended with the resurrection of Jesus on 25 March. This was, as we have seen, the commonly accepted resurrection date in the Eastern tradition.[37]

For Bede, shifting the resurrection from 27 to 25 March seems to have been an attractive alternative, especially in the light of his own chronology of creation, according to which Adam had been created on Friday, 23 March. In the preface to his *Greater Chronicle* he revisited this chronology and noted that it favoured a date of the crucifixion on Friday, 23 March, seeing how this would provide a neat typological connection between the first day (and fall) of mankind and the day of its salvation (and re-creation) through Christ's suffering.[38] However, as Bede's subsequent remarks go to show, not even such a reconsideration of the Passion date could save the believer from all the arising chronological discrepancies. The only instance of a Friday falling on 23 March, which occurred in the relevant range of years, was in

[36] Ibid.: "Quod autem VIII kal. apr. Crucifixus, VI kal. earundem die resurrexerit, multorum late doctorum ecclesiasticorum constat sententia vulgatum; quamvis Theophilus caesariensis, antiquus videlicet vicinusque apostolicorum temporum doctor, in epistola synodica quam adversus eos qui decima quarta luna cum Iudeis pascha celebrabant una cum caeteris Palestinae episcopis scripsit, ita dicit."

[37] *Acts of the Council of Caesarea*, ed. Krusch, *Studien* (I), 306–10. German translation and commentary: Strobel, *Texte*, 80–95. On the various recensions of this text, see Warntjes, *Munich Computus*, lxv, n 167. See also Jones, *Bedae Opera*, 87–89. The text was apparently based on Eusebius's account of the Synod of Caesarea in his *Historia ecclesiastica* (5.23.3), GCS 9.1:489.

[38] Bede, *De temporum ratione* (66), CCSL 123B:464–65. The passage is repeated nearly verbatim in the ninth century chronicle of Ado of Vienne, *Chronicon in aetates sex divisum*, PL 123:24.

AD 31. As Bede realized, this could not be reconciled with the 'long' chronology of the life of Jesus that he himself had defended:

> But if you are looking for such a year, and you are unable to find it in that place you thought [it would be,] blame the carelessness of the chronographers, or better yet, your own slowness, being very wary lest in defending the text of the chronicles you do not appear boldly to impugn the testimony of the Law and the Gospel by saying that our Lord and Saviour underwent the most holy mystery of the Cross in either the 15th or the 16th year of the Emperor Tiberius Caesar, or in either the 29th or 30th year of his age, when the Gospels plainly indicate that the forerunner of our Lord began to preach in the 15th year of Tiberius, and that this same [forerunner] subsequently baptized (among others) Jesus, *who was then at about the beginning of his 30th year.*[39]

As a result, the chronological discussion concerning the dates of Jesus's birth and death had reached a painful cul-de-sac, a fact which was grimly emphasized by Bede's own ironic remark that anyone who managed to find an appropriate set of data in AD 34 should "give thanks to God, for He has granted that you find what you were looking for, just as He promised."[40] For subsequent computists, this impasse with regard to one of the central questions of Christian chronology was to become both a serious burden and an incentive to new research.

3. *Carolingian Reactions: The 'Seven-Book-Computus' and the Chronicle of Claudius of Turin*

The influence of Bede's DTR on Western computistics reached an early peak on the continent during the reign of Charles the Great, of

[39] Bede, *De temporum ratione* (47), CCSL 123B:433: "Sin vero annum qualem quaerebas, in loco quem putabas invenire non poteris, vel chronographorum incuriae vel tuae potius tarditati culpam adscribe, tantum diligentissime cavens ne chronicorum scripta defensando intemerabile legis vel evangelii testimonium videaris impugnare, dicendo dominum salvatorem vel XV aut XVI imperii Tiberii Caesaris vel XXVIIII aut XXX suae aetatis anno sacrosanctum crucis subisse mysterium, cum evangelia manifeste significent XV anno Tiberii praecursorem domini praedicare coepisse, ipsumque mox inter alios baptizasse Iesum *incipien* tem iam fieri *quasi XXX annorum.*" Translation according to Wallis, *Bede*, 129.

[40] Bede, *De temporum ratione* (47), CCSL 123B:431–32: "Et ideo circulis beati Dionysii apertis, si quingentesimum sexagesimum sextum ab incarnatione domini contingens annum, quartam decimam lunam in eo VIII kal. Apr. quinta feria repereris, et diem paschae dominicum VI kal. Apr. luna decima septima, age Deo gratias quia quod quaerebas, sicut ipse promisit, te invenire donavit."

whom Einhard wrote that he "learned the art of the computus and with wise intention and great curiosity investigated the movements of the stars."[41] That Charles considered the computus a discipline of special importance for clerical education in his realm can already be seen from an injunction in his *Admonitio generalis* (789), according to which monastic and cathedral schools were required to

> carefully correct the psalms, notation, chant, computus, grammar, and the Catholic books in every monastery and diocese, because often some desire to pray to God properly, but they pray badly because of uncorrected books.[42]

During the following decades, Carolingian scholars produced massive collections of computistical material, the most important of which are known as the 'Seven-Book-Computus' of 809/12 and the 'Three-Book-Computus' of 818. In these convoluted anthologies, which were often only selectively copied by later scribes, readers could find a wealth of knowledge, encompassing chronography, arithmetic, astronomy, meteorology, and metrology, making them an early medieval *summa* of the natural and numerical sciences. In terms of computistical theory, the Carolingian anthologies drew heavily on Bede's DTR, which was itself widely copied during the ninth century.[43]

Two chapters, which can be found in the second book of the gigantic 'Seven-Book-Computus' are particularly noteworthy for the way in

---

[41] Einhard, *Vita Karoli Magni* (25), ed. Georg Waitz, 6th ed. (Hannover: Hahn, 1911), 30: "Discebat artem conputandi et intentione sagaci siderum cursum curiosissime rimabatur." The expression *rimabatur* recalls the Hiberno-Latin *rimarius*, meaning 'computist'. See Maura Walsh and Dáibhí Ó Cróinín, *Cummian's Letter* De controversia paschali *and the* De ratione conputandi (Toronto: Pontifical Institute of Medieval Studies, 1988), 33n126. Einhard may have known the word from the Irish scholar Dúngal of St. Denis. I owe this observation to Leofranc Holford-Strevens.

[42] *Admonitio generalis* (72), in *Capitularia regum Francorum*, vol. 1, ed. Alfred Boretius (Hannover: Hahn, 1883), 60: "Et ut scolae legentium puerorum fiant. Psalmos, notas, cantus, compotum, grammaticam per singula monasteria vel episcopia et libros catholicos bene emendate; quia saepe, dum bene aliqui Deum rogare cupiunt, sed per inemendatos libros male rogant."

[43] On the background, see Arno Borst, "Alkuin und die Enzyklopädie von 809," in *Science in Western and Eastern Civilzation in Carolingian Times*, ed. Paul Leo Butzer and Dietrich Lohrmann (Basel: Birkhäuser, 1993), 53–75; Borst, *Das Buch der Naturgeschichte*, 2nd ed. (Heidelberg: Winter, 1995), 121–76; Kerstin Springsfeld, *Alkuins Einfluß auf die Komputistik zur Zeit Karls des Großen* (Stuttgart: Steiner, 2002); Nadja Germann, *De temporum ratione* (Leiden: Brill, 2006), 29–35, 88–94. On the reception of Bede, see also Charles W. Jones, "Bede's Place in Medieval Schools," in *Famulus Christi*, ed. Gerald Bonner (London: SPCK, 1976), 261–85.

which they reflect Carolingian attempts to grapple with the chronological challenges posed by Bede's writings. The first of these chapters is headed *Annus magnus de concurrentibus compositus ad ostendendam diversitatem Nativitatis et Passionis Domini* ('The Great Year made up of concurrents in order to demonstrate the [chronological] difference in [the years of] the nativity and the Passion of the Lord') and begins, as the title indicates, with a 532-year list of concurrents, which are arranged according to the 19 different Easter full moons of the Alexandrian cycle.[44] The concurrents are essentially weekday numbers (from I to VII), which correspond to the weekday of 24 March in a particular year, and which had originally been designed to aid computists in finding the date of Easter Sunday. The sequence of 532 concurrents, however, which constitutes the *Annus magnus* in the present text, differs from that of the regular cycle by starting out in the 54th year of the 532-year Easter cycle. Although the anonymous author never clearly acknowledged this discrepancy, he made clear that it was motivated by his intention to begin the sequence in what he considered to have been the historical year of the creation of the world. The parameters for the latter could be found in chapter six of Bede's DTR, according to which the sun and the moon had been created on Wednesday, 21 March, and on the 14th day of the Easter lunation. In principle, there would have been four different years to choose from, i.e. years 54, 149, 369, and 491 of the cycle, each of which had a different position in the Julian order of leap years. Our author decided on year 54, which was the first year after a leap year. This was a rational choice, seeing that the additional quarter day of the Julian year could only be expected to have accumulated to a full leap day in the fourth year after the sun's creation.

Based on this new fixed point for his calculations, the anonymous author tried to establish the Easter full moon for the year of Christ's Passion by simply counting forward the years of his *Annus magnus*. On the basis of Bede's chronicles, where the nativity of Jesus was assigned to AM 3952, as well as the view that Jesus was 33½ years old when he

[44] Edited in Borst, *Schriften*, 3:1119–23. Of the 'Seven-Book Computus', only five complete copies survive, but parts and excerpts from this collection have found their way into no less than 200 codices from the ninth to fourteenth centuries. For a list of MSS, see ibid., 3:1070–86. The *Annus magnus* was later also used by Marianus Scottus. See Peter Verbist, "Reconstructing the Past: The Chronicle of Marianus Scottus (d. 1082)," *Peritia* 16 (2002): 291–94; Verbist, *Duelling*, 93–96, who wrongly ascribes the text to Dúngal of St. Denis.

died, he assumed that the Passion had taken place in AM 3952 + 34 = 3986. For this particular year, the *Annus magnus* yielded an Easter full moon on Thursday, 15 April.[45] At first glance, this was a satisfactory result, as the synoptic Gospels dated the crucifixion to the 15th day of the lunar month, which from a computistical vantage point required *luna* 14 to fall on a Thursday. Doubts could arise, however, considering that the Easter full moon in the present example fell extremely late, on 15 April, and thus ran afoul of the dating of the crucifixion to 23 or 25 March which had been advocated by various patristic authorities. Yet there was a graver and more insistent problem: since our author had begun his count from year 54/532, his Passion year was bound to correspond to 315/532 (3985 = 261 mod 532 and 54 + 261 = 315). Since the regular 532-year cycle began in 1 BC, this year was hence equivalent to 219 BC, AD 314, AD 846, etc. and no computistical trick in the world could bring the historical date of the Passion anywhere near its vicinity.

We cannot be sure if and how this glaring inconsistency was perceived by our author, but the fact that he refrained from any commentary on his results may be telling—as telling as the fact that he proceeded to repeat his calculation twice, this time swapping the Bedan interval of 3952 years for the 5199 years found in the chronicle of Eusebius and once again for the 5500 years of the 'Greeks' (i.e. the Alexandrian system). The results, however, were once again less than compelling: in the Eusebian case, *luna* 14 of the Passion year fell on Wednesday, 2 April, in what was really year 498 of the 532-year cycle (5198 = 410 mod 532 and 54 + 410 + 34 = 498); the Greek scenario, on the other hand, led to Thursday, 5 April, in year 267/532 (5499 = 179 mod 532 and 54 + 179 + 34 = 267).[46] It seems that the anonymous author of the *Annus magnus* had worked in the sincere hope that he could bypass the overt conflict between the Dionysiac era and the

---

[45] Borst, *Schriften*, 3:1122: "Ab Adam usque ad nativitatem Christi conputantur anni tria millia nongenti quinquaginta duo, inde usque ad passionem eius anni triginta tres et dimidius. Si in primo anno, quo Adam factus est, posueris terminum Paschae in XII. Kalendas Aprilis et septem concurrentem, erit ipsa dies quarta feria, in qua sidera condita esse creduntur et plenilunium atque aequinoctium. Quodsi cum tali termino talique concurrente cyclum inchoaveris et usque ad Christi nativitatem vel passionem supputando praedictum annorum numerum perduxeris, erit anno nativitatis dominicae terminus Paschae Kalendas Aprilis et concurrens quattuor, annoque passionis terminus XVII. Kalendas Maias et concurrens quattuor."

[46] Ibid., 3:1123. See also Borst's remarks ibid., 3:1059–60, and Borst, *Das Buch*, 159, which fail to do justice to the computational problems involved.

Easter cycle, which had been uncovered by Bede, by calculating the date of Christ's Passion from a different starting point, namely the year of creation. In doing so, he piously followed Bede's remarks on the chronology of the hexaëmeron in chapter six of DTR, not realizing that his creation year 54/532 stood in no adequate chronographic relation to the year of the nativity (2/532).

This interesting example of computistical confusion, which was preserved for posterity as part of the 'Seven-Book-Computus', is not a singular and isolated accident. In several manuscript versions of the same anthology, the chapter on the *Annus magnus* is immediately followed by yet another calculation of the Passion date (*ratio quomodo feria, qua dominus passus est, invenitur*), which was later also incorporated into the 'Three-Book-Computus' of 818. This time, an attempt was made to prove that 25 March had really been a Friday in the year of Jesus's death, despite the fact that the sequence of concurrents in the Easter table seemed to tell otherwise.[47] Once again, the anonymous author proceeded from the year of creation, basing himself on Bede's view that the world had begun on Sunday, 18 March. Unlike the author of the *Annus magnus*, however, the composer of the present argument made no use of the 532-year Easter cycle, but instead proposed a solution based purely on the weekday-cycle.

He began with the basic insight that the duration of the Julian common year exceeded the length of 52 weeks by one day (365 = 52 × 7 + 1). Seizing on this fact, the computist proposed a hypothetical 'solar year' of 364 days, which began on the Sunday of creation, 18 March. Since 364 days comprised a full number of weeks, the first day of the year would always return on Sunday, but its date in the Julian calendar would recede by one day each year: the second year would begin on 17 March, the third year on 16 March, and so forth. The anonymous computist continued to calculate forward on this basis until he reached AM 3983, which he identified as the 17th year of Tiberius according to Bede's *Greater Chronicle*. In this year, the hypothetical 364-day calendar began on 20 April. At the end of the same year (i.e. in March of the 18th year of Tiberius), the Passion should have taken

---

[47] Edited in Borst, *Schriften*, 3:1124–27. This chapter enjoyed a far higher dissemination in medieval manuscripts than the preceding *Annus magnus*. Of the former, 21 versions are known, whereas the latter is only preserved in nine instances. In 14 out of 21 cases, the *ratio quomodo feria* is transmitted independently of the *Annus mangnus*. See the apparatus in ibid., 3:1119, 1124, 1395.

place. In order to complete the calculation, however, one had to take account of all the Julian leap days, which had been intercalated since creation and which by AM 3983 had supposedly accumulated to two additional 364-day years, bringing the beginning of the *Annus Passionis* from 20 April, AM 3983, to 18 April, AM 3895. Since 18 April was a Sunday, the following 25 March fell on a Friday, just as the Latin tradition presupposed.

While the computist presented this result as a veritable success to his readers, a closer look at his calculation reveals two serious problems. First of all, in taking account of the leap day increment for 3983 years, the anonymous author should have added 3983 × 0.25 = 995.5 days, which is equivalent to roughly 2.7 years, instead of the two years added in his calculation. An even more serious discrepancy arose from the fact that his calculation completely disregarded the difference between the hypothetical 364-day year and the Julian common year of 365 days. In the course of 3985 years, this difference would have accumulated to 3985 ÷ 365.25 = 10.91 years, which are unaccounted for in this computational 'proof'.[48] Contrary to what both chapters in the 'Seven-Book-Computus' sought to pretend, the date of the Passion remained a chronological problem, which these Carolingian writers could only solve by inflicting serious violence on the core principles of calendric arithmetic. For all other computists, who chose not to resort to such questionable methods, it was difficult to avoid the conclusion that they were faced with a discrepancy which to them seemed almost impossible to overcome. When a group of Frankish computists was required in 809 to demonstrate their knowledge to an unknown interrogator, they confidently dated Christ's Passion to 25 March, only to find themselves trapped in a catch-22 question. As the preserved protocol of the examination records:

> At this point they were told first to count back the years of the Lord from the present to the first, and then from it forward to His Passion, finding whether the day of His Passion agreed with their answer of March 25, *luna* 15. And so, when they studied the traditional cycles, they were not able to find what had been commanded them because of an inherent difficulty. They replied that they could not find a formula to make those calculations agree. Then they were asked which calculation they wished to accept. Response: The authority of the fathers, that is of

[48] See also Verbist, "Reconstructing," 312–14; Verbist, *Duelling*, 120–22.

> Augustine, Jerome, Dionysius, Bede, who preached the Lord's Passion on March 25.[49]

The awkwardness of their response points to a deep dilemma, posed by their desire to conform with patristic authority on the one hand and their acceptance of the 532-year Easter cycle on the other. Since the lunar age and weekday of the Passion, as found in the Gospels, were not negotiable, the only way out of this dead end would have been to take a stand against patristic authority and cherished calendar traditions. According to Arno Borst, whose three-volume edition of *Schriften zur Komputustik im Frankenreich* has been an epochal contribution to the study of Carolingian computistics, such an assault on tradition was made only a few years later by Claudius of Turin (died ca. 827), who composed a chronicle of the world in 814. Born in Spain, Claudius was one of the period's most copious biblical commentators before being named bishop of Turin in 816 by his benefactor, the emperor Louis the Pious. In the later years of his life, Claudius became a highly controversial figure, owing to some of his theological positions, which verged towards iconoclasm and earned him accusations of heresy from some of his opponents, which included Dúngal of St. Denis and Jonas of Orléans.[50] Borst seems to have kept this reputation in mind, when he claimed that, in his chronicle, Claudius had controversially held Christ to have died on 23 March, AD 31.

[49] *Capitula, de quibus convocati compotiste interrogati fuerint*, edited in Borst, *Schriften*, 3:1040–41. "Hinc eis iniunctum ut annos domini a praesenti sursum versus computando usque ad primam perducerent, et ab eo deorsum versus usque ad passionem eius inquirentes diem eiusdem passionis dominicae in VIII k. ap. luna XVma concordantem. Inspectis itaque maiorum ciclis cum invenire quod eis mandatum fuerat prae difficultate quadam nequirent, responderunt non posse se convenientem in hoc reperire rationem. Igitur quid sequi vellent quaesitum est. Responsio: Auctoritatem patrum, id est augustini, hieronimi, dyonisii, bedae, qui passionem domini VIII kl. ap. praedicant." Translation modified from Charles W. Jones, "An Early Medieval Licensing Examination," *History of Education Quarterly* 3 (1963): 26. See also Borst, *Schriften*, 3:1034–39; Bernhard Bischoff, *Mittelalterliche Studien*, 3 vols. (Stuttgart: Hiersemann, 1966–81), 3:228–29; Springsfeld, *Alkuins Einfluß*, 105–6; Contreni, "Counting," 65–67.

[50] See Michael Gorman, "The Commentary on Genesis of Claudius of Turin and Biblical Studies under Louis the Pious," *Speculum* 72 (1997): 279–329; Johannes Heil, "Claudius von Turin—eine Fallstudie zur Geschichte der Karolingerzeit," *Zeitschrift für Geschichtswissenschaft* 45 (1997): 385–412; Michael Idomir Allen, "The Chronicle of Claudius of Turin," in *After Rome's Fall*, ed. Alexander Callander Murray (Toronto: University Press of Toronto, 1998), 288–319; Elisabetta Bellagente, "La *Chronica de sex aetatibus* di Claudio vescovo di Torino," *Aevum* 73 (1999): 237–46; Pascal Boulhol, *Claude de Turin* (Paris: Institut d'Études Augustiniennes, 2002).

Assuming that such a date was prone to cause serious irritation among Claudius's contemporaries, Borst even went as far as speaking of a "revolution of the time scale," which shattered conventional views of the time Christ's crucifixion and resurrection, as well as an "explosive truth... bound to make him new enemies."[51]

The sole basis for these assertions is a passage from the beginning of the second part of Claudius's chronicle, which establishes a chronological parallelism between the six days of creation and the week of Christ's Passion. In the oldest of the three existing manuscripts (Paris, BN, lat. 5001), which was copied still in the ninth century and formed the basis for Philippe Labbé's fragmentary *editio princeps* of 1657, the passage appears thus:

> And so it follows by reckoning that the first man, Adam, was formed from the slime of the earth and fitted with a soul on 23 March, the seventeenth day of the moon. According to my calculation, the second Adam, that is Christ, the Son of God... is found to have risen from the dead at the same time, that is 23 March, but not on the same weekday.[52]

It would seem that the passage immediately falsifies at least part of Borst's claim. Clearly, 23 March is here regarded to be the date of the resurrection rather than that of the Passion. This can also be seen from Claudius's remark "not on the same weekday," which makes clear that, although Adam had been brought to life on the sixth day of creation

[51] Borst, *Schriften*, 1:61–62: "Doch rang sich Claudius bei seinen komputistischen Rechnungen... zu einer brisanten Wahrheit durch, die ihm neue Feinde einbringen mußte. Sie... erschütterte aber die Überzeugung des Gottesvolkes vom Zeitpunkt der Kreuzigung und Auferstehung Christi, somit die derzeit herrschende Osterberechnung und Osterfeier. Bewirkt wurde dieser Umsturz der Zeitskala auf einfachste Weise, durch komputistische und typologische Verknüpfung der ersten weltzeitlichen Datenreihe mit der zweiten." See also ibid., 3:1337, 1341. Borst's claim was obviously inspired by a suggestion found in Allen, "The Chronicle," 314n128: "Claudius' 'calculation' reflects the date for 'Easter' in A.D. 31 according to Bede." This assertion is evidently weaker than what Borst subsequently proposed.

[52] Philippe Labbé, *Novae Bibliothecae Manuscriptorum Librorum*, 2 vols. (Paris: Cramoisy, 1657), 1:309 = PL 104:917: "Et ita ratione deducta X Kal. Aprilis luna septima decima invenitur protoplastus Adam ex terrae limo esse formatus pariter atque animatus. Secundum hanc nostram supputationem, secundus Adam, id est Christus, Dei Filius... eodem tempore, sed non eadem feria, id est X Kal. Aprilis invenitur resurrexisse a mortuis." I have slightly modified punctuation. The three preserved MSS of the chronicle are: Paris, BN, lat. 5001, 1r–8v (s. IX$^{3/4}$); Monza, Bibl. Capit., c-9/69, 66ra–83vb (s. X^{1}); Madrid, BN, 9605, 103r–11r, 112r–16v (a. 1026). In addition, there is a short excerpt in MS Paris, BN, lat. 2341, 12v (Le Puy before 843), recently published in Borst, *Schriften*, 3:1335–49. A new and complete edition of the chronicle is being prepared by Michael I. Allen.

and hence on a Friday, the typological recurrence of the corresponding calendar date (23 March) in the year of the Passion did not apply to the Friday of Christ's crucifixion, but to his resurrection two days later. Claudius had thus adopted Bede's date for Adam's creation, but had obviously decided to put the crucifixion two days earlier, on 21 March, a fact which must have startled later readers and commentators. In another manuscript (Monza, Biblioteca Capitolare, c-9/69), which was copied in the tenth century, the scribe accordingly tried to correct the above passage by clumsily inserting the words *fuisse passum et VIII* between *invenitur* and *resurrexisse*. This brought Claudius's words in tune with Bede's, but at the same time rendered them incoherent, because the new date presupposed what Claudius explicitly denied: that during the weeks of creation and crucifixion, 23 March had each taken place on the same weekday. It is hence clear that Borst erroneously based himself on an interpolated manuscript when he made the claim that Claudius dated the Passion to AD 31, relying on the fact that 23 March fell on a Friday in this year.[53]

While Borst's specific claim must be therefore rejected, he was right in assuming that Claudius's dating of Christ's death deviated from the contemporary norm. In order to see more clearly on this matter, we have to turn to the intricate computistical calculations which pervade the second part of the chronicle. As Michael Allen has pointed out, these calculations served the purpose of assigning Julian calendar dates and weekdays to important biblical events (such as the flood, the Exodus, and the destruction of Solomon's Temple), whose Hebrew calendar dates were given in the Old Testament.[54] Claudius naturally assumed that these dates followed the ancient Hebrew lunisolar calendar, which was occasionally mentioned in Bede's DTR.[55] A key passage

[53] MS Monza, Bibl. Capit., c-9/69, 79rb–va: "Et ita ratione deducta X Kal. Aprilis luna septima decima invenitur protoplastus Adam ex terrae limo esse formatus pariter atque animatus. Secundum hanc nostram supputationem, secundus Adam, id est Christus, Dei filius…eodem tempore, sed non eadem feria, id est X Kal. Aprilis, invenitur fuisse passum et VIII resurrexisse a mortuis." See Allen, "The Chronicle," 314n128; Borst, *Schriften*, 1:62. See also von den Brincken, *Studien*, 117, who erroneously identifies 23 March as Claudius's date of the Passion despite basing herself on Labbé's edition.

[54] Allen, "The Chronicle," 313–18.

[55] Bede, *De temporum ratione* (11; 13; 45), CCSL 123B:312–315, 326–327, 420–22; *Computus Coloniensis* (5.11), edited in Borst, *Schriften*, 2:941–42. On early medieval views of the Hebrew calendar, see Wiesenbach, *Sigebert*, 119–22, and the sources assembled in Warntjes, *Munich Computus*, 242–43. See also the tenth-century *Libellus*

for Claudius's understanding of the biblical calendar appears to have been Bede's exposition of the chronology of Noah's flood:

> It is obvious that the Hebrews were accustomed to observe their months according to the course of the Moon. When in the book of Genesis Noah is said to have gone into the Ark with his family on the 17th day of the second month, and to have left it on the 27th day of the same month after the flood, the only correct way of understanding this is that a whole solar year, that is, 365 days, is signified. For the Moon, which this year, for instance, is in its 17th day on the nones of May [7 May] will next year be 27 days old on the day before the nones of May.[56]

Since Bede here implied that the Hebrew calendar was lunisolar without explicitly mentioning any differences from the 19-year cycle used by the church, Claudius naturally assumed that the lunar months of the Old Testament could be identified with the lunar months of the Dionysiac Easter cycle. In principle, there was thus one very straightforward way of locating the beginning of the 'Hebrew' year in the Julian calendar: since the Easter full moon was supposed to represent the 14th day of Nisan, any computist could easily look up the latter's current Julian date in his Easter table and then count back 13 days to arrive at 1 Nisan, the beginning of the first month of spring. Claudius, on the other hand, used a slightly more elaborate approach, which we can glimpse by turning to his treatment of the Exodus from Egypt, the only date for which he outlined the steps of his calculation in any detail.

According to Claudius's adaptation of Bede's chronology, the Exodus had taken place in AM 2453. Since the year of creation (AM 1 = 3952 BC) was year 2 in the 19-year lunisolar cycle, he first had to divide the year in question by 19 and increase the remainder by one. In the case of the Exodus, the remainder was 2 (2453 mod 19 = 2), meaning that the Israelites had left Egypt in the third year of the 19-year cycle.

---

de argumentis lunae, PL 90:723, where it is alleged that the 19-year cycle was revealed by Moses through divine inspiration.

[56] Bede, *De temporum ratione* (11), CCSL 123B:315: "Verum haec utcumque acta vel computata fuerint, claret tamen hebraeos ad lunae cursum suos menses observare consuesse. Nec aliter in genesi recte sentiendum, ubi Noe cum suis septimo decimo die secundi mensis arcam ingressus et septimo vicesimo eiusdem mensis die post diluvium egressus asseritur, quam annum solis integrum, hoc est CCCLXV dierum, esse descriptum. Quia videlicet luna, quae praesenti anno, verbi gratio, per nonas maias septima decima existit, anno sequente septima vicesima pridie nonas Maias occurret." Translation modified from Wallis, *Bede*, 43. See also Bede, *In Genesim* (2.8.15–18), CCSL 118A:126–27.

The regular lunar epact of 22 March for this year was 2 × 11 = 22, to which Claudius added 8 days to arrive at his own customized version of the epact, which indicated the lunar age on the preceeding 31st of December. The final result was 22 + 8 = 30, which, for computistical purposes, was equivalent to an epact of zero. This in turn meant that 1 January fell on *luna* 1 and hence coincided with the beginning of a Hebrew lunar month. In order to get from 1 January to 14 Nisan, Claudius had to count 30 + 29 + 30 + 14 = 103 days. If day 1 corresponds to 1 January, day 103 (in a common year) will naturally fall on 13 April (31 + 28 + 31 + 13 = 103), which is indeed the date of the Easter full moon in the third year of the Dionysiac cycle. For reasons unexplained, however, Claudius counted only 102 days and thus settled for 12 April, putting the Exodus (15 Nisan) on 13 April, the arrival in the wilderness of Sinai (1 Siwan; Exodus 19:1) on 27 May and the reception of the Law (4 Siwan; Exodus 19:16) on 31 May.[57]

In order to arrive at the corresponding day of the week, Claudius employed a method of calculation that kept track of the leap day increment since creation. First, the total number of leap days had to be found by dividing the present year of the world by 4. In the case of the Exodus, which took place in AM 2453, 613 days had to be added (2453 ÷ 4 = 613.25). A second step prescribed the addition of six *dies intercalares*, yielding a total sum of 3072, which he then divided by 7. The remainder of this division indicated the weekday on 1 January,

---

[57] Claudius, *Chronicle*, PL 104:922–23: "Iterum sume superiores annos 2453, hos partire per decimam nonam partem…adde unum et fiunt tria, et invenies quod tertius fuerit annus cycli solaris, qui est decemnovalis: subtrahe unum, et remanent duo: ipse est secundus annus epactae lunaris. Iterum dic: undecies bini fiunt 22, tantos enim dies habet in duobus annis solaris cursus amplius lunari. His adde 8, et fiunt 30. Et quia annus embolismus fuit, tanti exstiterunt in epacta. *Iterum sume summam numeri 132 annorum subtrahe triginta, qui fuerunt in epacta*, et remanent 102, et invenies quod secundum dies solares, pridie Idus Aprilis feria secunda luna quarta decima exstiterit ipsum Pascha. Juxta vero dies lunares III Kal. Aprilis fuit Neomenia, id est novae lunae principium. Secundum vero supputationem nostram, quae fit per dies solares, Idus Apriles, luna quinta decima, alter die post Pascha, inveniuntur exisse filii Israel de Ramesse, et venisse in Socoth et de Socoth VIII. Kal. Maii, luna sexta decima in Phyairoth, et VI Kal. Junii, luna prima, qui est dies primus tertii mensis lunaris, venerunt ad montem Sinai, et acceperunt legem Deo loquente ad se per angelum pridie Kalendas Junii, die quinquagesimo post agni occisionem, feria secunda, luna quarta, quo die et Pascha in Aegypto celebraverunt." The italicized passage is garbled in MS Paris, BN, lat. 5001, on which Labbé's edition was based. I have emended it on the basis of MS Monza, Bibl. Capit., c-9/69, 81va.

which, in Claudius's calculation for the year of the Exodus, was six, meaning Friday (3072 ÷ 7 = 438 R 6).[58]

In principle, this calculation followed the standard computistical algorithms (*argumenta*) for finding the 'solar epacts' or concurrents, which were modified according to the parameters used by Claudius.[59] According to his chronicle, the world had begun on Sunday, 18 March, 3952 BC. Since Sunday is the first day of the week, it was possible to keep track of the year-by-year increment of the weekday on 18 March by simply adding up the years of the world along with the number of leap days which had accrued since creation. By AM 4, for example, the weekday on 18 March had shifted from 1 = Sunday to 5 = Thursday, because a leap day had to be added in this year. For sums greater than seven, it was of course necessary first to subtract multiples of seven days and then turn to the remainder. Yet the calculation just outlined only applies to the weekday on 18 March or equivalent days of the Julian year. In order to adapt this method to *any* day of the year, it was much more convenient to first find the weekday on 1 January, which was identical with that of 18 March only in leap years. In common years, on the other hand, it differed by six days: in AM 1, for instance, 1 January fell on a Monday, while 18 March was a Sunday. Accordingly, in order to arrive at the weekday on 1 January in a common year, six days had to be subtracted from the total weekday increment before it could be divided by seven. This seems to be the reason behind the six *dies intercalares*, which Claudius adduces in his calculation of the Exodus date. By *adding* rather than subtracting these six days, however, he introduced an error of two days into his computation, which went (2453 + 613 + 6) ÷ 7 = 438 R 6 (= Friday), whereas the correct

[58] Claudius, *Chronicle*, PL 104:922: "Si vis hujus Paschae nosse tempus, vel diem, vel feriam, aut lunam, sume annos ab initio mundi, qui sunt usque ad id tempus 2453; hos partire per quartam partem, et dic: quater 600 fiunt 2400, et supersunt 53. Iterum dic: quater deni, 40, et supersunt 13; quater terni, 12, et superest 1; et invenies quod primus annus tunc exstiterit post bissextum. Sume igitur ipsos dies bissextiles, qui fuerunt in toto retro tempore 613, hos adde ad superiores annos, et invenies summam eorum 3066: his iterum adde sex dies, quos intercalares vocamus, et erunt 3072. Et ut feriam ipsius anni invenire possis, partire supradictam summam, et dic septies 400 fiunt 2800, et supersunt 272. Iterum dic septies 30 fiunt 210, et supersunt 62; et rursum septies octoni fiunt 56, et supersunt sex; et invenies quod feria sexta secundum dies solares exstitit principium anni quam nos more gentilium Kalendas Januarias nuncupamus."

[59] See, e.g., Dionysius Exiguus, *Argumenta titulorum paschalium* (4; 8), ed. Krusch, *Studien* (II), 76–77.

result should have been (2453 + 613 – 6) ÷ 7 = 437 R 1 (= Sunday). It is precisely this computational error which holds the key to an understanding of his dates for Christ's crucifixion and resurrection.

From a remark in the preface to his chronicle, it becomes apparent that Claudius probably followed Bede in dating the Passion to AD 34, which in his system was equivalent to AM 3986.[60] In this year, the Easter full moon fell on 21 March, which was a Sunday. Following Claudius's special reckoning rules, however, we instead get (3986 + 996 + 6) ÷ 7 = 712 R 4, indicating that the year in question began on a Wednesday. If 1 January fell on a Wednesday, 18 March was a Tuesday, with the next Sunday falling on 23 March. This is precisely the combination presupposed by Claudius's own comments at the beginning of the second part of the chronicle, where it is stated that "according to my calculation" the resurrection fell on 23 March. As it turns out, Claudius's strange computational error had the attractive side-effect of bringing his Passion date in harmony with the Gospels. Interestingly, he appears to have sided with John, because he put the Passion on the date of the Easter full moon (*luna* 14). Bede, by contrast, had clearly favoured the synoptic chronology, writing that "no Catholic may doubt that the Lord mounted the Cross on Friday, on the 15th day of the Moon."[61] Even more daring than the Johannine lunar age was Claudius's calendar date of 21 March, which unlike Bede's 23 March or the more orthodox 25 March, had no backing from patristic tradition. It is worth pondering, however, whether Claudius may have found some reassurance of his view in chapter 30 of DTR, where the dates of the equinoxes and solstices in the Julian calendar are discussed. Bede begins by detailing an old Roman tradition, which put the four cardinal points of the year on the 8th day before the kalends [i.e. the first day] of April, July, October and January. His comments on this tradition are worth quoting in full:

---

[60] Edited in Mirella Ferrari, "Note su Claudio di Torino 'Episcopus ab ecclesia damnatus'," *Italia medioevale e umanistica* 16 (1973): 307: "Nos vero, secundum divinos quos in manu tenemus codices ex hebrea auctoritate per beatum Iheronimum christianum interpretem, non plus a conditione mundi usque ad passionem domini invenire possumus quam III. DCCCC LXXXVI."

[61] Bede, *De temporum ratione* (47), CCSL 123B:432. One may speculate if Claudius, in calculating the lunar age at the Passion, committed the same mistake as in the case of the year of the Exodus, where his lunar data turn out one day lower than expected. In this case, 21 March would have corresponded to 15 Nisan, thus bringing the whole calculation in line with the synoptic chronology.

> This is what some of the pagans say; and very many of the church's teachers recount things which are not dissimilar to these about time, saying that our Lord was conceived and suffered on the 8th before the kalends of April [25 March], at the spring equinox, and that he was born at the winter solstice on the 8th before the kalends of January [25 December]. And again, that the Lord's blessed precursor and Baptist was conceived at the autumn equinox on the 8th before the kalends of October [24 September] and born at the summer solstice on the 8th before the kalends of July [24 June]. To this they add the explanation that it was fitting that the Creator of eternal light should be conceived and born along with the increase of temporal light, and that the herald of penance, who must decrease, should be engendered and born at a time when the light is diminishing. But because, as we have learned in connection with the calculation of Easter, the judgment of all men of the East (and especially of the Egyptians, who it is agreed, were the most skilled in calculation) is in particular agreement that the spring equinox is on the 12th before the kalends of April [21 March], we think that the three other turning-points of the seasons ought to be observed a little before [the date] given in the popular treatises.[62]

The mentioned connection between the annual course of the sun and the conception and birth of Jesus and John the Baptist could be justified by recourse to John 3:30, where the Baptist is quoted as saying "he must increase, but I must decrease," which was sometimes interpreted as referring to the change of day length that was associated with the equinoxes and solstices.[63] According to Bede, however, the

[62] Bede, *De temporum ratione* (30), CCSL 123B:374: "Haec quidem gentiles, quibus non dissimilia de tempora etiam perplures ecclesiae tradidere magistri, dicentes: VIII kl. Apriles in aequinoctio verno dominum conceptum et passum, eundem in solsitio brumali VIII kl. Ianuarias natum; item beatum praecursorem et baptistam domini VIII kl. Octobres in aequinoctio autumnali conceptum, et in aestivo solstitio VIII kl. Iulias natum—addita insuper expositione quod auctorem lucis aeternae cum cremento lucis temporariae concipi simul et nasci deceret, poenitentiae vero praeconem quem oportebat minui cum inchoata minoratione lucis generari pariter et concipi. Verum quia, sicut in ratione paschali didicimus, aequinoctium vernale duodecimo kalendarum Aprilium die, cunctorum orientalium sententiis et maxime Aegyptiorum quos calculandi esse peritissimos constat, specialiter adnotatur, caeteros quoque tres temporum articulos putamus aliquanto priusquam vulgaria scripta continent esse notandos." Translation modified from Wallis, *Bede*, 87–88.

[63] Augustine, *De diversis quaestionibus* (58.1), CCSL 44A:104; Augustine, *Enarrationes in Psalmos* (132.11), CCSL 40:1934; Augustine, *In Iohannis Evangelium* (14.5), CCSL 36:144; ps.-Chrysostom, *De solstitiis et aequinoctiis*, PLS 1:561; Caesarius of Arles, *Sermones* (216), CCSL 104:859; ps.-Bede, *Libellus de argumentis lunae*, PL 90:724; Bede, *In Lucae Evangelium expositio* (1.1.24), CCSL 120:28. See also Rosalind Love, "Bede and John Chrysostom," *Journal of Medieval Latin* 17 (2007): 76–78; Warntjes, *Munich Computus*, 106–7. On medieval calendar traditions assigning the crucifixion to 25 March see, e.g., *Martyrologium Hieronymianum*, AASS 63:36–37;

traditional dates to which the conception and birth of Jesus and John had been assigned, did not really correspond to the true dates of the equinoxes and solstices. Indeed, his words could be taken to imply that the four calendar dates commemorating Jesus and John should be shifted by four days, from the 8th day to the 12th day before the kalends of April, July, etc.[64]

Such a shift could also be taken to affect the date of the Passion, which was generally assumed to have taken place on the same calendar date as Christ's conception (i.e. the Annunciation on 25 March). One important authority for this notion was St. Augustine, who, in his *Quaestiones Exodi*, had claimed that the typological link between the date of the conception and the Passion was already hinted at in the book of Exodus (23:19: "Do not cook a young goat in its mother's milk").[65] Claudius was in all likelihood aware of this idea, as he inserted a longer passage from the *Quaestiones* into his chronicle, and may have taken it very seriously. It is thus conceivable that he extended Bede's criticism of the traditional date of the vernal equinox to the date of the crucifixion, thereby finding an important justification to move it from 25 March to 21 March.[66] In doing so, Claudius was—at least by his own opinion—able to solve the most nagging chronological problem which Bede's DTR had posed to subsequent generations of computists: to find a satisfactory date of the crucifixion, which would simultaneously fall on a Friday and a full moon. Naturally, this solution could only be achieved on the price of bad arithmetic. As we have seen from the examples preserved in the 'Seven-Book-Computus', he was no exception in this respect.

64:159–60, 164; Polemius Silvius, *Kalendarium*, PL 13:678; Rabanus Maurus, *Martyrologium*, CCCM 44:31–32; Piper, *Die Kalendarien*, 13, 17–21; Arno Borst, *Die karolingische Kalenderreform* (Hannover: Hahn, 1998), 417–20; Borst, *Der karolingische Reichskalender und seine Überlieferung bis ins 12. Jahrhundert*, 3 vols. (Hannover: Hahn, 2001), 1:713–26; Warntjes, *Munich Computus*, 150–53.

[64] Such conclusions were indeed drawn at the beginning of the eleventh century by Heriger von Lobbes, *Epistola ad quendam Hugonem Monachum*, PL 139:1133–34. On Heriger see chapter four below.

[65] Augustine, *Quaestiones Exodi* (90), CCSL 33:114–16. See also Augustine, *De diversis quaestionibus* (56), CCSL 44A:96; *De Trinitate* (4.5), CCSL 50:172; ps.-Chrysostom, *De solstitiis et aequinoctiis*, PLS 1:562–63.

[66] See Claudius, *Chronicle*, PL 104:922, and Augustine, *Quaestiones Exodi* (47.6), CCSL 33:92.

CHAPTER FOUR

ALL COHERENCE RESTORED? THE AGE OF THE CRITICAL COMPUTISTS

The technical problems faced by Western computists when it came to dating the day of Christ's Passion and the year of the world's creation were so persistent and perplexing that only a radical solution could restore the lost coherence of computus and chronography. Since the basic data provided by the Gospels (Friday + *luna* 14 or 15) were hardly the subject of negotiation, the only available way out of the dilemma was to abandon either the Easter cycle, which failed to provide agreeable data, or some of the traditional chronological assumptions about the life of Jesus. The former option, reasonable as it may sound to us today, was beyond the grasp of early medieval computists. The centuries during which the Dionysiac Easter table had been promoted in the face of stiff competition from Victorius and the 84-year Irish *latercus* had led to a state in which the 19-year cycle of the Alexandrian church was invested with quasi-sacral authority. This development had already begun with Dionysius Exiguus himself, who made the doubtful claim that his cycle had been formally adopted by the fathers of the Council of Nicaea in 325. Subsequent computists would repeat this legend countless times and add to it the idea that the Alexandrian sequence of 19 Easter full moons had been dictated to the Egyptian monk Pachomius by an angel, thereby contributing to the growing idealization of the 19-year cycle as an example of both scientific and divine wisdom.[1]

Since they readily accepted the Dionysiac Easter cycle as a faithful depiction of the natural astronomical order, medieval computists instead began to turn to the attached *Annus Domini* as the root of all

[1] On the background, see Charles W. Jones, "A Legend of St. Pachomius," *Speculum* 18 (1943): 198–210; Wallis, *Bede*, 331–34; Stephen C. McCluskey, "Changing Contexts and Criteria for the Justification of Computistical Knowledge and Practice," *Journal for the History of Astronomy* 34 (2003): 201–17; Max Lejbowicz, "Les Pâques baptismales d'Augustin d'Hippone, une étape contournée dans l'unification des partiques computistes latines," in Warntjes and Ó Cróinín, *Computus*, 1–39.

chronological confusion. From the late tenth to the early decades of the twelfth century, a small but outspoken group of monastic scholars decided to tackle the problem of the Passion date by shifting around the dates of Jesus's life in relation to ancient chronology. Two of these scholars, Heriger of Lobbes (died 1007) and Gerland the Computist (died after 1093), opted for 23 March, AD 42, thereby effectively reinventing the very late Passion date which had been a pivotal element in the chronological system of Annianus of Alexandria. All other known authors, Abbo of Fleury (died 1004), Marianus Scottus (died 1082), Sigebert of Gembloux (died 1112), Hezelo of Cluny (died 1123), Heimo of Bamberg (died 1139) as well as an early twelfth-century anonymous from Limoges, preferred to stick with the more orthodox 25 March as the date of the crucifixion and decided to *ante*-date the life of Jesus, arriving at a Passion in either AD 12 or—in the singular case of Heimo of Bamberg—in AD 1. Since they all upheld conventional views on the lifespan of Jesus Christ (32 or 33 years + roughly three months), each of the authors named also had to re-date the year of Christ's birth, forcing them to abandon the *Annus Domini* of Dionysius. The critical acumen with which some of these authors treated received tradition and their readiness to reform Christian chronology by proposing a new 'true' incarnation era inspired Johannes Wiesenbach, whose 1986 edition of the *Liber decennalis* of Sigebert of Gembloux included the first comprehensive survey of the subject, to jointly refer to them as the 'critical computists' (*kritischen Komputisten*), a label that I too shall adopt for the purpose of the following discussion.[2]

The earliest scholar to openly reject the accuracy of the Dionysiac era was Abbo of Fleury, who served as the abbot of Saint-Benoît-sur-Loire and has remained famous as one of the foremost computistical and mathematical scholars of his day. In an admirable analysis of the relevant manuscript material, Peter Verbist was able to show that Abbo's thought on the Passion problem and related chronological issues went through at least four stages from 982 to 1004.[3] In an initial step, Abbo

[2] Wiesenbach, *Sigebert*, 63. The standard monography on this movement is now Verbist, *Duelling*. See also Alfred Cordoliani, "Abbon de Fleury, Hériger de Lobbes et Gerland de Besançon sur l'ère de l'incarnation de Denys le Petit," *Revue d'histoire ecclésiastique* 44 (1949): 463–87.

[3] Peter Verbist, "Abbo of Fleury and the Computational Accuracy of the Christian Era," in Jaritz and Moreno-Riaño, *Time and Eternity*, 63–80; Verbist, "Abbon de Fleury et l'ère chrétienne vers l'an mil: un esprit critique vis-à-vis d'une tradition erronée," in *Abbon de Fleury*, ed. Barbara Obrist (Paris: Publications universitaires

merely considered a prolongation of the Dionysiac era by three years, dating the birth of Christ to 3 BC. This move was supported by the chronicle of Eusebius, which counted 287 years between the reign of Diocletian and the birth of Christ, whereas Dionysius Exiguus had hinted at an interval of merely 284 years. As a result of this shift, the Passion could be dated to AD 31, in which *luna* 15 fell on 25 March. Yet in this year, 25 March was a Sunday instead of a Friday. Abbo solved the problem by changing the alignment of the solar cycle with the 19-year cycle in such a way that the concurrent of AD 31 moved from VII to V, making 25 March a Friday. The justification for such tinkering came from Abbo's erroneous assumption that the Dionysiac table did not match up correctly with its predecessor, the table of Cyril of Alexandria, because the sequence of concurrents had been disturbed in late antiquity. Needless to say, this step was highly problematic, since it technically implied that the Passion had taken place in a year which was actually equivalent to year 450/532—an additional 114-year shift compared to the three years admitted by Abbo. Luckily, he did not stick to this incoherent theory, but eventually proposed a different solution, according to which the Passion had taken place on Friday, 25 March, AD 12, thereby also creating a new incarnation era that started in 21 BC.

Besides looking for an adequate solution to the Passion problem, Abbo was equally interested in the date of death of St. Benedict of Nursia, whose remains had been transferred from Montecassino to Fleury-sur-Loire in the tenth century.[4] From the available hagiographic sources, Abbo concluded that Benedict had passed away on Holy Saturday, 21 March, but soon had to find that no such combination of dates had occurred within the possible range of years, which he established as AD 529 to 604. Thanks to his change of the Passion year, however, Abbo was now able to reinterpret AD 509, in which Holy Saturday did in fact fall on 21 March, as the year 531 of his newly corrected incarnation era. As the example of St. Benedict goes to show,

Denis Diderot, 2004), 61–93; Verbist, *Duelling*, 35–84. See also Nadja Germann, "À la recherche de la structure du temps: Abbon de Fleury et le comput," in *Abbon, un abbé de l'an mil*, ed. Annie Dufour and Gillette Labory (Turnhout: Brepols, 2008), 153–76; Germann, "Zwischen *veritas naturae* und *fides historiae*: Zeit und Dauer bei Abbo von Fleury," in *Das Sein der Dauer*, ed. Andreas Speer and David Wirmer (Berlin: de Gruyter, 2008), 171–95.

[4] On the background, see Elizabeth Dachowski, *First among Abbots* (Washington: The Catholic University of America Press, 2008), 1, 182–83.

the methods of computistical chronography could be extended and applied to problems which lay outside the narrow scope of biblical chronology. Abbo was by no means the only medieval scholar to explore the connections between hagiography and chronology in this fashion, as can be seen from the work of Heimo of Bamberg, who used the Easter cycle to investigate the dates of death of other saintly figures, such as Mark the Apostle.[5]

While Abbo and most of the other critical computists relied on the Latin tradition, exemplified by Augustine, who put the crucifixion on 25 March, two authors from this group instead opted for 23 March. As we have seen in the case of Annianus of Alexandria, this choice could motivate a shift of the Passion to AD 42, in which 23 March fell both on a Friday and on *luna* 15. The same date was espoused only a few years after Abbo by the learned abbot Heriger of Lobbes (died 1007), whose *Epistola ad Hugonem* (ca. 995) contained some acute observations on chronological questions.[6] Having realized that the Dionysiac era led to serious contradictions in the chronology of Christ's earthly life, Heriger turned to the traditions of the 'Greeks', who counted eight years less (*viii annos ab incarnatione Domini minus quam circulus Dionysii supputant*) from the incarnation of the Lord than Dionysius Exiguus.[7] This was an obvious nod to the Alexandrian chronological

---

[5] Hans Martin Weikmann, *Heimo von Bamberg: De decursu temporum* (Hannover: Hahn, 2004), 42, 339–40, 347, 372, 374–75. See also Arno Borst, "Ein Forschungsbericht Hermanns des Lahmen," *Deutsches Archiv* 40 (1984): 398–406, 432–34; Verbist, *Duelling*, 137, 227–28, 303–4, 307–8.

[6] Heriger von Lobbes, *Epistola ad quendam Hugonem Monachum*, PL 139:1129–36. A re-edition can be found in Cordoliani, "Abbon," 480–84. On this work, see now Peter Verbist, "De Epistola ad Hugonem van Heriger van Lobbes († 1007): Een belangrijk chronologisch traktaat," *Millennium* 12 (1998): 30–42; Verbist, *Duelling*, 15–33. On Heriger, see further Oskar Hirzel, *Abt Heriger von Lobbes 990–1007* (Leipzig: Teubner, 1910).

[7] Heriger, *Epistola*, PL 129:1131: "Sed Graecis, auctore Nicaeno concilio et beato Cyrillo Alexandrino episcopo, magis assentiendum, qui et finitimi Orientalibus et vicini apostolicorum virorum fuere temporibus.... Hi itaque VIII annos ab Incarnatione Domini minus quam circulus Dionysii supputant; ergo, secundum illos, primus ab incarnatione Domini annus is est qui in circulo Dionysii nonus. Is autem est XXXIIII qui est in circulo Dionysii XLII, habens VII concurrentes, XIIII epacatas, terminum in XI Kalend. Aprilis, feria scilicet V, quando et Caena factus est, et vespere Pascha celebrari coepta est, sicut dicit Dominus, die XIIII ad vesperum Pascha Domini est, et in XV sub luna solemnitatem celebrabilis, quando et passus est, id est X Kalendas Aprilis, et Sabbatum quando requievit in sepulchruo, luna XVI, IX Kalendas Aprilis, et Dominicum diem Resurrectionis VIII Kalendas Aprilis luna XVII." See also Verbist, *Duelling*, 25n50: "Heriger refers here not to the so-called Alexandrian Creation era (AM 5493), but to the Byzantine Creation era (AM 5509)." This is clearly false, since

system created by Annianus, who had dated the birth of Christ to AD 9. Although Heriger's indications do not suffice to identify his sources for this 'Greek' tradition, it is possible that he knew of Annianus's dates from the Byzantine chronicle of Theophanes Confessor, which had been partially translated into Latin by Anastasius Bibliothecarius at the end of the ninth century.[8]

Unfortunately, this connection has been missed by previous scholars such as Peter Verbist, who instead proposes that the abbot of Lobbes used the spurious *Acts of the Council of Caesarea* of pseudo-Theophilus as his source of inspiration.[9] This assumption, which is not very plausible in the light of Heriger's own remarks, is obviously motivated by the comparison of Heriger's correction with that of Gerland the Computist (died after 1093), who likewise opted for a Passion on 23 March, AD 42. In setting the crucifixion on the tenth instead of the eighth day before the calends of April, Gerland (like Bede before him) relied on the testimony of pseudo-Theophilus, which seemed to confer a very high age on this dating tradition. As a skilled computist, Gerland could easily see that the chronological shift required by a Passion on 23 March (i.e. from AD 34 to AD 42) was less extreme than the one presupposed by the 'Latin' 25 March (i.e. from AD 34 back to AD 12). From his newly found Passion date, Gerland subtracted 33 years and 3 months of Christ's life to arrive at a nativity on 25 December, AD 8.[10]

---

the Byzantine era is usually associated with a Passion in AD 31 and nativity dates that *precede* the Dionysiac era by several years.

[8] Theophanes Confessor, *Chronographia*, ed. Carl de Boor, 2 vols. (Leipzig: Teubner, 1883–85), 2:61–63. See also Anna-Dorothee von den Brincken, "Welt- und Inkarnationsära bei Heimo von St. Jakob: Kritik an der christlichen Zeitrechnung durch Bamberger Komputisten in der ersten Hälfte des 12. Jahrhunderts," *Deutsches Archiv* 16 (1960): 156n9.

[9] Verbist, *Duelling*, 24n48.

[10] See MS Oxford, Bodleian Library, Auct. F.1.9, 19ra–20ra, 21vb, and the partial edition in Cordoliani, "Abbon," 484–87. See also Alfred Cordoliani, "Le comput de Gerland de Besançon," *Revue du moyen âge latin* 2 (1946): 309–13; Lambert Marie de Rijk, ed., *Garlandus Compotista: Dialectica* (Assen: Van Gorcum, 1959); Jennifer Moreton, "Before Grosseteste: Roger of Hereford and Calendar Reform in Eleventh- and Twelfth-Century England," *Isis* 86 (1995): 565–69; Borst, *Kalenderreform*, 336–37; Borst, *Reichskalender*, 1:265–27; Verbist, *Duelling*, 147–71. Verbist refers to the Passion on 23 March as the "Greek tradition" and treats both Gerland and Heriger as adherents to the same view. As we have seen, however, it would be more appropriate to say that only Heriger truly followed the Greek tradition (namely the system of Annianus), whereas Gerland simply based himself the Pseudo-Theophilan *Acts*, which is more correctly defined as an alternative date with the Latin tradition. See ibid., 11–12, 24–25, 167–68.

Of all the medieval attempts to correct conventional chronology by computistical means, the one produced by the monk Marianus Scottus (died ca. 1082) stands out at as the most influential and well-received in his own period. Marianus hailed from Ireland, but undertook the *peregrinatio pro Christi* to the continent, where he spent most of his life in various German monasteries. His re-dating of the life of Jesus took place in the context of a world chronicle, which he completed in 1076 in Mainz, where he spent the last thirteen years of his life in voluntary enclosure in his cell. One may presume that this form of life provided Marianus with the necessary time and solitude to ponder chronological problems, which had been left unsolved by previous computists, such as the question after the year in which the world had been created. As we have seen, Bedan doctrine had prescribed a full moon on the fourth day of creation, which was also the day of the vernal equinox, i.e. 21 March. Yet neither Bede's own world era nor the Septuagint-based era found in the chronicle of Eusebius (which put the birth of Christ in AM 5199) could be computistically reconciled to this theory. Where the author of the *Annus magnus* in the 'Seven-Book-Computus' of 809/12 had merely been able to spread further confusion, Abbo of Fleury had at least recognized the problem and acknowledged its existence several times in his writings, but without providing any definite solution. When Marianus, who was aware of the failed attempts of the computist of 809/12, turned to the problem at the beginning of his chronicle, he did the only reasonable thing and added further years to the Hebrew world era, re-setting its start to 4184 BC, which corresponds to year 54/532. Since he also re-dated the incarnation of the Lord to 22 BC, this left a total interval of 4162 years between the world's creation and Christ, 230 years more than what Bede had proposed. How could this extension be justified on the basis of the Hebrew Old Testament? In a brilliant move, Marianus noticed that the Gospel of Luke (3:36) mentioned an additional patriarch, Cainan, in its list of generations between Noah and Abraham, for whom there was no equivalent in the Hebrew and Vulgate versions of Genesis. By assuming that Cainan had begotten his son Shelah at the age of 230, he was able to account for precisely the time span needed to support his own re-dating of the creation of the world.

Things were far less straightforward in the case of the post-biblical period. Although the shift of Christ's incarnation by 22 years was a relatively minor change compared to the insertion of 230 years, it was more difficult to account for, owing to the way in which the life of

Jesus Christ was embedded in the regnal chronology of the Roman Empire. Unlike Annianus of Alexandria, who seems to have unwaveringly accepted the idea that the 19th year of Tiberius corresponded to AD 42, Marianus Scottus understood that an addition of 22 years to the incarnation era could be justified only if these years could somehow be extracted from the duration of the reigns of the Roman emperors which had intervened between the Passion of Christ and the present. His methods consisted in the collation of the Eusebian chronicle with a (seemingly corrupt) copy of the chronicle of Isidore of Seville as well as a careful re-calculation for the number of regnal years between Julius Caesar and Vespasian, which seemed to reveal that all in all 21 years had gone missing in Eusebius's scheme of counting the years by Olympiads.[11]

Marianus's chronological inventiveness was closely mirrored by that of the chronicler Heimo of Bamberg, whose monumental work *De decursu temporum* took the extreme measure of dating the Passion of Christ to *luna* 14 on 25 March, AD 1, making his incarnation occur in 33 BC.[12] Contrary to what one might think, the rationale behind this shift was not an acceptance of the Johannine chronology of the Passion, but Heimo's theory, according to which the lunar dates of the biblical Hebrew and those of the Alexandrian 19-year lunisolar cycle differed systematically by one day. In the introductory portion of his work, he states:

> Let it be known that the computus of the Hebrews, which the old patriarchs and prophets, but also the evangelists, have used, is of a different form than the computus of the Greeks and Latins, which Christians today use. The Hebrews calculated the course of time according to lunar years, the Greeks and the Latins, however, use solar years; and the lunations of April, June, August, and of all other months, which we

[11] For details, see Verbist, "Reconstructing," 306–10, 319–22; Verbist, *Duelling*, 111–32. See further Anna-Dorothee von den Brincken, "Marianus Scottus: Unter besonderer Berücksichtigung der nicht veröffentlichten Teile seiner Chronik," *Deutsches Archiv* 17 (1961): 191–238; von den Brincken, *Studien*, 166–73; von den Brincken, "Marianus Scottus als Universalhistoriker iuxta veritatem Evangelii," in *Die Iren und Europa im früheren Mittelalter*, ed. Heinz Löwe, 2 vols. (Stuttgart: Klett-Cotta, 1982), 2:970–1009. For an edition and commentary on the *Vita Mariani Scotti*, see now Stefan Weber, *Iren auf dem Kontinent* (Heidelberg: Mattes, 2010).

[12] See in general Otto Meyer, "Weltchronistik und Computus im hochmittelalterlichen Bamberg," *Jahrbuch für fränkische Landesforschung* 19 (1959): 241–60; von den Brincken, "Welt- und Inkarnationsära," 155–94; von den Brincken, *Studien*, 175–81; Wiesenbach, *Sigebert*, 104–7; Weikmann, *Heimo*, 30–69; Verbist, *Duelling*, 251–339.

> count as comprising 29 days, they count as months of 30 days; and those months, which have 30 days in our calendar, have 29 days in theirs.... They use the same epacts and the same concurrents as we do today, but the seat of the epacts, which we put on 22 March, they have on 21 March.[13]

Heimo's peculiar views on the Jewish calendar are best understood in light of the information, which Bede had provided on the same subject in chapter 11 of *De temporum ratione*:

> The ancients customarily calculated their months not by the course of the Sun, but by that of the Moon. Hence, whenever Holy Scripture (whether [speaking of a time] under the Law or before the Law) indicates a day of the month on which something was said or done, it signifies nothing other than the age of the Moon. The Hebrews, to whom the oracles of God are entrusted, have never ceased to observe the months after the ancient custom of their fathers. They call the first month of the new fruits, which is dedicated to the rites of Passover, Nisan. Because of the fluctuating course of the Moon, it sometimes falls in March, sometimes in April, and occasionally encroaches on some days of May. But it is properly assigned to April, and always either begins, ends, or is totally included within it, provided that the rule we discussed above is observed.... By a similar calculation, their second month, Iyyar, corresponds to May...and the twelfth, Adar, to March.[14]

[13] Weikmann, *Heimo*, 135: "Sciendum est autem, quia aliter se habet compotus Hebreorum, quo usi sunt antiqui patriarche et prophetae, sed et evangelistae, et aliter se habet compotus Grecorum et Latinorum, quo nunc maxime utuntur Christiani. Hebrei quippe seriem temporum per lunares annos estimabant, sed Greci et Latini per solares, et lunationes Aprilis, Iunii, Augusti et caeterorum mensium, quos nos numeramus XXIX dierum, illi numerabant XXX, quos nos XXX, illi XXIX, embolismos omnes XXX dierum sicut et nos preter ultimum; illi XXIX dies tantum evenerant. Eisdem epactis, eisdem quoque concurrentibus usi sunt, quibus et nos utimur, sed sedem epactarum, quam nos in XI Kal. Aprilis ponimus, ipsi posuerunt in XII Kal. Aprilis." Verbist, *Duelling*, 260, errs in attributing to Heimo the view that "in the last cyclical year [the Jews] reckoned one lunar day fewer than the Christians (instead of 384 days)." The *saltus* is evidently the same in both calendars.

[14] Bede, *De temporum ratione* (11), CCSL 123B:313–14: "Antiqui enim menses suos non a solis sed a lunae cursu computare solebant; unde quoties in scriptura sacra sive in lege seu ante legem, quota die mensis quid factum dictumque sit indicatur, non aliud quam lunae aetas significatur; a qua semper Hebraei, quibus credita sunt eloquia Dei, antiquo patrum more menses observare non cessant, primum mensem novorum, qui paschae caeremoniis sacratus est, Nisan appellantes, qui propter multivagum lunae discursum nunc in Martium mensem, nunc incidit in Aprilem, nunc aliquot dies Maii mensis occupat. Sed rectius Aprili deputatur quia semper in ipso vel incipit vel desinit vel totus includitur, ea dumtaxat regular, cuius et supra meminimus, observata.... Secundus eorum mensis Iar Maio...duodecimus Adar Martio simili ratione comparatur." Translation according to Wallis, *Bede*, 42.

Given these and other explanation found in Bede's work, readers could be easily lead into assuming that the Hebrews, just like the Christians, had used a regular sequence of 'full' and 'hollow' months as well as an ordered cycle of intercalation.[15] Since Bede at one point mentioned the ancient Hebrews in connection with the Alexandrian 19-year cycle, it was tempting to assume that the Hebrews had already used this same cycle during Old Testament days.[16] An attentive computist, who closely followed the doctrines of the Venerable Bede could thus easily come to the conclusion that the 19-year cycle of the Alexandrian church and the Hebrew calendar used in the biblical scriptures were basically one and the same. In fact, there was only one small hint which suggested a structural difference: in medieval computistics, it was an established convention to count the lunar months of the 19-year cycle as alternating between 'full' months of 30 days and 'hollow' months of 29 days, starting with the lunation that belonged to the month of January.[17] In this order, the lunation of April, which contained the Easter full moon, was bound to be 'hollow' or 29 days long. According to the above-cited quotation from Bede's *De temporum ratione*, however, the April lunation was also identical with Nisan, the first month in the Hebrew calendar. If the Hebrew lunar year *began* with the lunation of April, it was reasonable to suppose that it was counted as a 'full' month of 30 days. Counting from January, the Christian pattern was hence 30 + 29 + 30 + 29, etc., while the corresponding Hebrew pattern was 29 + 30 + 29 + 30, etc. As a result of this deviation, the Hebrew age of the moon on a particular day of the Julian calendar was usually one day ahead of its Christian equivalent. Judging from his own words, it was precisely this line of thought, which Heimo of Bamberg followed when he claimed that the Hebrew *luna* 15 was equivalent to *luna* 14 in the Christian 19-year cycle. In making this interpretation, he became the first computist to at least implicitly acknowledge that Nisan, the month in which Christ had been crucified, belonged to a different calendar from the Christian Easter lunation, namely the calendar of first-century Judaism.[18]

[15] Bede, *De temporum ratione* (11; 13), CCSL 123B:312–17, 326–27.

[16] Ibid. (45), CCSL 123B:420–22.

[17] Ibid. (20), CCSL 123B:346–49.

[18] According to Verbist, *Duelling*, 335n540, Heimo "argued that the moon at sunset on the evening of Christ's Passion had changed from luna xiv to luna xv, so that Christ according to the Jews was crucified on luna xv and according to the Greco-Roman chronology on luna xiv." As we have seen, this is simply not the case. The same

For all their well-meaning intentions, the impact of the critical computists on the chronological customs of Latin Christianity remained minimal. Although it is true that we know of at least one papal letter, which provides a double dating that may have been inspired by Marianus Scottus's corrected incarnation era,[19] the *Annus Domini* of Dionysius Exiguus had by that time been so firmly entrenched in day-to-day practice that it was never in serious danger of being superseded by any of the alternative eras that were proposed by Abbo of Fleury and his successors. Yet it would be shortsighted to use the practical application of their ideas as the sole criterion by which to judge the project of the critical computists.[20] In their sometimes ambitious re-evaluations of the chronological problems surrounding the date of Christ's Passion, the critical computists effectively revived the tradition of computistical chronography, which had lain dormant in the West since the work of Victorius of Aquitaine (with the possible exception of the chronicle of Claudius of Turin). In order to propose new solutions, they had to overcome many of the strictures and limitations which had been set by ecclesiastical tradition and which had done much to confuse their predecessors in the eighth and ninth centuries. Eager to restore the lost coherence within Christian chronology, the critical computists bravely took it on themselves to adjudicate between conflicting traditions, reinventing the 19-year Easter cycle as a universal chronological tool, which was powerful enough to override received authority. As a result, they helped to considerably widen the boundaries of that which could be thought and written in medieval computistical literature.[21]

---

mistake is found in von den Brincken, "Welt- und Inkarnationsära," 178n119. On the lack of "Differenzierungsvermögen" between the Jewish/Hebrew calendar and the Easter cycle among the critical computists, see already Wiesenbach, *Sigebert*, 119–22.

[19] See Pope Urban II, *Epistolae et privilegia*, PL 151:501, where AD 1098 is simultaneously designated as 1121 "secundum certiorem Evangelii probationem."

[20] *Pace* Thomas Vogtherr, "100 Jahre Forschungen zur Chronologie des Mittelalters und der Neuzeit," *Archiv für Diplomatik* 54 (2008): 233: "Letztlich blieben die Kritiker des Dionysius in einer Außenseiterposition. Deswegen wird man ihre Berechnungen und Ergebnisse im nachhinein auch kaum als ernsthaft zu diskutierende Minderheitenmeinungen einiger Komputisten zu würdigen haben."

[21] See also the assessment in Verbist, *Duelling*, 345–54.

CHAPTER FIVE

NEW FOUNDATIONS: CHRONOLOGY AND THE TWELFTH-CENTURY RENAISSANCE

1. *The Roots of Reform: From the* 'Computus Naturalis' *to the Jewish Calendar*

As has been hinted at in the previous chapter, the chronicle of Marianus Scottus was the one work among the critical computists' writings which enjoyed the greatest recognition during the High Middle Ages. Soon after its completion, the chronicle was taken to England by Robert de Losinga, a Lotharingian by birth, who became bishop of Hereford in 1079. Robert also composed an abridged version of the text, by which he contributed significantly to the dissemination of the latter's chronological arguments. According to Marianus's revised system of Christian chronology, the birth of Christ had occurred in 22 BC, which necessitated a re-dating of important events in recent history such as the Battle of Hastings, which now fell in 1088 instead of 1066. These corrections were viewed favourably by some English chroniclers, including William of Malmesbury, who openly complained about his contemporaries' imperviousness to new ideas, which made them unwilling to abandon the false Dionysiac reckoning of the years of the Lord.[1]

Robert de Losinga was by no means the only continental scholar to make important contributions to the intellectual culture of the British Isles during the latter part of the eleventh century. Robert's native Lorraine was a region of particular importance in this respect, seeing how it had become a seedbed for Arabic learning, which swept across

[1] On the reception of Marianus Scottus, see W. H. Stevenson, "A Contemporary Description of the Domesday Survey," *English Historical Review* 22 (1907): 72–84; Alfred Cordoliani, "L'activité computistique de Robert, évêque de Hereford," in *Mélanges offerts à Réné Crozet*, ed. Pierre Gallais and Yves-Jean Riou, 2 vols. (Poitiers: Société d'études médiévales, 1966), 1:333–40; Wiesenbach, *Sigebert*, 99–102, 107–12; Verbist, *Duelling*, 143–46.

the Pyrenees at an increasing rate, long before the same could be said about other parts of Christian Europe. For instance, it seems to have been mainly through Lotharingian channels that knowledge of the plane astrolabe, which could be used for timekeeping and the measurement of the position of celestial objects, spread forth into the Latin West in the course of the eleventh century.[2] The first Western scholar of whom we have definite knowledge that he used the astrolabe for a specific quantitative astronomical observation is Walcher of Malvern. Like Robert de Losinga, Walcher was a Lorrainer who later migrated to England, where he held the position of Second Prior at the monastery of Great Malvern (Worcestershire) from ca. 1091 until his death in 1125 or 1135. Among his preserved writings, we find a personal report, written at some point after 1107, in which he describes two observations of lunar eclipses, carried out by him during the 1090s. The first of these observations had taken place during Walcher's stay in Italy, near a town called Romona, where he had become witness to an unexpected eclipse of the moon on 30 October 1091 (Oppolzer no. 3552). The eclipse "took place before dawn looking westwards," but, as Walcher writes, "I had no instrument that would give me the time of the full moon, nor could I see the moon clearly because of thick clouds." To his good fortune, he soon had the opportunity to experience another eclipse, this time near his priory in Great Malvern, which occurred on 18 October 1092 (Oppolzer no. 3554). "I at once seized my astrolabe and made a careful note of the time of full eclipse," which turned out to be 12:45 when measured in equal hours after sunset.[3]

---

[2] On the background, see James Westfall Thompson, "The Introduction of Arabic Science into Lorraine in the Tenth Century," *Isis* 12 (1929): 184–93; Mary Catherine Welborn, "Lotharingia as a Center of Arabic and Scientific Influence in the Eleventh Century," *Isis* 16 (1931): 188–99; Arno Borst, *Astrolab und Klosterreform an der Jahrtausendwende* (Heidelberg: Winter, 1989).

[3] English translation cited after Richard W. Southern, *Medieval Humanism and Other Studies* (Oxford: Basil Blackwell, 1970), 166–67. The original text is preserved in MS Oxford, Bodleian Library, Auct. F.1.9, 90r, and partially edited in Charles Homer Haskins, *Studies in the History of Mediaeval Science* (Cambridge, MA: Harvard University Press, 1924), 114–15. On the background, see also Stephen C. McCluskey, *Astronomies and Cultures in Early Medieval Europe* (Cambridge: Cambridge University Press, 1998), 171–84. On Walcher, see William Dugdale, *Monasticon Anglicanum*, ed. John Caley, Henry Ellis, and Bulkeley Bandinel, 6 vols. (London: March, 1817–30), 3:442b; Dom David Knowles, C. N. L. Brooke, and Vera C. M. London, *The Heads of Religious Houses, England and Wales*, 2nd ed., vol. 1, *940–1216* (Cambridge: Cambridge University Press, 2001), 1:90.

In order to gain a better understanding of the context and motivation behind Walcher's observations, it is worth returning to Bede's *De temporum ratione*, whose 43nd chapter tackles the question of "why the Moon sometimes appears older than its computed age" (*Quare luna aliquoties maior quam computatur pareat*). According to Bede, astute observers could regularly notice a two-day discrepancy between the calculated new moon of the first spring lunation, which fell on 4 April in year 19/19 of the Alexandrian cycle, and the observable moon, which then appeared to be two days older. In his—not fully stringent—attempt to explain this discrepancy, Bede turned to the sequence of full and hollow months in the 19-year cycle, which, as he knew, did not truthfully reflect the actual duration of the mean lunation, which was slightly above 29½ days. The resulting inequality reached its peak in the last year of the cycle, just before order was restored by omitting one day from the final embolismic month—a correction also known as the 'leap of the moon' or *saltus lunae*. As Bede was quick to point out, the overt difference between the calculated and the observable phases of the moon did not imply an actual imperfection on the part of the Easter cycle, but had been wisely taken into account by the fathers of the Council of Nicaea, to whom he (wrongly) attributed its institution.[4]

In reality, the problem was far deeper than the Northumbrian monk was ready to admit. The Alexandrian 19-year cycle, with its equation of 19 Julian years and 235 lunar months led to a mean lunation of 6393.75d ÷ 235 = 29.530851d. Although the deviation from the accurate value (which, during late antiquity, was ca. 29.530585d) was far from dramatic, it was still enough to cause the calculated moon to lag behind the observable moon by one day in about every 304 years. Hence, while the astronomical full moon (i.e. the opposition of sun and moon) used to closely match *luna* 15 during the decades around AD 300, by Bede's age it had moved back towards *luna* 14, sometimes even encroaching upon *luna* 13.[5] While this difference was not always

[4] Bede, *De temporum ratione* (43), CCSL 123B:415–16: "Et non potius intelligendum quia, cum lunam anni illius paschalem a pridie nonas Apr. incipere signabant, aliud maius periculum per hoc declinaverint ne videlicet, si aliter decernerent, indissolubilis ille communium annorum et emblismorum status solveretur, quem inviolabiliter observandum divinae legis auctoritate Hebraeis tradentibus agnoverant?" For a more detailed analysis of this chapter, see Wallis, *Bede*, 328–32.

[5] Kenneth Harrison, "Easter Cycles and the Equinox in the British Isles," *Anglo-Saxon England* 7 (1978): 2–3.

easy to notice by simple observation, it could quickly reveal itself during eclipses, which occurred only at the time of true conjunction (solar eclipses) or opposition (lunar eclipses). One noteworthy example is the solar eclipse, which could be seen over Northern England on 1 May 664 (Oppolzer no. 4465), the same year in which the Synod of Whitby in Northumberland decided on the adoption of the 'Roman', i.e. Alexandrian, style of Easter reckoning. According to the Alexandrian cycle, 1 May in that year was supposed to correspond to *luna* 28 of a 'hollow' (29-day) lunation, which was far enough removed from the calculated day of conjunction to cause potential consternation. As a matter fact, at some early stage in the historical transmission of said astronomical event, the date of the eclipse was changed to 3 May, which was the beginning of a new lunar month in the 19-year cycle, apparently in an effort to obfuscate the discrepancy. In Bede's works, this falsified date for the eclipse of 664 is recorded no less than three times, whereas the correct date has been preserved in Irish annals.[6]

Confusion of this sort was, in theory at least, bound to repeat itself whenever a solar eclipse was found to occur two days earlier than *luna* 1, as was the case no less than four times during the first half of the ninth century, in 809, 810, 812, and 841. The eclipse on 30 November, AD 810, even became the subject of inquiry on the part of Charles the Great, who wrote a letter to the Irish scholar Dúngal of St. Denis, in which he showed himself eager to know more about the causes and methods of prediction of such phenomena.[7] Owing to the slight excess in the duration of the Alexandrian mean lunation, overt discrepancies between the observable and the 'ecclesiastical' moon continued to increase as the centuries went by. While the critical computists seem

---

[6] Daniel McCarthy and Aidan Breen, "An Evaluation of Astronomical Observations in Irish Annals," *Vistas in Astronomy* 41 (1997): 128–31. See also Jennifer Moreton, "Doubts about the Calendar: Bede and the Eclipse of 664," *Isis* 89 (1998): 50–65. For Bede's doctrine on eclipses, see *De temporum ratione* (27), CCSL 123B:362–63.

[7] See McCluskey, "Changing Contexts," 205–11; Bruce S. Eastwood, "The Astronomy of Macrobius in Carolingian Europe: Dungal's Letter of 811 to Charles the Great," *Early Medieval Europe* 3 (1994): 117–34; Brigitte Englisch, "Karolingische Reformkalender und die Fixierung der christlichen Zeitrechnung," in Warntjes and Ó Cróinín, *Computus*, 238–58. See also Borst, *Kalenderreform*, 268–69, where a comment regarding the lunar eclipse of 2 November 784 (Oppolzer no. 3083) in a ninth-century calendar manuscript (MS Berlin, Staatsbibliothek, Codex Philipps 1869) is cited: "Anno DCC.LXXXIIII. IIII. Non. Nov. Aeclypsis lunae facta est quasi VIII. hora noctis lunae XIIIImae. Cui vero placeat scire, quomodo potuisset luna XIIII. aeclypsin pati, dum calculatores veteres non posse esse nisi in XV. dicant, computet horam lunae que fit in mense lunari."

to have ignored the problem, the eleventh century finally saw some burgeoning attempts to improve the accuracy of lunar computations. During the 1040s, the learned monk Hermann of Reichenau (1013–54) saw himself faced with the problem that the calculated lunar ages often lagged behind the observable ones by two days, a fact which he claimed was known even to common people and peasants in his time.[8]

In his *Epistola de quantitate mensis lunaris* (ca. 1040), Hermann traced the underlying problem to the unequal lengths of the lunar months in the 19-year cycle, where the sequence of 'full' (30-day) and 'hollow' (29-day) lunations was frequently disturbed by 30-day embolisms, bissextile days and the 'leap of the moon'. Since the natural lunar months were all of equal length, it was necessary to divide the 6939.75d of the 19-year cycle by its 235 lunations in order to obtain its mean value, which is 29.530851d in decimal notation. In Hermann's terminology, which was also common to other computists, this figure was expressed as 29 days 12 hours 29 moments and 348 atoms (where there are 40 moments to an hour and 564 atoms to a moment). In his *Abbreviatio compoti*, written in 1042, he went on to use this value to construct a new type of lunar calendar, whose new moons were all equally spaced by the duration of the mean synodic month.[9] Hermann referred to his lunar tables as a *computus naturalis* to contrast it with the established *computus usualis*, which only reckoned in units of days. The nucleus of this conceptual distinction can already be found in Bede's *De temporum ratione*, where it is clearly stated that there is no perfect correspondence between the ecclesiastical lunisolar cycle and the natural movements of sun and moon. As Bede himself put it,

---

[8] Hermann of Reichenau, *Abbreviatio compoti cuiusdam idiotae* (25), edited in Germann, *De ratione temporum*, 326: "Si quem autem fortassis curiosiorem et in talibus diligentiorem una mecum permoverit, que causa quisve error sit, ut lune etas compoto nostro regulisque antiquorum supradictis persepe non conveniat, sed plerumque pridie, nonnumquam biduo—ut ipse dominus Beda fatetur et visus noster affirmat—prius quam primam computemus, luna non gracilis in celo appareat, et absurdum putaverit regulas sequendo lunam necdum esse contendere, cum omnibus vel rusticis clare novam liceat cernere, sciat se quaestionem admodum scrupulosam querere et antiquam a pluribus quidem sepe temptatam, sed nondum ab aliquot quem mea parvitas attingere potuit cauta perscrutatione discussam."

[9] Alfred Cordoliani, "Le computiste Hermann de Reichenau," *Miscellanea storica ligure* 3 (1963): 165–90; Borst, "Forschungsbericht," 407–31; Werner Bergmann, "Chronographie und Komputistik bei Hermann von Reichenau," in *Historiographia medaevalis*, ed. Dieter Berg and Hans-Werner Goetz (Darmstadt: Wissenschaftliche Buchgesellschaft, 1988), 103–17; Germann, *De temporum ratione*, 177–232.

in the Christian computus "custom and authority or at least convenience of calculation prevails over nature."[10]

To construct an actual *computus naturalis*, which adequately reflected the natural sequence of lunations, was also the avowed aim of Walcher of Malvern, who professed an interest in the influence of the moon's waxings and wanings on "human activities" and "medical exercises," which necessitated the construction of more accurate lunar tables.[11] While his basic ideas were not too dissimilar to those of Hermann of Reichenau (whom he does not seem to have known), he substantially improved on his predecessor by providing his calculated lunar phases with an empirical footing. Using the lunar eclipse of 18 October, 1092, and his astrolabe to time the opposition of sun and moon precisely, Walcher was able to obtain an astronomically observed starting point for a new 76-year lunar table, which covered the years AD 1036 to 1111. Unfortunately, he subsequently weakened this promising approach by rounding his already inaccurate value for the mean lunation (which is the same as Hermann's) down to the quarter of an hour. Given such imprecision, it is hardly surprising that Walcher, who continued to check his calculations against observed eclipses, soon began to notice serious discrepancies, which seemed to indicate a lack of constancy in the moon's movement.[12]

For all its defects, Walcher's lunar cycle—just like a similar table, which was constructed only a few years later by the computist Gerland—gives a good indication of the enormous intellectual sea change that was about to unfold in the course of the twelfth century thanks to the reception of Arabic science. One of the results of this development was a serious improvement in the field of astronomical knowledge,

---

[10] Bede, *De temporum ratione* (11), CCSL 123B:317: "Sed in utroque mense computando, consuetudo vel auctoritas vel certe compendium calculandi naturae praevaluit; nam non solum lunae menses, quod calculandi necessitas cogit, tricenis undetricenisque diebus ordinant, sed et lunam superfluam, quae iuxta naturae rationem in fine anni debuerat intercalari, plerique ubilibet intercalant." See also ibid. (2; 43), CCSL 123B:274–75, 412–18, and Faith Wallis, "The Church, the World, and the Time: Prolegomena to a History of the Medieval Computus," in *Normes et pouvoir à la fin du moyen âge*, ed. Marie-Claude Déprez-Masson (Montreal: Editions CERES, 1990), 15–29.

[11] MS Oxford, Bodleian Library, Auct. F.1.9, 86va: "Si in humanis actibus, velut in exercitationibus medicinae, aliquot habet effectus lunaris incrementi sive detrimenti varietas, sic sapientes experti senserunt, necesse est ut accensionis lunae dies et hora semper ac deinde totius discursus eius dimensio ad purum dinoscatur."

[12] Ibid., 95vb–96ra.

which profoundly changed both the Western outlook on the computus and, as we shall see, the treatment of traditional chronological problems. Where eleventh-century scholars such as Hermann of Reichenau had already been aware of the defect of the Alexandrian 19-year cycle, but could only guess at the precise nature of its causes, a wealth of new astronomical sources, translated from Arabic and imported from the Iberian peninsula, now confronted Western computists with valuable new data and fresh insights into the calculation of celestial phenomena.[13]

In the wake of this development, the English West Country soon became a stronghold of astronomical and scientific knowledge. An important role in this was played by Petrus Alfonsi, a Jewish convert to Christianity, who emigrated from Spain to England in the early twelfth century, bringing with him the fruits of Arabic and Hebrew learning. Among his students were the famous translator and scientist Adelard of Bath and the aforementioned Walcher of Malvern, who later, in 1120, composed a treatise entitled *Sententia Petri Ebrei, cognomento Anphus, de dracone*, which purports to contain Petrus Alfonsi's instructions on the prediction of eclipses and stands as the first Latin treatise to use scientific quantifications based on degrees, minutes and seconds in the sexagesimal system.[14] Four years earlier, in 1116, Alfonsi

[13] On the general background, see Haskins, *Studies*, 82–140; Charles Burnett, "Mathematics and Astronomy in Hereford and its Region in the Twelfth Century," in *Medieval Art, Architecture and Archaeology at Hereford*, ed. David Whitehead (Leeds: British Archaeological Association, 1995), 50–59; Burnett, *The Introduction of Arabic Learning into England* (London: The British Library, 1997); McCluskey, *Astronomies*, 165–208; Raymond Mercier, "East and West Contrasted in Scientific Astronomy," in *Occident et Proche-Orient*, ed. Isabelle Draelants, Anne Tihon, and Baudouin van den Abeele (Turnhout: Brepols, 2000), 325–42. For Gerland's lunar table and its descendants, see Arvid Lindhagen, "Die Neumondtafel des Robertus Lincolniensis," *Arkiv för Matematik, Astronomi och Fysik* 11, no. 2 (1916): 1–41; van Wijk, *Le Nombre*, 38–45; Jennifer Moreton, "Before Grosseteste"; Moreton, "On Not Editing Grosseteste," in *Editing Robert Grosseteste*, ed. Evelyn A. Mackie and Joseph Goering (Toronto: University of Toronto Press, 2003), 176–79; Borst, *Reichskalender*, 1:265–67, 299–300, 302–6.

[14] MS Oxford, Boldeian Library, Auct. F.1.9, 96ra–99ra. On Alfonsi, see John Tolan, *Petrus Alfonsi and his Medieval Readers* (Gainesville: University Press of Florida, 1993); María Jesús Lacarra, ed., *Estudios Sobre Pedro Alfonso de Huesca* (Huesca: Instituto de Estudios Altoaragonenses, 1996); Charles Burnett, "The Works of Petrus Alfonsi: Questions of Authenticity," *Medium Aevum* 66 (1997): 42–79. For an edition of the *Sententia*, see José María Millás Vallicrosa, "La aportación astronómica de Pedro Alfonso," *Sefarad* 3 (1943): 87–97. For an updated English version of the same article, without the attached editions, see Millás Vallicrosa, "Pedro Alfonso's Contribution to Astronomy," with postscript by Charles Burnett, *Aleph* 10 (2010): 139–68.

had already produced a Latin redaction of the *zīj al-Sindhind*, a set of ninth-century astronomical tables ascribed to al-Khwārizmī, the Arabic version of which he seems to have brought with him from Spain. Aside from the *Preceptum Canonis Ptolemei*, a crude and obscure translation from Greek, based on Ptolemy's *Handy Tables*, the *zīj al-Sindhind* was the first collection of astronomical tables to make its way into the Latin West and thereby reestablish Western Christendom's contact with the tradition of Greek mathematical astronomy, which it had been sorely lacking for many centuries. Al-Khwārizmī's *zīj* was soon followed by the Toledan Tables, which had been compiled in Spain in the eleventh century. After their translation from Arabic into Latin, they were adapted by Western astronomers for their local purposes, which resulted in the circulation, during the twelfth and thirteenth century, of astronomical tables re-calculated for the meridian of places such as Toulouse, Marseilles, Paris, London, Hereford, and Novara.[15]

That Christian monastic scholars, whose outlook on astronomy was in many ways shaped by the exigencies of the Easter computus, should have eagerly embraced the new knowledge offered by Arabic sources is not entirely self-explanatory. Charles H. Haskins, the pioneer of studies in the twelfth-century scientific 'Renaissance', held Latin computistics in such low esteem that he ascribed to it a retarding effect on the reception of Arabic astronomy.[16] A very different conclusion was reached in more recent times by Faith Wallis, who justifiably called computus "the door through which ancient and Arabic astronomy and mathematics entered the West."[17] The progressive role of computistics in the quick integration of Arabic astronomy into the Western canon of

---

[15] For editions, see David Pingree, *Preceptum Canonis Ptolomei* (Louvain-la-Neuve: Academia-Bruylant, 1997); Heinrich Suter, ed., *Die astronomischen Tafeln des Muhammed ibn Mūsā al-Khwārizmī* (Kopenhagen: Andr. Fred. Høst & Søn, 1914); Otto Eduard Neugebauer, *The Astronomical Tables of Al-Khwārizmī* (Kopenhagen: Munksgaard, 1962); Fritz S. Pedersen, *The Toledan Tables*, 4 vols. (Kopenhagen: Reitzel, 2002). See further G. J. Toomer, "A Survey of the Toledan Tables," *Osiris* 15 (1968): 5–174; Raymond Mercier, "Astronomical Tables in the Twelfth Century," in *Adelard of Bath*, ed. Charles Burnett (London: The Warburg Institute, 1987), 87–118.

[16] Haskins, *Studies*, 87: "In such conservative circles it was natural that Arabic astronomy should penetrate slowly."

[17] Wallis, "The Church," 25. See also Jennifer Moreton, "The *Compotus* of 'Constabularius' (1175): A Preliminary Study," in *Langage, Sciences, Philosophie au XIIe siècle*, ed. Joël Biard (Paris: Vrin, 1999), 62: "*Computus* seems to have been at the 'cutting edge' of the new science of the later middle ages." On the wider context see further McCluskey, *Astronomies*, 77–96, 198–202; John North, *Cosmos* (Chicago: University of Chicago Press, 2008), 237–40. For a short survey of later medieval computistical literature, see Lynn Thorndike, "Computus," *Speculum* 29 (1954): 223–38.

knowledge has also been emphasized by other contemporary scholars, most recently by Max Lejbowicz, who points to an interesting detail in vocabulary: even though the West has a well-known track record in borrowing technical astronomical terms from Arabic, it never came to use the word *zīj* to refer to astronomical tables, instead staying content with the Latin word *tabula*, as in *tabula paschalis* or Easter table.[18]

There are many ways to frame the historical affinity between computistics and mathematical astronomy, one of which would be to say that Arabic sources provided new answers to questions which Western computists had been wrestling with for a long time. These questions partly pertained to the length of the mean lunation, whose inaccurate estimation by the creators of the Alexandrian 19-year cycle had caused much of the chronological confusion noted in the previous chapters. By contrast, the newly imported astronomical tables provided Christian computists with first-hand knowledge of the Muslim calendar, thereby confronting them with a completely different approach to lunar computation. In order to keep the lunar months of the calendar (which alternated between 29 and 30 days) in step with the phases of the moon, Islamic astronomers intercalated 11 days over the course of 30 lunar years. Expressed in modern notation, this implied a mean value of 29.53055d, which is reasonably close to the 29.53059d proposed by modern methods of measurement. In addition to a great wealth of astronomical data, the *zījāt* also provided their Western users with valuable information on historical chronology, some of which had previously been unavailable. A typical *zīj* began with tables comparing different chronological eras, which were aimed at enabling astronomers to base their calculations on different epochs (tied to different calendars) by converting one era into another. These lists usually included the eras of Nabonassar and Philip Arrhidaeus, which had already been used by Claudius Ptolemy in the *Almagest* and the *Handy Tables*, as well as the Islamic Hijra and the Persian era of the Sassanid king Yazdegerd III.[19]

---

[18] Max Lejbowicz, "Des tables pascales aux tables astronomiques et retour: Formation et réception du comput patristique," *Methodos: Savoirs et textes* 6 (2006), 1–2, http://methodos.revues.org/documents538.html. In fact, it was neither incorporated nor loan-translated, as could have been done by using the word *linea*. I owe this observation to Leofranc Holford-Strevens.

[19] See, e.g., Neugebauer, *Astronomical Tables*, 82–84; Al-Bīrūnī, *Chronology*, 27–36; David Pingree, *The Thousands of Abū Maʿshar* (London: The Warburg Institute, 1968), 37–40.

One early exemplar of such a chronological list can be found in Petrus Alfonsi's redaction of al-Khwārizmī's *zīj al-Sindhind*, as it appears in a manuscript from Oxford, Corpus Christi College (cod. 283, fol. 114r).[20] The list is very unusual in so far as it expresses each era by giving its distance in years, months, and days from the same epoch date, 1 October, AD 1116. This date, which must have been reasonably close to the time the table was composed, still falls within the lifetime of Petrus Alfonsi. The same list of eras makes a surprising second appearance as part of an astronomical appendage to the *Liber ysagogarum Alchorismi* (before 1143), a mathematical treatise, which contains one of the earliest known Latin expositions of Hindu-Arabic numerals and the corresponding algebraic methods of calculation.[21]

Scholars have long inconclusively speculated about the author or redactor of the *Liber ysagogarum*, which in one manuscript is attributed to a certain *Magister A.* This would match the names of either Petrus Alfonsi or his student and translator Adelard of Bath, who are both linked to the *Liber ysagogarum* via the aforementioned chronological table. Although both names have been suggested, more recent research by André Allard has made it seem likely that the work was composed in Spain in the milieu of Toledo, with Abraham ibn Daud as a possible contributor.[22] That the work stems from an author with a Jewish background is strongly suggested by another striking feature of the *Liber ysagogarum*, which has to date been largely overlooked. The chronological table is preceded by a short set of conversion rules for

---

[20] See Neugebauer, *Astronomical Tables*, 137, 143–45.

[21] The table is found in MSS Paris, BN, lat. 16208, 71va; Munich, BSB, Clm 13021, 30v; Milan, Biblioteca Ambrosiana, A.3.sup, 18r. See the edition in Bruce George Dickey, "Adelard of Bath: An Examination Based on heretofore Unexamined Manuscripts" (PhD diss., University of Toronto, 1982), 318.

[22] MS Paris, BN, lat. 16208, 67ra: "Incipit liber ysagogarum alchorismi in artem astronomicam a Magistro A compositus." On the problem of authorship, see André Allard, "The Arabic Origins and Development of Latin Algorisms in the Twelfth Century," *Arabic Sciences and Philosophy* 1 (1991): 242–49; Allard, *Muhammad Ibn Mūsā Al-Khwārizmī: Le calcul indien (Algorismus)* (Paris: Blanchard, 1992), viii–xxi, xxxvi–xxxviii; Burnett, "The Works," 51–52, and Burnett, "Catalogue: The Writings of Adelard of Bath and Closely Associated Works, together with the Manuscripts in Which They Occur," in Burnett, *Adelard of Bath*, 173–74. For further opinions, see Haskins, *Studies*, 24; Millás Vallicrosa, "La aportacíon," 83; Richard Lemay, "The Hispanic Origin of our Present Numeral Forms," *Viator* 8 (1977): 446n46; Dickey, *Adelard of Bath*, 77–111; Menso Folkerts, "Adelard's Version of Euclid's *Elements*," in Burnett, *Adelard of Bath*, 63; Louise Cochrane, *Adelard of Bath* (London: British Museum Press, 1994), 81, 84n32; Shlomo Sela, *Abraham Ibn Ezra and the Rise of Medieval Hebrew Science* (Leiden: Brill, 2003), 21n15.

dates in the Muslim and Jewish calendars, which represent the earliest discussion of the medieval conjunction-based Jewish calendar in a Latin text. *Magister A.*'s remarks on the Jewish calendar are frustratingly terse (to the point of being incomprehensible) and mostly consist of instructions on how to convert Arabic lunar dates into Jewish lunisolar dates and vice versa. These instructions are followed by Latin transcriptions of the Jewish month names and a short introduction to the numerical values of the letters of the Hebrew alphabet. In one of the preserved manuscripts (Paris, lat. 16208), this latter passage was copied by a different hand from the rest of the text. It would seem that the scribe initially responsible for copying the work was daunted by the confrontation with exotic Hebrew letters in this section and left its completion to a more experienced colleague. A similar situation seems to have occurred in the case of another manuscript, which was copied during the 1160s in the Bavarian monastery of Prüfening near Regensburg (Munich, Clm 13021). This manuscript omits the Hebrew alphabet altogether and leaves three lines of blank space in its place, indicating that the scribe intended the passage to be inserted by someone else at a later stage.[23]

The *Liber ysagogarum* provides us with an early glimpse at the burgeoning reception of the medieval Jewish calendar in the Latin West. After centuries during which Western computists had been labouring under the misapprehension that the Hebrew/Jewish lunisolar calendar and the ecclesiastical 19-year cycle were basically interchangeable systems of computation, twelfth-century Christian scholars finally began to realize that the Jews of their own time had long reverted to a far more sophisticated scheme of calculation, which is still in use today. The computational basis of this new scheme was the time of mean conjunction, also known as *molad* (מולד) in Hebrew (indicating the 'birth' of the moon). All *moledot* occur at intervals of exactly 29 days, 12 hours, and 793 *chalakim*, with the 'chelek' corresponding to 1080th of an hour or 3⅓ seconds. The resulting value of 29; 31, 50, 8, 20d (in sexagesimal notation), which is very close to modern estimates of the mean synodic month, has an ancient pedigree, as it can be found in Ptolemy's *Almagest* and even in earlier Babylonian sources. In terms of accuracy, the Jewish *molad*-system was a great improvement over

[23] MSS Paris, BN, lat. 16208, 71va; Munich, BSB, Clm 13021, 30va.

previous schemes, especially over the Christian 19-year cycle with its error of one day in three centuries.[24]

It is difficult to tell with any certainty when this medieval Jewish calendar, with all its intricate rules of operation, came into being, but the remaining evidence suggests that the transition occurred in several steps between the fourth and the tenth century AD.[25] It is only from the early twelfth century onwards, however, that a number of comprehensive scholarly tractates, explaining the calendar's astronomical and legal basis, have been preserved in Hebrew manuscripts. The most famous of these, Maimonides's *Laws of the Sanctification of the New Moon* (ca. 1170/78), which was part of his *Mishneh Torah*, was preceded earlier in the century by Abraham bar Ḥiyya Savasorda's *Sefer ha-ibbur* (ca. 1123) and a work of the same name by Abraham Ibn Ezra (1146). In addition to his *Sefer*, Ibn Ezra also penned a 'Letter of the Sabbath' (*Igeret ha-Shabbat*, 1158), in which he turned to some specific problems of calendar reckoning. At one point in his 'letter', he criticized fellow Jews who communicated the technical details of their calendar to Gentiles in an inexact manner. This passage is of some interest, as it hints towards a lively atmosphere of exchange between Jews and Christians on calendrical matters.[26] In fact, Ibn Ezra himself, in spite of his criticism of other scholars, included some remarks on the Jewish 19-year cycle in his Latin *Liber de rationibus tabularum* (1154), which he specially composed for a Christian audience. Comparing the

---

[24] For useful introductions to the fixed Jewish calendar, see Feldman, *Rabbinical Mathematics*, 185–210; Arthur Spier, *The Comprehensive Hebrew Calendar* (New York: Behrman House, 1952), 217–27.

[25] This development, however, was not necessarily a linear or progressive one. The best account is Stern, *Calendar*, 164–210. See now also Sacha Stern and Piergabriele Mancuso, "An Astronomical Table by Shabbetai Donnolo and the Jewish Calendar in Tenth-Century Italy," *Aleph* 7 (2007): 13–41, with corrigenda in *Aleph* 8 (2008): 343–44.

[26] See Michael Friedländer, "Ibn Ezra in England," *Transactions of the Jewish Historical Society of England* 2 (1894–95): 71; Shlomo Sela, "Contactos científicos entre judíos y cristianos en el siglo XII: El caso del *Libro de las Tablas Astronómicas* de Abraham Ibn Ezra en su versión latina y hebrea," *Miscelánea de estudios árabes y hebraicos*, Sección de Hebreo, 45 (1996): 216–17; Sela, *Abraham Ibn Ezra*, 49–57. Disputes with Christian priests about the computus are also recorded in Abraham bar Ḥiyya, *Sefer ha-Ibbur* (2.5; 3.10), ed. Herschell Filipowski (London: Longman, Brown, Green, and Longmans, 1851), 44–45, 109. An even earlier Jewish-Christian exchange, centering on the birth year of Jesus, is documented in Simcha Emanuel, "A Jewish-Christian Debate—France, 1110" [in Hebrew], *Zion* 63 (1998): 143–55. On the twelfth-century *Sifrei ha-Ibbur*, see now Elisheva Carlebach, *Palaces of Time* (Cambridge, MA: Harvard University Press, 2011), 11–27.

19-year cycle used by the Jews with that of the Alexandrian Easter computus, Ibn Ezra observed that the

> Christians celebrate their Easter according to this cycle, which is why the Hebrew and the Christian Pasch both occur around the same time. But in two years out of the aforementioned cycle they differ by one month. This happens because the Christians begin their cycle three years earlier than the Jews do, which can be seen from the present year, which is the 1154th [of the Christian era]. According to the Christians, it is the 15th year of the 19-year cycle, but according to the Jews it is the 12th. Consequentially, when it is the fifth year of the cycle according to the Jews, it is the eighth according to the Christians, at which time they have an intercalary month, whereas the fifth year according to the Jews is not an intercalary year, which is why they diverge in this year.... For the same reason, the 16th year according to the Jews is the 19th year according to the Christians. And this is the reason why the Christians, in what is the 5th and 16th year according to the Jews, would have the Pasch before 21 March, when the sun has not yet entered the head of Aries, unless they insert an intercalary month.... And in each year in which the beginning of the lunar month in which the Pasch occurs falls on a Sabbath or a Sunday, the Jews celebrate seven days earlier than the Christians. The reason for this is that the lunar month according to the Jews begins either with the conjunction of sun and moon or with [the moon's] first visibility. The Christians, on the other hand, reason that one cannot define a day as the first day of the moon except the one on which it was first sighted by Adam, who was created on the third day after the creation of the moon itself; and this is why the Christians refer to that day as the 'first', which comes about two days after the conjunction of sun and moon, in accordance with the [difference between] the ages of the moon at its creation and its first sighting by man.[27]

[27] José María Millás Vallicrosa, *El libro de los fundamentos de las Tablas astronómicas de R. Abraham Ibn Ezra* (Madrid: Casa Provincial de Caridad de Barcelona, 1947), 99–100: "Et secundum hunc circulum feriant christiani suum pascha. Ideoque pascha hebreorum et christianorum fere eodem tempore contingunt. In circulo tamen predicto in duobus locis a se dissident fere per spatium unius mensis. Et hec res contingit quia christiani inchoant circulum per tres annos antequam iudei. Quod possumus cognoscere per hunc annum presentem qui est millesimus centesimus quinquagesimus quartus: est secundum christianos quintus decimus in ciclo decemnovenali; secundum iudeos vero 12mus. Ut quando est quintus circuli secundum iudeos, est octavus secundum christianos, qui bisextilis secundum eos esse debet, cum quintus secundum iudeos non sit bisextilis, et hic est annus in quo dissident. Quoniam est sextum secundum iudeos est nonus secundum christianos, iudei faciunt bisextum tunc christiani non faciunt, tamen consentiunt in termino paschali. Secundum hanc rationem quando est 16tus secundum iudeos est 19 mus secundum christianos. Et hec est ratio christianorum quod in quinto et 16o anno secundum iudeos, nisi faciant bisextum in his, eveniet pascha ante vicesimum primum diem Marcii cum sol adhuc non intervenit caput arietis. Ratio vero iudeorum est quod ipsi non curant considerare

Ibn Ezra's remarks are worth quoting in full, because they reveal to us some of the reasons why Christians could develop a strong interest in certain details of the Jewish *molad*-calendar. Unlike the Arabic calendar, which was purely lunar, the Jews used a 19-year lunisolar cycle with the same order of intercalation (with embolismic months in years 3, 6, 8, 11, 14, 17, and 19) that also underlay the Alexandrian cycle, although the Jewish cycle started three years later, making it synchronous with both the cycle of the Byzantine world era and what was known as the 'circle of the moon' among Western computists.[28] Owing both to this structural similarity and to the obvious calendrical link between Easter and Passover, the Jewish calendar was particularly suited as a template for Christian computists, to which they could compare their own method of lunisolar reckoning. As a result, it could not take long until they realized that the calendar of their Jewish neighbours was technically superior. By basing all their lunar month on the day of mean conjunction, which was calculated using an equal value that included fractions of days and hours, the Jews had actually successfully put into practice the kind of *computus naturalis* that some Christian computists had been searching for since the eleventh century. When it came to estimating the duration of the mean lunation, the Jewish value of 29d 12h 793ch or 29.53059d was considerably shorter than that of the Alexandrian 19-year cycle, coming closer to the 29.53055d of the Arabic 30-year lunar cycle. This latter point was nicely demonstrated by the astronomer Roger of Hereford, who systematically compared the Muslim (Chaldaean), Jewish (Hebrew) and Christian lunations in a table included in his *Computus*, written in 1176. As Roger duly noted, the number of 'atoms' by which the value of the mean lunation exceeded 29 days 12 hours and 29 moments, was 188 in the case of the 'Chaldaeans' and 208 8/9 in the case of the

ingressum solis in caput arietis secundum figuram animalis, nisi secundum punctum equationis diei et noctis. Et in omni anno in quo kalende mensis lunaris in quo et pascha est, in quo inquam kalende sunt in sabbato vel in dominica die, 7 dies ante christianos celebrant iudei pascha. Et ratio est huius rei quod kalende secundum iudeos sunt in ipsa adunatione solis et lune vel secundum quod ipsa apparet visui. Ratio vero christianorum est quod non dicenda est prima nisi secundum hoc quod primo visa est ab Adam, qui tertia die creatus est post creationem lune, et ita christiani circiter spatium duarum dierum ab adunatione sua cum sole primam esse dicunt secundum etatem lune creationis et humane visionis." A similar passage, with regard to the comparison between the Jewish and Christian cycles, is already found in Abraham bar Ḥiyya, *Sefer ha-Ibbur* (2.5), ed. Filipowski, 45.

[28] On the 'circle of the moon', see Mosshammer, *Easter Computus*, 85–95.

'Hebrews', whereas the Christian calendar implied a staggering excess of 348 atoms. Since it was the ecclesiastical calendar that was burdened with the greatest discrepancy between observed and calculated moons, there could be little doubt that both Muslims and Jews had done a better job in estimating the mean synodic month.[29]

The fact remained that the beginning of the lunar month, as calculated by Christian computists, was often lagging two days behind the actual day of conjunction, as it could be found by using the Jewish *molad*-computation. From Ibn Ezra's remarks, it can be inferred that some Christian scholars tried to justify or explain away this glaring discrepancy by evoking not the greater precision of the Jewish calculation, but the chronology of creation. Since Adam had been created on the sixth day, whereas the heavenly luminaries had begun their course on the fourth day, they concluded that, from Adam's point of view, the first day of the lunar month had actually corresponded to the second day after conjunction. This peculiar explanation of the observable error in the calculated lunar phases seems to have been rather common among medieval apologists of the ecclesiastical calendar. Aside from being mentioned in Ibn Ezra's treatise, it can also be encountered in a computus-text ascribed to a certain *Magister Chonrad*, first written in 1200, but only preserved in a later redaction made in 1396.[30]

In the meantime, the problem was noticed not only in the Latin West, but also in the Byzantine Empire, where Greek computists began to devise new formulae that added two further days to the customary calculation of the lunar epact (the Byzantine equivalent to which was known as the *themélios*, indicating the lunar age on 1 January).[31] In his computus treatise, written in 1372, the mathematician Isaac Argyros tried to explain this augmentation by once again appealing to the chronology of the hexaëmeron. According to him, previous computists had ignored the fact that the moon had only been created on the

[29] MS Cambridge, University Library, K.K.1.1, 238r. On Roger of Hereford and his *Computus*, see further Haskins, *Studies*, 124–26; Josiah C. Russell, "Hereford and Arabic Science in England about 1175–1200," *Isis* 18 (1932): 14–25; Jennifer Moreton, "Before Grosseteste," 58, and the introduction to Roger Bacon, *Compotus*, ed. Robert Steele (Oxford: Clarendon Press, 1926), xix–xx.

[30] See Ferdinand Kaltenbrunner, "Die Vorgeschichte der gregorianischen Kalendereform," *Sitzungsberichte der kaiserlichen Akademie der Wissenschaften (Wien)*, phil.-hist. Kl., 82 (1876): 295.

[31] Mentz, *Beiträge*, 24–50.

fourth day, instead reckoning with the epact of the first day.[32] It is important to note that this theoretical 'reform' of the Byzantine lunar epact or *themélios* never had any effect on the actual date of Easter in the Greek Church. As far as can be ascertained, the only genuine advance towards calendar reform in the Byzantine world was undertaken in 1324, when Argyros's teacher Nicephorus Gregoras brought the matter before Emperor Andronicus II. If we can trust Nicephorus's own testimony, Andronicus was sympathetic towards his proposal of fixing the problem with the aid of Ptolemaic astronomy, but he eventually refrained from implementing the necessary reform, because he believed that his subjects would be unwilling to accept any changes to their traditional calendar.[33] When, in 1582, the Roman Church finally brought itself to implement the necessary reform, which had been discussed by Western computists for four long centuries, the Byzantine Empire was already a thing of the past.

2. *Reinher of Paderborn*

While the prehistory of the Gregorian calendar reform has been relatively well studied, modern accounts have tended to assign the beginnings of this discussion to the thirteenth century, based on the pioneering works of John of Sacrobosco and Robert Grosseteste (see next chapter). In doing so, they often ignored the fact that the first call for a reformation of the calendar dates from as early as the 1170s.[34]

---

[32] Isaac Argyrus, *Computus ecclesiasticus* (7), edited in Dionysius Petavius, *Uranologion* (Paris, 1630), 368 = PG 19:1292–93. See also Mentz, *Beiträge*, 35. Contrary to what Mentz claims, Argyrus, in his explanation of the discrepancy, did not appeal to the day of Adam's creation.

[33] Nicephorus Gregoras, *Byzantina Historia* (8.13), ed. Schopen, 1:364–73. See also Rodolphe Guilland, *Essai sur Nicéphore Grégoras* (Paris: Geuthner, 1926), 10, 282–85; Stefan Bezdechi, "Un projet de réforme du calendrier par Nicéphore Grégoras," in *Mélanges d'Histoire Générale*, ed. Constantin Marinescu, 2 vols. (Cluj: Cartea românească, 1927–38), 1:68–74; Kaltenbrunner, "Die Vorgeschichte," 324–26.

[34] The fundamental study on the prehistory of the Gregorian reform remains Ferdinand Kaltenbrunner's "Die Vorgeschichte," which has not been fully superseded. See further Pierre Duhem, *Le système du monde*, 10 vols. (Paris: Hermann, 1913–59), 4:42–60; John North, "The Western Calendar—'Intolerabilis, Horribilis, et Derisibilis': Four Centuries of Discontent," in *Gregorian Reform of the Calendar* ed. George V. Coyne, Michael A. Hoskin, and Olaf Pedersen (Vatican City: Specola Vaticana, 1983), 75–113; Christine Gack-Scheiding, *Johannes de Muris: Epistola super reformatione antiqui kalendarii* (Hannover: Hahn, 1995), 3–46; Peter Aufgebauer, "'Ein vielfaches

It can be found in the *Compotus emendatus*, written by Reinher of Paderborn in 1170 or 1171. Little is known about the author himself, who served as a dean (*decanus*) at Paderborn cathedral in Westphalia and as a schoolmaster (*magister*) at the local cathedral school, but his only preserved work testifies to a mind of exceptional erudition and mathematical skill, a fact which is duly acknowledged by the incipit of the two oldest known manuscripts, where Reinher is praised as a *perspicacissimus calculator*.[35] In later manuscripts, the ascription was instead changed to Albert the Great, which has confused some modern scholars.[36]

Reinher's work, which has been regrettably overlooked in much previous scholarship, is of great interest for the history of science. Aside from being the first complete description of the Jewish conjunction-based

---

Ärgernis in der Kirche': Probleme der christlichen Zeitrechnung im Mittelalter," in *Herrschaftspraxis und soziale Ordnungen im Mittelalter und in der frühen Neuzeit*, ed. Peter Aufgebauer and Christine van den Heuvel (Hannover: Hahn, 2006), 73–85. Only few modern accounts have properly acknowledged the twelfth-century roots of Western calendar reform. See, e.g., Steele's introduction to Bacon, *Compotus*, xix–xxi, and Jennifer Moreton's articles "Before Grosseteste," and "The *Compotus*."

[35] See Walter Émile van Wijk, *Le comput emendéde Reinherus de Paderborn (1171)* (Amsterdam: North-Holland Publishing Company, 1951), 10, whose edition is based on MS Leiden, Universiteitsbibliotheek, B.P.L. 191E (s. XII), 129r–40r: "Incipit praefatio magistri Reinheri decani Patherbornensis, perspicacissimi calculatoris, in compotum emendatum." See also MS Hannover, Niedersächsische Landesbibliothek, IV 373 (s. XIII[1]), 3v–10v: "Incipit praefatio magistri Reinheri decani Patherburnensis perspicacissimi calculatoris in compotum emendatum." On Reinher and his *Compotus*, see further Walter Émile van Wijk, "Un Comput de la fin du douzième siècle," *Archives Internationales d'Histoire des Sciences* 4 (1951): 867–73; Klemens Honselmann, "Magister Reinher: Schrittmacher für die Kalenderreform und die moderne Rechenkunst," in *Von der Domschule zum Gymnasium Theodorianum in Paderborn*, ed. Klemens Honselmann (Paderborn: Verein für Geschichte und Altertumskunde Westfalens, 1962), 107–26; Werner Herold, "Der *computus emendatus* des Reinher von Paderborn," in *Kulturarbeit und Kirche*, ed. Werner Chobrak and Karl Hausberger (Regensburg: Verlag des Vereins für Regensburger Bistumsgeschichte, 2005), 39–47.

[36] MS Admont, Stiftsbibliothek, 442 (s. XIII[4]/XIV[1]), 38r–51v: "Incipit correctio antiqui computi per fratrem Albertum facta et edita." MS Prague, Národní Knihovna, XIII.F.8 (s. XIII[4]/XIV[1]), 105r–27r: "Rectificatio computi secundum Albertum fratrem ordinis praedicatorum Ratisponensem episcopum." See Martin Grabmann, *Mittelalterliches Geistesleben*, 3 vols. (Munich: Hueber, 1926–56), 2:342; Nicholas of Cusa, *Die Kalenderverbesserung*, ed. Viktor Stegemann (Heidelberg: Kerle, 1955), lxxi; Tom Müller, *"Ut reiecto paschali errore veritati insistamus": Nikolaus von Kues und seine Konzilsschrift* De reparatione kalendarii (Münster: Aschendorff, 2010), 140, and Aufgebauer, "'Ein vielfaches Ärgernis'," 75, who erroneously reiterates the attribution to Albert. An anonymously transmitted version of the *Compotus emendatus* is found in MS Oxford, Bodleian Library, Can. Misc. 561 (s. XV), 82r–94v: "Incipit computus emendatus Rubrica." The latter version is incompete and breaks off in the middle of chaper II.14.

calendar from a Christian author, it is also among the earliest Latin texts to employ the Western form of Hindu-Arabic numerals (including zero) for a specific scientific purpose.[37] In the preface to his work, Reinher briefly acknowledges the novelty of this step, noting that he relied on the unusual numerals "for the sake of economy in writing and calculating."[38] Meanwhile, his opening words evoke the dramatic picture of an ecclesiastical calendar under siege:

> Those who attack the Catholic faith rejoice in finding errors in the computus employed by the church. For they believe and affirm that we also commit errors in other points. Eager to put an end to disgrace and mockery, not only on the part of our adversaries, but also of many believers, some of my friends have asked me to ponder how this scandal and this state of confusion could be removed.[39]

As Reinher goes on to admit, he initially hesitated to follow this request, because he knew only too well that the church was not easily persuaded to abandon long-standing customs. Yet seeing how the error of the calendar was growing in magnitude over time, he finally saw himself compelled to write down his ideas for posterity, knowing that a correction would become inevitable in the foreseeable future.[40] As he was quick to stress, it was not his intention to introduce a completely new calendar, but instead to suggest a return to the calendar of the ancient Hebrews, which had once been accepted by the early fathers of the church, before it was unwittingly abandoned by their successors.[41] In making this judgment, Reinher took several cues from the notes on

---

[37] See van Wijk, *Le comput*, 5, 72–73, and Honselmann, "Magister Reinher," 117–23. On the transmission and impact of Hindu-Arabic numerals, see J. Lennart Berggren, "Medieval Arithmetic: Arabic Texts and European Motivations," in Contreni and Casciani, *Word, Image, Number*, 351–65; Charles Burnett, "The Semantics of Indian Numerals in Arabic, Greek and Latin," *Journal of Indian Philosophy* 34 (2006): 15–30.

[38] van Wijk, *Le comput*, 10: "In designatione numerorum, figuris plerumque utimur aliis quam latinis, propter scribendi et computandi compendium."

[39] Ibid.: "Qui fidem catholicam impugnant gaudent quod errorem inveniunt in compoto quem tenet ecclesia. Putant enim et affirmant etiam in aliis nos errare. Hoc autem obprobium et hanc subsannationem non tantum adversariorum sed etiam multorum fidelium amputare cupientes, quidam amicorum meorum rogaverunt ut intenderem qualiter scandalum hoc et confusio tolleretur."

[40] Ibid.: "Verumtamen errorem magis ac magis augere videns, animum tandem apposui ut vel posteris prodessem cum cogens necessitas correctionem exposceret."

[41] Ibid.: "Compotum secundum quem annorum antiquorum quantitates et revolutiones cognoscuntur, quem primitivae ecclesiae patres non abjecerunt, a quo etiam se non declinare vel aberrare putabant qui moderni temporis errorem condiderunt et induxerunt, in medium proferre intendimus, quanto brevius et facilius possumus."

the history of Easter reckoning found in Bede's *De temporum ratione*. After the Council of Nicaea had formally decided on celebrating Easter on the Sunday that followed the first spring full moon, it had (so Bede asserts) become the special prerogative of the bishops of Alexandria, whose expertise in the quantitative sciences was highly esteemed, to notify the Roman bishop and the churches in the East every year of the date of Easter. After a while, however, the other churches became weary of this procedure and started to develop their own rules, which resulted in a period of confusion, during which Easter was celebrated on different dates in different regions of the Empire. This variety, Reinher claimed, persisted until the introduction of the 532-year cycle attributed to Dionysius Exiguus, which once again ensured ecclesiastical unity.[42]

The relative simplicity of this cycle, however, was paid for with a loss of astronomical accuracy. Two problems in particular, according to Reinher, plagued the Dionysiac cycle and hence made it impossible to celebrate Easter in accordance with the Law. For one thing, Dionysius Exiguus and his followers had defined the Easter full moon as the first full moon falling on or after the vernal equinox, even though no such rule existed in the Pentateuch. Worse still, they had been wrong in assuming that the equinox would always fall on the same date in the Julian calendar, namely 21 March. Secondly, they had been mistaken in believing that the new and full moons would return on the exact same Julian calendar dates after 19 Julian years. Since the full moon in present times often fell after the astronomical equinox, but before 21 March, Easter was in reality frequently celebrated one month later than prescribed by the Alexandrian rules.[43] In order to remedy the defects of the Dionysiac reckoning, Reinher proposed a return to the ancient roots of Christian computistics, which to him were identical with the Jewish conjunction-based calendar. *Compotus*, in Reinher's definition of the word, was a science (*scientia*) established for the determination of religious feast days, which could be traced back directly to the time of the Exodus and the divine revelation to Moses and his people.[44] In contrast to Dionysius Exiguus, Moses had been instructed

---

[42] Ibid., 24. See Bede, *De temporum ratione* (44), CCSL 123B:418–19.

[43] van Wijk, *Le comput*, 26, 28.

[44] Ibid., 16: "Compotus est scientia inveniendi quibus diebus et feriis annua festa redeant. Est autem haec scientia ad cultum divinum necessaria. Cum enim per Moysen filiis Israel Dominus praeceperit ut per annos singulos quarto decimo die primi mensis

in the astronomical and mathematical wisdom of the Chaldaeans and Egyptians and would never have accepted an inaccurate computus of the kind which the Latin Church presently had in use. "In order to return to ancient wisdom, that is in order to calculate the lunar phases in accordance with nature," it was hence essential to learn more about the methods by which the Jews calculated the 14th day of Nisan.[45]

The bulk of Reinher's *Compotus emendatus* was accordingly dedicated to an extensive and astoundingly competent explanation of the present-day Jewish calendar and its computational methods. As Reinher went on to show, one Jewish 19-year cycle comprised 6939d 16h 595ch, as opposed to the 6393.75d of the Christian cycle. As a result, the mean lunation of the Dionysiac cycle exceeded that of the Jewish calendar by what Reinher designated as 313/50760th of an hour, a difference which accrued to a whole day in every 314 years, 275 days, 2 hours, and 112 parts of an hour (where 235 parts = 1 hour). This meant that since the beginning of the Dionysiac calculation in the sixth century, the calculated new moons had started to lag roughly two days behind the astronomically accurate dates.[46] In order to better demonstrate the admirable accuracy of the Jewish calendar, Reinher furnished his *Compotus* with a set of conversion tables and extensive instructions for their use.[47] Since he knew that not all of his readers would be able "to gain such proficiency in reckoning" that they would be able to put these tables to their proper use, he generously added another table "for the more simpleminded" (*simpliciores*), which indicated the exact day of the Jewish lunation for the first day of each Julian month in the years 1171 through to 1270. In addition, he promised that "anyone capable of working with the tables described above will be able to extend this table for any future period."[48]

---

ad vesperem Pascha initiarent et ceteras solemnitates suas certorum mensium certis diebus agerent, quibusdam computationum regulis fieri oportuit ut mensium et annorum principia et quantitates cognoscerent."

[45] Ibid., 28: "Quod si ab errore ad veritatem redire placuerit, scilicet ut lunae aetatem verius computemus et ut mense primo Hebraeorum proxima (prima) feria post XIIII diem eiusdem mensis solemnitatem Paschae inchoemus sicut etiam hi qui errorem induxerunt se instituisse putabant, oportet ut Hebraeorum computationibus invenire sciamus, qui sit mensis primi dies XIIII-us."

[46] Ibid., 28, 48, 50.

[47] Ibid., 30–40.

[48] Ibid., 40: "Quoniam autem numerandi facilitatem non omnes assequi valent ut secundum praecepta supraposita annorum et mensium principia invenire queant, idcirco etiam simplicioribus providentes tabulam subjunximus in qua ab anno incarnationis

Since his main purpose was to make the Jewish calendar applicable to the calculation of Easter, Reinher refrained from basing his calculations on the Jewish world era, but instead chose the Christian era (beginning on 31 December, 1 BC, at noon, according to astronomical convention) as his epoch or 'root-date' (*radix*). According to his own calculation, this date was separated by an interval of 3760 years, 3 months, 15 days, and 21h 952ch from the beginning of the Jewish calendar.[49] This result is somewhat puzzling. Using the present-day Jewish calendar, we find that the *molad Tishri* of year 3761 of the Jewish world era (JE) fell on a Friday, 11h 989ch, which is equivalent to 17 September, 1 BC, in the Julian calendar. The fact that Reinher added 104d 12h 91ch (= 3 × 29d 12h 793ch + 15d 21h 952ch) to this figure in order to reach the noon of 31 December, 1 BC, shows that he assumed that his Jewish *molad*-dates were calculated from a noon epoch, rather than from sundown, as is actually the case. Since Reinher was clearly aware of the Jewish calendar's evening epoch, this seems to have been an intentional alteration, presumably aimed at facilitating conversion between Jewish and Christian dates.[50]

In a further cumbersome calculation, Reinher showed that by the beginning of 13 September, 1170—a date evidently close to the composition of the *Compotus emendatus*—1,800,657 days as well as 18h 876ch had elapsed since the start of the Jewish calendar. He could thus extrapolate that the latter had begun on Monday, 7 October, 5h 204ch.[51] This is the value of the so-called *molad baharad* (2.5.204 = בהר"ד), whose civil time corresponds to Sunday, 6 October, 3761 BC, 23 hours, 11 minutes, and 20 seconds after midnight. As Reinher went on to point out, there were reasons to be dissatisfied with this Jewish epoch, which purported to indicate the time of the creation of the world. After all, most Christians agreed that the world had been

Domini 1171 usque ad 1270 facile sit invenire cuilibet quo die anni romani incipiat annus antiquae computationis et quilibet mensis eius ut ubi incipit mensis quem Dominus primum mensem anni appellari jubet, cuius die XIIII ad vesperam Pascha initiare praecipit. Tabulam autem hanc ad quantumlibet prolixi temporis spatium extendere poterit, qui in supradictis inveniendis expeditus fuerit."

[49] Ibid., 32, 36.

[50] Ibid., 36: "Sciri autem oportet quod in hac computatione dies singulos a meridie incipimus, id est sex horis ante usualem eius inceptionem." Ibid., 52: "Principium ergo annorum Hebraeorum fuit post meridiem secundae feriae, transactis horis 5 et partibus 204." In his commentary (ibid., 75–76), van Wijk unnecessarily complicates the issue.

[51] Ibid., 52–54.

created not in autumn, but in spring, and on a Sunday rather than on Monday, which seemed like a bewildering choice for the beginning of a lunar calendar, seeing how the moon was only created on the fourth day.[52]

Reinher's criticism shows that he was unaware of the actual rationale behind the epoch of the Jewish calendar. The year of Adam's creation, as it can be inferred from the *Seder Olam Rabbah* and other works of Rabbinic chronography, was 3760 BC, whose *molad Tishri*, according to the present-day Jewish calendar, fell on Friday, 26 September, at 14:00, which is why it is also known as *molad vayad* (ו״יד = 6.14). The fact that this is a round value (without any fractions of hours) indicates that *molad vayad* used to be the original calendar epoch, on which the Jewish creation era was based. However, since *molad vayad* marks the day of Adam's creation, the first five days of the hexaëmeron necessarily belonged to the preceding calendar year. For these reasons, the era that came to be favoured by Jews in this period begins with the previous 1 Tishri, which by consequence falls nearly a whole year before the creation of the world. Accordingly, the *molad baharad* is sometimes referred to as *molad Tohu*, in allusion to the time of the primordial *Tohuwabohu* in Genesis 1:1.[53] It almost seems that the Jewish expert, who presumably instructed Reinher on matters of the Jewish calendar, was himself somewhat confused about these matters. In the absence of any evidence to indicate that Reinher had been educated in Muslim Spain—as Honselmann conjectured—the theory that he received his knowledge through personal contacts with Jews in his native Westphalia or the Rhineland area seems by far the likeliest explanation for the impressive display of calendar lore in the *Compotus emendatus*.[54] That said, it is worth pointing out that Reinher was also familiar with Arabic astronomical tables, as becomes clear from another calculation for Sunday, 13 September, 1170, which according to him was the day of a conjunction of Mars and Jupiter as well as the first day of year 566 since the Hijra in the Muslim calendar.[55]

[52] Ibid., 56.

[53] On the background, see Sylvie Anne Goldberg, *La Clepsydre* (Paris: Michel, 2000), 246–47; Stern, *Calendar*, 192, 272–73.

[54] Honselmann, "Magister Reinher," 107, 123–25; van Wijk, *Le comput*, 73: "Reinherus peut très bien avoir recueilli ses données sur les chiffres en même temps que celles sur le comput juif, dans des discours avec un Juif instruit et je me plais à me représenter mon doyen s'entretenant avec un vénérable rabbin ou moré."

[55] van Wijk, *Le comput*, 38.

As close-lipped as Reinher may have been with regards to his sources, he became very outspoken when it came to applying his knowledge to a long-standing chronological problem: that of Christ's Passion. The latter took up the remaining eight chapters (II.8–15) of the *Compotus emendatus*, making them by far the lengthiest part of his work. As Reinher recapitulated at the beginning, the system of Dionysius equivocated the year 34 of the Christian era with the 16th year of the 19-year cycle, in which *luna* 15 of the Easter lunation fell on 22 March. The concurrent of AD 34 was IV, meaning that 24 March was a Wednesday. However, neither the weekday of the resurrection nor any of the corresponding Julian dates found in church tradition (25, 27, or 28 March) could be reconciled with these data. Since Reinher took the Gospel narratives of Christ's public ministry to be unanimous in indicating that Jesus died in AD 34, the Dionysiac parameters caused a classic trilemma, where acceptance of two elements compelled rejection of the third. The three propositions in question were: (1) One of the three traditional dates for Christ's Passion (23, 25, or 26 March) is the historical date of the crucifixion; (2) The Dionysiac era of incarnation designates the historical year of Christ's nativity; (3) Jesus died in the 34th year since his birth.[56]

Some of his predecessors, such as "Gerland, a certain computist who claimed that the number of years of the Lord was different from the one held by the entire church," had dropped proposition (2) in order to move the Passion to a more computistically convenient year.[57] As Reinher realized, the mere fact that the deficient rules of the Dionysiac cycle led into such an awkward situation provided enough reason to abandon this cycle and instead re-examine the date of the Passion on the basis of the Jewish calendar. Before this could be done, however, it was necessary to consider another chronological problem, which his predecessors had largely ignored. As a matter of fact, the 'critical computists' had been unanimous in assuming that Jesus was crucified on the 15th day of the lunar month, thus acting as if all four evangelists had been in perfect harmony on this question. Reinher, by contrast, was able to report the opinion of 'certain' (*quidam*) exegetes who dated the death of Jesus to the day on which the Paschal lamb was slain,

[56] Ibid., 56, 58, 60.

[57] Ibid., 60: "Huiusmodo rationibus seductus Gerlandus, quidam compotista, numerum annorum Domini alium esse dixit quam universalis teneat ecclesia."

i.e. on 14 Nisan. Three reasons could be given in favour of this view: first of all, there was the testimony of John 18:28, where it was stated that the Jews, on the day of the crucifixion, refrained from entering the praetorium, "lest they should be defiled, but that they might eat the Passover." The second reason concerned the status of 15 Nisan as a high Jewish feast day, on which physical labour and other kinds of activities were prohibited. This conflicted with the scriptural depiction of the events surrounding the trial of Jesus. As Reinher pointed out, not only did the Jews

> stay up all night, but they were also busy from the morning until the hour of the crucifixion, restlessly dragging [Jesus] now to Annias, now to Caiaphas, now to Pilate, now to Herod and then back to Pilate. All of this would have been forbidden on a great feast day.[58]

The third reason pertained to a particular rule in the Jewish calendar, according to which 15 Nisan could never fall on a Friday. This rule seemed to clearly contradict the synoptic depiction of the Last Supper as a Passover meal, leading some to assume "that the Lord anticipated the Passover with his disciples, because he knew he was going to suffer before the Passover legally began."[59] The calendrical rule in question is known as *lo BaDU Pesaḥ* (לא בד״ו פסח) and belongs to the complex of *deḥiyyot* or postponement rules of the Jewish calendar. According to said rule, the first day of unleavened bread (15 Nisan) can never fall on a Monday, Wednesday or Friday (hence the term *BaDU*, a mnemonic word composed of the letters in the Hebrew alphabet that correspond to the respective numbers of the weekdays, i.e. ב = 2 = Monday, ד = 4 = Wednesday, and ו = 6 = Friday). As Reinher himself knew, this rule was a mere corollary of another postponement rule, called *lo ADU Rosh* (לא אד״ו ראש), which prevented *Rosh Hashanah* (1 Tishri) from falling on a Sunday, Wednesday, or Friday (hence *ADU*, from א = 1 = Sunday, ד = 4 = Wednesday, and ו = 6 = Friday). Since

---

[58] Ibid., 60–62: "Secundo dicunt quod si dies ille magnus solemnitatis dies fuisset, circa condemnandum Dominum ita impediri eos non licuisset, quia non tamen tota nocte sed etiam a mane usque ad horam passionis non quieverunt et nunc ad Annam nunc ad Caïpham, nunc ad Pilatum, nunc ad Herodem et iterum ad Pilatum eum trahentes, requiem non habebant, quod magno die festivitatis non licuisset."

[59] Ibid., 62: "Tertio dicunt quod magnum Paschae diem numquam secunda vel IIII vel VI feria esse contingit, licet lunaris aetatis ratio supradicta his feriis sicut et aliis Pascha initiari debere exigat. Dicunt autem quod Dominus Pascha cum discipulis suis anticipaverit pro eo, quia se passurum noverat antequam, secundum legem, Pascha initiaretur."

the interval between Nisan and Tishri is a constant of 177 days in the Jewish calendar, *BaDU* follows logically from *ADU*.[60]

Early traces of these *deḥiyyot* appear in the Babylonian Talmud (*Rosh Hashanah* 20a), where a prohibition of having *Yom Kippur* (10 Tishri) fall next to a Sabbath (i.e. on a Friday or Sunday) is discussed. Such a constellation, in which two days of strict sabbatical rest directly follow each other, would have had disadvantageous consequences for the preparation of fresh food and the burial of the dead, both serious problems in hot climates. In order to forestall these scenarios, it was decided that 1 Tishri can never fall on a Wednesday or Friday. The additional prohibition pertaining to Sunday instead derived from the feast day *Hoshana Rabbah* (21 Tishri), which is accompanied by a ritual beating of willow branches (Reinher mistakenly speaks of carrying 'palms'). Since this custom conflicted with the duties of sabbatical rest, it was deemed necessary to prevent *Hoshana Rabbah* from falling on a Saturday, making it impossible for 1 Tishri to occur on a Sunday (*Sukkah* 43b). The Mishnah shows no traces of these postponement rules, which means that their existence in the first century can be excluded. Furthermore, their first mention in the amoraitic period need not imply that the postponement rules were already a fixed part of the calendar as it was actually in use among late antique Jewry. Only from the mid-ninth century onwards is there decisive evidence that *lo ADU Rosh* had made its way into the conjunction-based calendar.[61]

As the presence of the *deḥiyyot* in Reinher's discussion of the Passion date shows, some commentators, probably both Christian and Jewish, in his time believed otherwise and projected their use back to the time of Jesus. Some evidence from twelfth-century sources indicates that the postponement rule *BaDU* had become an object of serious discussion in Christian circles even before the corresponding calendar became fully known thanks to the work of Reinher and others. The earliest

[60] Ibid., 44: "Inde est quod mensem in quo Pascha celebrant, non incipiunt IIa vel IIIIa, vel VIa feria, propter decimum et vicesimum primum diem septimi mensis. Hoc enim observant quod a primo die mensis paschalis usque ad VII-i mensis primum diem, semper habent dies 177. Sed decimum diem VII mensis non licet esse in VI vel prima feria, ergo mensis paschalis non potest incipiere in IIa vel Iva feria. In VIa veria non licet eum incipiere ne VII mensis dies XXI, in quo palmas portant, in sabbatum veniat in quo non liceret palmas portare."

[61] Stern, *Calendar and Community*, 166–67, 194–95. See also Zuckermann, *Materialien*, 60–61; Mahler, *Handbuch*, 492–97; Ari Belenkiy, "A Unique Feature of the Jewish Calendar—Deĥiyot," *Culture and Cosmos* 6 (2002): 3–22; Feldman, *Rabbinical Mathematics*, 191–94.

reference to this rule in a Latin text can be found in the *Dialogue against the Jews* (ca. 1109), written by the aforementioned Petrus Alfonsi, who had converted to Christianity in the town of Huesca (Aragon) in 1106. This work, which became a popular source for Christian apologists and missionaries, pitted Alfonsi's new Christian self (*Petrus*) against his former Jewish self (*Moses*) in a series of polemical conversations. In the context of a prolonged exchange of opinion about the legitimacy of Christian vs. Jewish festivals and forms of worship, Petrus inquires as to the reason why Moses and his coreligionists sometimes postpone the Passover:

> Petrus: In reality, sometimes you even change the day of the Passover and defer it until the day following, because you never celebrate it on Monday, or Wednesday, or Friday. I want you to explain to me why you do this.
>
> Moses: I do not know why, other than that our sages ordained it so, and Gamaliel above all.
>
> Petrus: And do you know why Gamaliel did so?
>
> Moses: No.
>
> Petrus: Indeed, Gamaliel was a holy man and a faithful Christian. And because he knew that on Monday the Jews initiated a plan by which Christ could be condemned, and, moreover, on Wednesday the silver was given for the betrayal of Christ, and that on Friday Christ was fixed to the Cross, because, I say, he knew this and did not want any joy to be expressed on those days, for this reason he forbade them from celebrating the Passover on those days and enjoined that it be deferred until the day following. He did not want to reveal this secret, however, to everyone.[62]

Petrus Alfonsi's fanciful explanation of rule *BaDU* was apparently based on a conflation of Rabban Gamaliel II (ben Simeon), who is often associated with astronomy and the calendar in Jewish tradition, and his grandfather Gamaliel 'the Elder' (fl. AD 20–50), whose name is mentioned in the Acts of the Apostles (5:34–39; 22:3). The younger Gamaliel flourished at the turn of the second century (fl. AD 80–120) and is credited in medieval manuscript versions of the Babylonian Talmud (*Rosh Hashanah* 25a) with formulating the present-day length of the mean month in the Jewish calendar: 29d 12 2/3h 73ch = 29d 12h 793ch. In addition, he played a central role in the development of

[62] Petrus Alfonsi, *Dialogue against the Jews*, trans. Irven M. Resnick (Washington: The Catholic University of America Press, 2006), 263–64. On the background, see Tolan, *Petrus Alfonsi*, 12–41.

the Passover *Haggadah*.[63] Given Alfonsi's literary agenda, which aimed at the de-legitimization of Jewish custom and the demonstration that the Jewish texts contained evidence for the truth of Christianity, it is possible that this conflation of two important rabbinic sages was intentional rather than accidental. As his *Dialogue* indicates, the twelfth century was a time of change not only for secular learning, but also for the Christian attitude towards Judaism, which was viewed with increasing hostility by scholars and laymen alike. Christian polemicists such as Alfonsi himself began to develop a highly negative view of Jews as the wilful slayers of the Son of God, who not only refused to recognize Jesus as the Messiah, but who had also abandoned the pristine Law in favour of the 'heretical' Talmud.[64] At the same time, however, the apologetic need to engage intellectually with Judaism increased the amount of knowledge about Jewish religion that was available in Christian quarters. In fact, only a few years after the composition of Alfonsi's *Dialogue*, the *deḥiyyot* reappear in the treatise *De sancta trinitate et operibus suis*, written between 1112 and 1116 by the theologian Rupert of Deutz (ca. 1075–1129):

> It is their custom to avoid wherever possible two consecutive Sabbaths, i.e. two feast days on which work is not allowed. "In such a case," they say, "much hardship and many inconveniences would arise, for example in caring for the dead: if two Sabbath days intervene, their bodies would start to spread a troubling smell on the fourth day." For these reasons, whenever the 14th day of the moon fell on a Thursday, they ate the Passover lamb on the same evening [!], but they left the feast to be postponed to the following Sabbath. Knowing this, no one needs to wonder why, when the Lord suffered... the Jews were free to hold court, even though it was the 15th day of the moon, which should have been a holy feast day according to the Law.[65]

---

[63] Shamai Kanter, *Rabban Gamaliel II: The Legal Traditions* (Chico, CA: Scholars Press, 1980); Stern, *Calendar*, 201–2.

[64] Anna Sapir Abulafia, *Christians and Jews in the Twelfth-Century Renaissance* (London: Routledge, 1995).

[65] Rupert of Deutz, *De sancta Trinitate et operibus eius* (15), CCCM 22:900: "Mos illis est ut, ubicumque vitare praevalent, numquam duo sabbata, id est duos dies festos, quibus operari non liceat continuent. Multa enim, aiunt, nobis importuna fierent, atque incommoda, verbi gratia, in curandis funeribus, quae interdum a prima vespera prioris sabbati servata, in quartum diem fetoris molestiam generarent. Idcirco ubi quinta feria Paschalis terminus, id est decima quarta luna evenisset, agnum quidem eadem nocte comedebant, sed sequentis diei solemnitatem relinquebant in sabbatum transferendam. Hoc praescito nemo miratur cur, quando Dominus passus est, cum praeterita vespera Iudaei agnum comedissent, quem et ipse Dominus cum discipulis

Rupert later repeated this explanation in his *Commentary on Matthew* (1125/27), adding that

> knowledge of this tradition immediately solves this question, because it was their custom to postpone the feast whenever the 15th day of the month would legally fall before a Sabbath and instead of having two Sabbaths, they would celebrate one Great Sabbath, thus having a double feast-day. And this is the reason why the evangelist [John 19:31] says: "The Jews, because it was the day of Preparation, [asked Pilate to have his legs broken,] so the bodies would not remain on the cross during Sabbath." And he adds as their reason: "For this was a great Sabbath day."[66]

Rupert's commentaries clearly show that the existence of the *deḥiyyot* had become known among Christian scholars at a relatively early stage, owing to their usefulness in clarifying the chronology of Christ's Passion. Naturally, this knowledge must have come into circulation thanks to contacts to learned Jews. These contacts, it would seem, were fueled by polemical disputations, as they were frequently conducted by Rupert with Jews in Cologne and the neighbouring Rhineland area.[67] As the above quotations indicate, Rupert's Jewish informants got him to accept that the rule *lo BaDU Pesaḥ* could not be reconciled with a crucifixion on 15 Nisan. This conclusion, however, was quite problematic, for it directly conflicted with the verdict of Bede, who had prohibited moving the crucifixion away from the 15th of the moon as being a "great danger to the Catholic faith."[68] Rupert apparently believed that

manducaverat, Iudaeis tamen iudicia habere vacaverit, cum esset luna decima quinta, quae dies ex lege sancta esse debuerat atque solemnis."

[66] Rupert of Deutz, *De gloria et honore filii hominis super Mattheum* (10.69–91), CCCM 29:300–301: "Hanc, inquam, quaestionem cognitio cito solvit illius iam dictae traditionis, quia traditum eius fuerat differre solemnitatem, cum evenisset ante sabbatum legale quinta decima dies mensis et pro duobus sabbatis unum celebrare sabbatum magnum solemnitate duplici. Unde cum dixisset evangelista: *Iudaei ergo, quoniam parasceve erat, ut non remanerent in cruce sabbato corpora*, causam istam subiunxit: *Erat enim magnus dies ille sabbati.*"

[67] On the background, see Maria Lodovica Arduini, *Ruperto di Deutz e la controversia tra Cristiani ed Ebrei nel secolo XII* (Rome: Istituto storico italiano per il medio evo, 1979); John H. Van Engen, *Rupert of Deutz* (Berkeley: University of California Press, 1983), 241–48; David E. Timmer, "Biblical Exegesis and the Jewish-Christian Controversy in the Early Twelfth Century," *Church History* 58 (1989): 309–21; Anna Sapir Abulafia, "The Ideology of Reform and Changing Ideas Concerning Jews in the Works of Rupert of Deutz and Hermannus quondam Iudeus," *Jewish History* 7 (1993): 3–23.

[68] Bede, *De temporum ratione* (61), CCSL 123B:452: "Tantum diligentissime cavendum ne hanc sexta decima luna, ut quidam, patratam confirmando, non solum

the problem could be easily defused by assuming that the postponement of 15 Nisan left the actual Passover meal (which lay technically at the beginning of 15 Nisan) in its initial place. This way, he could claim that Jesus and his disciples had eaten lamb around the same time as all other Jews, on Thursday evening, whereas the postponement of 15 Nisan merely made sure that the ensuing Friday was not a day of rest. Needless to say, this argument failed to address the fact that the Gospel of John indicated (18:28) that the Passover meal still lay ahead from the point of view of those Jews, who refused to enter the Roman praetorium on the morning of the crucifixion.

Unlike Rupert of Deutz, Reinher of Paderborn acknowledged this point in his discussion of the Passion chronology, but none of the different arguments that he enumerated in favour of the Johannine chronology made him prepared to admit that the evangelists were in any disagreement. To assume that such a blatant chronological contradiction could exist between the four Gospels was sheer audacity in Reinher's pious mind.[69] In search for a viable exegetical pathway, he instead turned to the three synoptic accounts of the Last Supper, which began with the disciples coming to Jesus, asking him where to prepare the Passover meal. Matthew (26:17), Mark (14:12), and Luke (22:7) all referred to the day in question as the *primus dies azymorum* or "first day [of the feast] of unleavened bread." Their terminology is imprecise, since the first day of unleavened bread did not begin until the evening which marked the transition from 14 to 15 Nisan. The fact that the day on which the disciples came to Jesus was at the same time supposed to be the day on which "the Passover" was slain (Mark 14:12; Luke 22:7), likewise strongly indicates that the synoptic evangelists were really meaning to describe events which had taken place one day earlier, on the afternoon of 14 Nisan.[70]

Reinher, by contrast, came to the opposite conclusion, for he assumed that the synoptic Gospels, when talking about the "first day of unleavened bread," were to be taken literally in saying that the Last Supper had been eaten at the evening of 15/16 Nisan. This assumption naturally put the crucifixion on the afternoon of 16 Nisan. In

---

inevitabile nostrae calculationis dispendium sed et gravissimum catholicae fidei incurramus periculum."

[69] van Wijk, *Le comput*, 64: "Nos autem evangelistas contraria dixisse nec audemus nec debemus dicere."

[70] See Ogg, *The Chronology*, 221–22, for further details.

support of this scenario, Reinher pointed out that the sacrifice of the Passover lamb and the ensuing meal were only the first in a line of several sacrifices and sacred meals that took place during the week of unleavened bread according to the Mosaic Law. Numbers 28:19, for instance, spoke of "a burnt offering unto the Lord; two young bullocks, and one ram, and seven lambs of the first year." Referring to this passage, Reinher assumed that the slaying of the *pascha* on the "day of unleavened bread," as mentioned in Mark and Luke, was not the Passover lamb of 14 Nisan, but one of the further sacrifices made on the following day. Given these presuppositions, it could be concluded that the Last Supper had actually taken place one day later than commonly supposed:

> And thus we can say that, after they had begun eating unleavened bread on the evening of the 14th day, [Jesus] came to Jerusalem on the following day, i.e. on the 15th day, which is the feast day...and so there is neither a contradiction to be found between the evangelists, nor is it right to say that the three [synoptic] evangelists referred to the 14th day as the day of the unleavened against all duty and custom, as if they were ignorant of the Law.[71]

Some interesting light on Reinher's exegetical arguments is shed by situating them within the contemporary disputes over the form of bread used during the Eucharist that raged between Roman and Greek Christianity during the High Middle Ages. In the Latin and Armenian Churches, it was customary to consecrate unleavened bread (*azyma*), a liturgical convention which reflected the synoptic depiction of the Last Supper as a Passover meal. Since Jesus had come to fulfil the Law and not to destroy it (Matthew 5:17), it was reasonable to suppose that he had celebrated the meal in accordance with all divine precepts, including the use of matsot. By contrast, the Greek Church insisted on consecrating ordinary fermented bread, which could be justified by recourse to the fourth Gospel, which put the Passion on the "day of preparation" and hence on the day *before* the beginning of the feast

[71] van Wijk, *Le comput*, 66: "Sic itaque dicere possumus, quod postquam XIIIIa die mensis ad vesperam azyma comedere incoeperant, quod sequenti die, id est XVa die mensis, scilicet in die solemni, Iherosolimam venerit et ibi, coena facta, traditus sit et sic nec evangelistae inter se contrarii invenientur, nec dicere oportet quod tres evangelistae contra debitum et consuetudinem, tamquam legis ignari, appellaverint XIIIIum diem mensis primum diem azymorum." An attempt at harmonization that is quite similar to Reinher's is discussed in Beckwith, *Calendar and Chronology*, 295–96.

of unleavened bread (John 13:1, 18:28, 19:14, 31). True discontent over these ritual differences had first begun to arise in mid-eleventh-century southern Italy, where Greek and Latin Christians lived side by side. The existing Greek misgivings about the Roman rite culminated in a letter written in 1053 by Leo, the archbishop of Ochrid in Macedonia, to John, the Greek archbishop of Trani in Apulia, in which he condemned the use of unfermented bread as a grave and 'judaizing' error. Leo's attacks were among the causes that triggered the famous Roman legation to Constantinople (1054), led by Cardinal Bishop Humbert of Silva Candida, who ended up excommunicating the letter's instigator, the patriarch Michael Cerularius, in an act that has often been construed as the starting point of the Great Schism between East and West.[72]

Humbert and his colleagues responded to the Greek charges in their written *Dialogues* between a Roman and a Byzantine cleric, which contained a plethora of theological arguments in favour of the Latin rite. From a chronological point of view, the most interesting idea presented in this text concerns the interpretation of John 18:28, which, taken by itself, seemed to indicate that Jesus and his disciples had already convened for the Last Supper on 13/14 Nisan, thereby vindicating the Greek position. According to the Latin rebuttal offered in the *Dialogues*, the expressions *pascha* and *azyma* or *dies azymorum*, as used by the evangelists, were essentially interchangeable synonyms. Matthew 26:2, for instance, quoted Jesus as saying that "the Passover [*pascha*] is two days away," whereas Matthew 26:17 apparently referred to the same day as "the first day of unleavened bread" (*primus dies azymorum*). Likewise, Luke expressly stated that the feast of unleavened bread was also "called the Passover" (22:1) and referred to the day of preparation, on which the lamb was slain as "the day of unleavened bread" (22:7). Similar ambiguities could be found in the Gospel of Mark (14:1, 12) and in the Acts of the Apostles (12:3–4). These passages could be jointly used to re-construe the Jews' mention

---

[72] On the background, see John H. Erickson, "Leavened and Unleavened: Some Theological Implications of the Schism of 1054," *St. Vladimir's Theologial Quarterly* 14 (1970): 155–76; Mahlon H. Smith III, *And Taking Bread*... (Paris: Éditions Beauchesne, 1978); Georgij Avvakumov, *Die Entstehung des Unionsgedankens* (Berlin: Akademie Verlag, 2002), 29–159; Axel Bayer, *Spaltung der Christenheit* (Cologne: Böhlau, 2002), 214–21; Henry Chadwick, *East and West* (Oxford: Oxford University Press, 2003), 200–32; Brett Edward Whalen, "Rethinking the Schism of 1054: Authority, Heresy, and the Latin Rite," *Traditio* 62 (2007): 1–24.

of the *pascha* in John 18:28 as a reference to the matsot or *panes azymi*, which were consumed throughout the week of unleavened bread. As a result, the scene at the praetorium and the ensuing crucifixion of Jesus could be understood to have taken place on 15 Nisan, the first day of unleavened bread. This interpretation naturally also presupposed that the "day of preparation for Passover" (*parasceve paschae*) mentioned in John 19:14 was a reference to the approaching Sabbath day (19:31) and not to the preparation of the Passover meal. In this scenario, Jesus and his disciples had indeed celebrated the Passover in line with the Mosaic Law on the evening of 14/15 Nisan, which entailed that they had only used unfermented bread, as was assumed by the Roman rite during the Eucharist.[73]

That concern over this subject persisted into the twelfth century and also captivated scholars north of the Alps is once again witnessed by Rupert of Deutz, who defended the Roman use of azymes in a chapter of his *Book on the Divine Offices*, composed around 1110.[74] It would be good to know whether Reinher of Paderborn was likewise aware of these debates and hence of the broader ritual and theological implications of his understanding of the Passion chronology. His efforts to deflect the claim that Jesus had anticipated the Passover by celebrating the Last Supper on the evening of 13/14 are indeed similar to those made by previous Latin authors in defence of the Roman rite, albeit Reinher saw himself compelled by the *deḥiyyot* to move the Last Supper to 15/16 Nisan, while the Humbertian *Dialogues* had been content with the traditional synoptic date of 14/15 Nisan. In this regard, it is striking to see that the idea, set forth in the *Dialogues*, according to which the *pascha* in John 18:28 should be understood not as a reference to the paschal lamb, but to the unleavened bread (*azyma*), also occurs in Reinher's *Compotus emendatus* in addition to his previous argument, which hinged upon the sacrificial animals mentioned in Numbers 28:19.[75] The 'certain' exegetes, who according to Reinher defended a crucifixion on 14 Nisan on the basis of the Johannine

[73] Humbert of Silva Candida (et al.), *Adversus Graecorum calumnias* (16), PL 143:943–45. See Avvakumov, *Entstehung*, 140–42.

[74] Rupert of Deutz, *Liber de divinis officiis* (2.22), CCCM 7:52–56. For another, roughly contemporary, take on the same topic, see Anselm of Canterbury's *Epistola de sacrificio azimi et fermentati*, in his *Opera omnia*, ed. Franciscus S. Schmitt, 6 vols. (Stuttgart-Bad Cannstatt: Frommann, 1968), 2:223–32.

[75] van Wijk, *Le comput*, 66.

account, might thus have been a veiled reference to the Greek side in the debate over unleavened vs. fermented bread.

In any case, thanks to his clever juggling of different passages from the Old and New Testament, Reinher had managed to shift the lunar date of the crucifixion from the Johannine 14 Nisan to 16 Nisan, whilst retaining the assumed harmony between all four Gospel accounts. What was now left to do was to show by means of calculation that 16 Nisan in AD 34 had really been a Friday, as presupposed by his own argument. Using his conversion tables, Reinher concluded that the beginning of Wednesday, 10 March, AD 34, was separated by an interval of 3793y 6m 20h 109ch from the epoch of the Jewish calendar. A look at the present-day Jewish calendar shows that this was essentially correct: the Jewish year 3794 began on Monday, 14 September, at 23h 533ch. Adding the value of six months or 6 × 29d 12h 793ch leads to a *molad Nisan* on Tuesday, 9 March, 3h 971ch. By adding 20h 109ch, one indeed reaches the beginning of Wednesday, 10 March. As this surplus value indicates, however, the 1st day of Nisan in AD 34 should have been 9 March instead of 10 March. In Reinher's case, his oversight that 1 Nisan fell one day earlier even caused a mistake of two days, as 10 March was a Wednesday, thus necessitating a postponement to Thursday according to rule *BaDU* (the weekday of 1 Nisan is the same as that of 15 Nisan). This had the convenient side effect of making 16 Nisan in AD 34 fall on Friday, 26 March, which was the date to which Victorius of Aquitaine had assigned the Passion in the fifth century.[76]

For Reinher of Paderborn, this result was a triumph: thanks to his advocacy of the Jewish *molad*-reckoning, he had seemingly managed to reconcile patristic tradition with the Gospel data and even with the Dionysiac era of the incarnation. The key to this success was his unorthodox and, with hindsight, rather doubtful assumption that the crucifixion had taken place on 16 Nisan. It seems likely that this assumption did not flow straightforwardly just from his reading of the Gospel texts, but had in fact been forced upon him by the calendrical

[76] Ibid., 68–70. Gack-Scheiding, *Johannes de Muris*, 10–11, makes the claim that Reinher believed that the Jewish calendar was inaccurate and could not be used for chronological calculations ("Da aber bereits bei der Errechnung des Datums der Weltschöpfung Fehler entstanden seien, könne selbst das Datum von Christi Geburt mittels des jüdischen Kalenders nicht exakt berechnet werden"). As the above discussion shows, this is not the case. A similarly poor understanding of Reinher's text mars the account in Borst, "Computus," 45; Borst, *Schriften*, 1:104.

situation, once he had settled for AD 34 as the year of the crucifixion. Be that as it may, Reinher of Paderborn, "the most perspicacious calculator," had managed to show one thing with great clarity: the 19-year lunisolar cycle, which had been relied upon so much by the critical computists, was no more than a blunt weapon. For future attempts to date the death of Jesus Christ there would be no way past the Jewish calendar and the application of astronomy.

3. *The* 'Compotus Constabularii'

One question, which Reinher left untouched, was how Augustine, Bede, and other ecclesiastical authorities had managed to go astray in putting the date of the Passion on 25 March instead of 26 March, even though the former was a Thursday in AD 34. Only four years later, another astronomically minded author, this time writing in England, went on to tackle these sensitive issues in a computistical tractate of his own, which was wholly independent of Reinher's *Compotus emendatus*. This remarkable work is preserved in a single manuscript, partially damaged by fire, which is kept in the Cotton Collection of the British Library (Vitellius A.XII). Hints that the bulk of the work was written in 1175 can be gathered from scattered references to the *annus praesens*, for instance when the author mentions the current Jewish year as 4935.[77]

An important secondary witness to the text and its presumed author can be found in a manuscript from Christ Church Library, Canterbury, copied in ca. 1185 by the monk Salomon. In a short note, included in the preface to a collection of computistical texts and materials, Solomon mentions that he chose to include some remarks on the work of a *Magister Cunestabulus*, in which was "argued convincingly and efficiently with astronomical arguments against Gerland and Marianus and

---

[77] MS London, British Library, Cotton Vitellius A.XII, 87ra–97va (hereafter cited as *Constabularius*). Hints on the *annus praesens* appear on fols. 90v (epact 26), 93r, 94rb (4935 JE). The English origin is alluded to on fol. 96: "Item, quandoque luna distans a sole Paulo minus quam XXIX gradibus, in Anglia non apparebit, maxime circa equinoctium autumpnale." For descriptions of the MS, see René Derolez, *Runica Manuscripta: The English Tradition* (Bruges: De Tempel, 1954), 222–25; Walsh and Ó Cróinín, *Cummian's Letter*, 51–52.

in favour of the church about the years of the Lord."[78] This description perfectly matches the 39th and final chapter of the computus found in the Cotton manuscript, making it possible to identify its author as *Magister Cunestabulus*. In addition, we know of three copies of a *Compotus Constabularii*, mentioned in a fourteenth-century catalogue from Christ Church Library. It is likely that one of them was the copy now found in the Cotton Collection. Following Jennifer Moreton's helpful 'preliminary study', I shall hereafter refer to the author of the Cottonian computus as *Constabularius*.[79]

In the preface to his work, *Constabularius* is found complaining about those "hunters of novelties and immodest despisers of antiquity" among his contemporaries who tried to introduce new ideas into the ecclesiastical calendar by claiming that there was a difference between the opinion of the church and the true astronomical state of things.[80] As he explains to the anonymous addressee of his words, whom he only refers to as *dilectissime*, his own work was written with the intention of following established authority wherever possible. In particular, he claimed to have based himself on the computist Gerland, who was to be "imitated in all things, except in those where he goes against ecclesiastical use. For where he speaks well, there is no one better."[81] These reaffirmations of orthodoxy, however, should not distract us from the

---

[78] MS Canterbury, Christ Church Library, Egerton 3314, 1b, cited according to P. J. Willets, "A Reconstructed Astronomical MS from Christchurch Library Canterbury," *The British Museum Quarterly* 30 (1965): 23: "Sed & in calce huius scedule quiddam de magestri Cunestabuli scriptis inserui, in quo sane contra Marianum & Gerlandum pro ecclesia de annis domini astronomicis rationibus efficaciter disputatur."

[79] See Montague Rhodes James, *The Ancient Libraries of Canterbury and Dover* (Cambridge: Cambridge University Press, 1903), 49 (nos. 288–90); Willets, "A Reconstructed Astronomical MS," 24; Moreton, "The *Compotus*," 65–66.

[80] *Constabularius*, 87ra: "His quaedam nostrorum modernorum applaudentes nuper ausi sunt cartulis pascalibus suas novitates inscribere et suorum vestigia pervertire. Sunt enim quidam novitatis venatores et antiquitates inprobi calumpniatores, qui in doctrina christiana locum ab auctoritate tan quam inartificiosum superciliose repudiant et desuo confidentes ingenio aliter quam tota ecclesia soli sentire volunt soli scire videantur. Sed quod deterius est vidi equidem dolui qui videre scripto quod commendatum quemdam aliter se habere secundum ecclesiam aliter secundum veritatem." See also Haskins, *Studies*, 87, where part of the preface is cited.

[81] *Constabularius*, 87ra: "Te quoque dilectissime timore domini et reverentia fidi catholice vehementer abhorrere feceritur veritatem et ecclesiam.... At tamen scias me in toto hoc opere ad modum pauca dicere quem non haberam ex auctore, nichilque penitus de meo adicere nisi quod nota dubitationis insigniam ubi auctoritatibus sanctorum patrum in quibus spiritus sanctus locutus est adquiescent qui sapiunt.... Universorum appositus percurrere compotum quatenus singula que mihi dubitabilia visa sunt explanaverem. Notus est praeter ceteros auctores Gerlandus quam imitatus et

fact that the new astronomy, which had swept into Western Europe during the twelfth century, had left palpable traces in the author's own learning. This already becomes clear from his introductory definition of the computus, which is thoroughly progressive by the standards of the time:

> Computus is the science of measuring time using the mean motions of the sun and the moon. It is partly called 'natural', partly 'artificial'. The natural computus divides time into equal portions according to equal motions. To this purpose, time is divided into very small units. The artificial computus observes equality only in the computation of whole days. For it uses solar and lunar years and months which sometimes have more, sometimes fewer days.[82]

When talking about the length of the lunar month, his outlook was similar to that of his contemporary Roger of Hereford, who wrote his *Compotus* in the following year. Accordingly, he knew that

> the Arabs, with whom Alfraganus [al-Farghānī] concurs, make one lunation smaller by 160 atoms than we do. The Jews, on the other hand, only differ by 139 atoms and 1/9 of an atom. With them, Azarchel [al-Zarqālī] agrees, in whose tables [sc. The Toledan Tables] the modern astronomers put great trust. Accordingly, if our value for the lunation is too large, the lunar phases will be indicated later than they should be.[83]

Elsewhere, he claims to have frequently observed the calculated moon to be two days behind, as had already been mentioned by Bede.[84] Some remarkable erudition is also displayed in *Constabularius*'s repeated references to ancient astronomers and their observations. In 432 BC, Meton and Euctemon (*Mitan et Actimon*) supposedly found the vernal equinox to fall on 25 March; in 146 BC, Hipparchus (*Abrachiz*)

---

imitandum in omnibus exceptis his in quibus obviat usui ecclesie. Nam ubi bene dicit nemo melius."

[82] Ibid., 87rb: "Compotus est scientia mensurandi tempora mediis motibus solis et lunae. Hic partim naturalis dicitur, partim artificialis. Naturalis: equis motibus equas temporum portiones distribuit. Cuius rei gratia secatur tempus in minimas particulas. Artificialis: solum in dies integros computationem equalitatem obseruat. Facit enim et annos et menses iam solis et lunae, nunc plurium, nunc pauciorum dierum."

[83] Ibid., 96ra: "Nam arabes quibus assentit Alpharganus minorem faciunt unam lunationem quam nos 160 athomis. Iudei vero 139 athomis et IXa parte unius athomi. His consentit Azarchel in cuius tabulis moderni astronomi maxime confidunt. Si ergo nostra lunatio productior est quam oporteret, tardius quoque luna prima dicetur quam oporteret."

[84] Ibid., 97b: "Immo etiam duobus multi multitotiens teste Beda lunam novam viderunt antequam prima diceretur. Nos quoque idem frequenter accidere videmus."

found it on 24 March, in AD 140, Ptolemy observed it on 22 March, and, finally, in AD 831, Thābit ibn Qurra is said to have found it on 17 March.[85] While the vernal equinox of Meton seems to be a spurious derivation (in ancient tradition, Meton and Euctemon are only credited with observing the summer solstice on 27 June), the following two observations can be found in the third book of Ptolemy's *Almagest* (3.1). Since the most widely used Latin translation of this work was only produced by Gerhard of Cremona in 1175, the same year *Constabularius* wrote his *Compotus*, it is far more likely that he used Thābit's *De anno solis*, a work which *Constabularius* mentions several time and which also contains the equinox cited for AD 831.[86] Another source, on which *Constabularius* appears to have partly relied, was Abraham Ibn Ezra's *Liber de rationibus tabularum*, in which he was able to find some information on the 19-year cycle used by the Jews.[87]

But not all of his ostensive knowledge about the Jewish calendar can be explained from his acquaintance with Ibn Ezra. Judging from what he told his readers at one point in his treatise, he seems to have had some personal contacts with learned Jews, with whom he conversed on calendrical matters.[88] This becomes particularly likely in the case of the *tequfot*, the equinoxes and solstices in the Jewish calendar, which *Constabularius* is the first Christian author to write about at any length. According to the doctrine ascribed in the Babylonian Talmud

---

[85] Ibid., 95rb. This chapter was edited in Moreton, "The *Compotus*," 79–82.

[86] *Constabularius*, 95rb–va. See Thābit ibn Qurra, *De anno solis*, edited in Francis J. Carmody, *The Astronomical Works of Thabit b. Qurra* (Berkeley: University of California Press, 1960), 67–68; Otto Neugebauer, "Thâbit ben Qurra 'On the Solar Year' and 'On the Motion of the Eighth Sphere'," *Proceedings of the American Philosophical Society* 106 (1962): 268–70; Kristian Peder Moesgaard, "Thābit ibn Qurra between Ptolemy and Copernicus: Analysis of Thābit's Solar Theory," *Archive for the History of Exact Sciences* 12 (1974): 204. That Thābit was a source of the *Computus Constabularii* is also assumed by Cecilia Panti, *Moti, virtù e motori celesti nella cosmologia di Roberto Grossatesta* (Florence: SISMEL, 2001), 77. Moreton, "The *Compotus*," 73–74, suspects Ibn Ezra as a source for the observations, but there appear to be no parallels in the *Liber de rationibus tabularum* to support this conjecture.

[87] See *Constabularius*, 90rb, and Moreton, "The *Compotus*," 73, 77; Millás Vallicrosa, *El libro*, 99.

[88] *Constabularius*, 91va: "In omni enim ciclo lune, sive Romano teste Beda, sive Pascali teste Dionisio, sive Iudeorum ipsos interrogate... secundum omnes XVII annus embolismalis est." See Moreton, "Before Grosseteste," 585; Moreton, "The *Compotus*," 72–73. On the background of Jewish-Christian contacts in late twelfth-century England, see Robert C. Stacey, "Jews and Christians in Twelfth-Century England: Some Dynamics of a Changing Relationship," in *Jews and Christians in Twelfth-Century Europe*, ed. Michael A. Signer and John Van Engen (Notre Dame: University of Notre Dame Press, 2001), 340–54.

(*Eruvin* 56a) to Mar Samuel (died ca. AD 250)—or *Samuel Iudeus*, as our author dubbed him—the length of the solar year is 365.25 days, just as in the Julian calendar. By consequence, the dates of the four *tequfot* fall roughly on the same day as the old Roman equinoxes and solstices in Julian calendar, being separated each by 91 days and 7½ hours, which means that their precise time shifts by six hours each year.[89] *Constabularius* elucidated this doctrine by using the example of the *tequfat Nisan*, i.e. the date of the vernal equinox. In what corresponds to the first year in a leap year-cycle, the *tequfat Nisan* was thought to have occurred at sunset or 18:00, on 25 March. By the following year it had shifted to midnight, reaching noon or 12:00 on 26 March in the fourth year before returning to its original position.

With this dating of the equinox to 25/26 March, Mar Samuel was in agreement with other ancient authorities known to *Constabularius*, such as pseudo-Hippocrates and pseudo-Theophilus, both of whom had been cited by Bede. As is well known, later computists rejected this tradition in favour of a date on 21 March, appealing to the wisdom of the 'Egyptians', i.e. the Alexandrian Church. *Constabularius* cited several witnesses to the latter opinion (such as the Alexandrian bishops Proterius and Cyril, Helperich of Auxerre, and Victor of Capua), only to add that the summer solstice, as it could presently be measured "by any layman" (*quilibet laicus*), actually fell as early as 16 June.[90] As he was quick to stress, however, this did not mean that "the ancients were so dormant that they would err in a subject matter which even simple laymen are able to investigate."[91] From the aforementioned observations of the vernal equinox, made by famous ancient astronomers, it could be inferred that the true length of the solar year was slightly less than 365.25, from which it followed that the equinoxes and solstices had moved closer and closer towards the beginning of the Julian year since early antiquity. At first glance, the available data suggested that, at the time of Jesus, the equinox had already fallen on 23 March, having moved away from the traditional

[89] *Constabularius*, 91vb, 95ra, 95va, 96vb. On the background, see Sacha Stern, "Fictitious Calendars: Early Rabbinic Notions of Time, Astronomy and Reality," *Jewish Quarterly Review*, n.s., 87 (1996): 105–7.

[90] *Constabularius*, 94vb–95rb. See Bede, *De temporum ratione* (30), CCSL 123B:371–76; Moreton, "The *Compotus*," 74–76.

[91] *Constabularius*, 95rb: "Neminem tamen arbitror esse tam insani capitis ut opinetur vigilantium antiquorum ignorantie sompni adeo subrepsisse ut fallerentur in eam rationem quam proprie indagare potest laicalis simplicitas. Absit."

Roman date. Yet *Constabularius* did not assume that the length of the solar year had been a constant throughout the centuries. Instead, he assumed that 25 March had still been the seat of the vernal equinox at the time of Christ, claiming, somewhat disingenuously perhaps, that "this accords with the rules of Ptolemy and Hipparchus and indeed, with all the tables I have encountered so far."[92]

Despite the erroneousness of this assertion, his observations concerning the changing date of the vernal equinox make him a pioneer in the history of calendar reform with regard to the length of the solar year, just as Reinher of Paderborn was a pioneering thinker with regard to the reform of the lunar calendar. Moreover, his assumption that the vernal equinox fell on 25 March in the first century provided the key to a completely fresh approach to the problem of the Passion date, to which *Constabularius* devoted the final chapter of his work.[93] Just like Reinher before him, *Constabularius* was aware that Latin scholars had previously considered three possible dates—23 March, 25 March, and 26 March—all of which failed when confronted with the calendrical parameters of the Dionysiac cycle for AD 34. He also knew that Gerland had tried to solve the problem by choosing 23 March and shifting the Passion to AD 42, whereas Marianus Scottus had opted for a crucifixion on 25 March, AD 12. *Constabularius* explicitly criticized the approach of these 'critical computists', noting its methodological flaws, especially when it came to the history of calendars. From the historical account in Bede's *De temporum ratione*, he took the idea that the 19-year lunisolar cycle had first been used by Eusebius of Caesarea, from which it could be inferred that this cycle had not yet been around in the first century of the Christian era. Before its introduction, the Alexandrians had supposedly determined the date of Easter "year by year with great labour, and for the entire world."[94] Commenting on this piece of information, *Constabularius* wrote:

---

[92] *Constabularius*, 96vb: "Constat vernum equinoctium secundum naturalem compotum... fuisse in tempore Christi singulis annis vel in die VIII kal aprilis, scilicet in anno bisextili et in primo post bisextum, vel in nocte sequenti, scilicet in anno IIIo et IIIIo; et hoc secundum regulas Ptolemei et Abrachaz, immo etiam secundum omnes tabulas quas adhuc invenerim."

[93] Ibid., 96rb–97va.

[94] Bede, *De temporum ratione* (44), CCSL 123B:418: "Non quod Aegyptus vel caetera per orientem Christi ecclesia verum eatenus lunae cursum vel diem nescierit rite invenire paschalem, sed quia ea quae ipsis temporibus annuatim cum labore investigata et per orbem mandata saepius in questionem venire solebant, facilius praefixa semel regula circulari semper observari et sine scrupulo ambigendi poterant edisci."

> This "laborious investigation" of the Egyptians was, I believe, a kind of astronomical speculation, which is why I find the methods of those modern authors to be lacking in efficiency, who try to find the year of the Passion by using the epacts [of the Dionysiac cycle]. For there is nothing that precludes the possibility that the Jews at this time used a computus, in which *luna* 17 fell on a Sunday and on one of the three dates [25, 27, or 28 March]...even though the [Dionysiac] epact produces different results.... I also would like to ask from where they take the idea that the Jews in those days calculated the age of the moon with the [same] computus that we use today, instead of the one that they currently have in use, especially given the fact that they often already note the fourth day of the moon...when it is still the first day according to our own calculation.[95]

Reinher of Paderborn and the English *Constabularius* were clearly thinking alike in many points, but the latter's awareness of the historical dimension of the Jewish calendar evidently went a step further, for he also doubted whether the calendar used by the Jews of his own day could be projected back by 1100 years in all its intricate details. In particular, he denied that the rule *BaDU* had already been established practice back in Jesus's day, seeing how the "Passover of the Jews fell on a Friday according to the testimony of the Gospels.... This can never happen according to their [present] computus."[96] Moreover, the Jewish conjunction-based calendar seemed to resemble the calendar of the Arabs, at least when it came to estimating the mean lunation, since both the Jews and Arabs used a shorter value than the one implied by the Alexandrian 19-year cycle. *Constabularius* found it unlikely that such an estimate had already existed at the time of Christ, "because one cannot read about any astronomer before the time of Christ who makes the mean lunation as small as the Muslims do."[97]

---

[95] *Constabularius*, 96va: "Fuitque illa laboriosa egiptiorum investigatio ut opinor astronomica speculatio, quapropter minus efficax mihi modernorum eorum ratio qui per epactas annum passionis inquirunt. Nichil enim prohibet iudeos illis temporibus habuisse compotum secundum quem luna fuit XVIIa die dominica in una trium datarum quas super posuimus eo anno quam ecclesia dicit fuisse passionis, licet secundum epactas secus accidit.... Quero etiam unde hoc habeant quod iudei tunc etatem lune quesiverint per compotum lune quem nos habemus potius quam per eum quem ipsi habent, utique interdum luna invenitur etiam IIIIm secundum eos...quando ipsam est prima secundum nos."

[96] Ibid.: "At tamen quod iudei tunc non habuint illum compotum quod nunc habent ex hoc elicio quod pascha iudeorum tunc teste evangelio fuit VIa feria.... Secundum compotum eorum hoc numquam potest accidere."

[97] Ibid.: "Saraceni quoque lune compotum habent secunda quam luna quando est IIa vel IIIa quando secundum nos prima, sed ipsum non fuisse in temporibus

In the end, the only viable method of finding the day on which Jesus had been crucified was to compare as many lunar computations as possible. In doing so, *Constabularius* assumed that not only the Christians, but also the Jews, used the vernal equinox as a lower limit for the Paschal full moon, i.e. the 14th day of the first spring lunation. Judging from the Julian date of the *tequfat Nisan* in the system of Samuel, this meant that the 15th of Nisan on which the crucifixion took place according to the synoptic Gospels cannot have occurred earlier than 26 March, an assumption which was seemingly in line with the Passion date suggested by Victorius of Aquitaine.[98] Naturally, this also meant that the corresponding new moon of Nisan should have fallen on 12 March. Yet a comparison of different lunar computations showed that the *molad* of AD 34 fell on 9 March according to the present-day Jewish calendar, whereas Ptolemy's data, as quoted by our author, indicated lunar first visibility on 11 March. In his view, this discrepancy was no reason to despair, as it merely confirmed his thesis that the Jewish calendar in the first century looked different from the one used by the Jews in his own day.

> For who can exclude that the Jews of those times had a computus according to which the new moon fell on 12 March…which is only one day later than it was [first] seen?[99]

These deliberations went to confirm our author's foregone conclusion that Christ's Passion had occurred on Friday, 26 March. Yet this result also raised an important question. If 26 March was the correct date, why did the fathers of the church instead claim that the crucifixion had taken place one day earlier, on 25 March, as could be read in the

---

passionis domini.… Per hoc conicio quod ante temporibus christi nullus legitur astronomus qui unam lunationem tam parvam fecit quam Saraceni, immo maiorem de multis athomis."

[98] Ibid., 96vb: "Samuel quoque qui multis annis posterius fuit qui et iudeorum privatissimus magister fuit illud idem voluit scilicet ut vernum equinoctium esset vel VIIIo kal aprilis in diem vel in nocte sequenti. Sed quecumque compotum haberunt iudei in tempore christi sive rationes suorum contemporaneos sive antiquorum secuti fuerint, nequaquam computaverunt equinoctium esse citius quam VIII kal aprilis."

[99] Ibid., 97r: "In anno autem passionis quem ponit ecclesia fuit luna prima secundum victorinum in nonis martii, secundum epactas VIIIo idus, secundum compotum qui est iudeorum VIIo idus; et invenit dionisius lunam fuisse primam in anno passionis IIIIo idus, sed secundum veritatem et equationies ptolomei et abrachaz fuit primus dies visionis nove lune Vo idus. At tamen qui prohibet iudeos tunc temporis habuisse compotum secundum quod luna dicitur prima IIIIo idus ut dicit dionisius, scilicet uno solo die tardius quam videtur?"

works of Bede, Augustine and many others? *Constabularius* was able to give an original and startling answer to this question by once again turning to the date of the *tequfat Nisan*. From both Samuel and Bede, he inferred that the date of the vernal equinox in the Julian calendar oscillated between two different days within the course of a four-year cycle. As a result, it was no more than a convenient fiction to ascribe the vernal equinox to a particular calendar date, as was customarily done when it was taught that it fell on either 25 March or 21 March. Since the equinox could fall on the evening of 25 March in one year and on the morning of 26 March two years later, there were in fact two possible ways of dating an event relative to the solar year. One could assign it to a particular day in the Julian calendar or to the day of the equinox (or solstice), which would lead to slightly different results.[100] *Constabularius* used this insight to explain away the 'error' of the church fathers in dating the Passion. If the patristic authorities stated that Christ had been incarnated and crucified on the same day, they were really meaning to say that both events had taken place on the day of the equinox. The conception had been on 25 March, whereas the Passion was on 26 March—different days in the calendar, but in both cases the true day of the vernal equinox.[101] The conclusion to all of this was that the kind of corrections that Marianus Scottus and Gerland had tried to apply to the Dionysiac era had to be rejected outright. As it turned out, "the Church has reckoned the year from the incarnation absolutely correctly."[102] The age of the 'critical computists', and to have clearly demonstrated this remains the undisputed achievement of the *Compotus Constabularii*, had definitely come to an end.

[100] Ibid., 97rb–va. See Bede, *Epistola ad Wicthedum* (1–8), CCSL 123C:635–40.

[101] *Constabularius*, 97rb: "Ex his ergo colligitur dominum fuisse incarnatum VIIIo kal aprilis ut fuit equinoctium in anno incarnationis et passus VIIo ut fuit equinoctium in anno passionis. Et tamen non sine ratione auctores latinorum precipui eidem date scilicet VIIIo kal aprilis incarnationem et passionem ascripserunt. Ipsi habebisse kalendas nostras non curantibus didicerant dominum in eodem die incarnatum et passum, scilicet in uno equinoctio."

[102] Ibid., 97va: "Constat ergo per premissas rationes immo longe potiores, licet a me ignorentur, ecclesiam rectissime annos incarnationis computasse."

CHAPTER SIX

A SCIENCE OF TIME: ROGER BACON AND HIS SUCCESSORS

1. *Bacon, the Calendar, and the Passion of Christ*

The acute criticism of the ecclesiastical calendar voiced in the *Compotus emendatus* of Reinher of Paderborn and—in somewhat more subdued form—in the works of *Constabularius* and Roger of Hereford fell on fertile ground in the following century, during which three well-known English scholars went on to pen influential treatises on the reform of the calendar, which took account of the deficiencies both of the Julian calendar with respect to the sun and the 19-year lunisolar cycle with respect to the moon. The three men in question were the astronomer John of Sacrobosco (ca. 1195–ca. 1256), who wrote a treatise *De anni ratione* (ca. 1232/35),[1] Robert Grosseteste (ca. 1170–1253), chancellor of Oxford University and bishop of Hereford, who wrote a *Compotus correctorius* (ca. 1220/30), and—most importantly—the Franciscan polymath Roger Bacon (ca. 1214/20–ca. 1292).[2]

Bacon's views on the calendar have gained particular fame for the strident tone in which they were expressed and for the fact that he was the first to present this kind of criticism to the only person in Latin Christendom who could have possibly authorized the desired change: Pope Clement IV (1265–68), to whom he addressed both his *Opus*

[1] The *editio princeps* of this text can be found in John of Sacrobosco, *Libellus de Sphaera*, ed. Philipp Melanchthon (Wittenberg, 1538), Br–H3r. On the background, see Olaf Pedersen, "In Quest of Sacrobosco," *Journal for the History of Astronomy* 16 (1985): 175–221; Jennifer Moreton, "John of Sacrobosco and the Calendar," *Viator* 25 (1994): 229–44.

[2] The *Compotus correctorius* of Robert Grosseteste was edited by Steele as an appendix to Bacon, *Compotus*, 232–40. On the background, see Duhem, *Le système*, 3:411–13; Richard C. Dales, "The Computistical Works Ascribed to Robert Grosseteste," *Isis* 80 (1989): 74–79; Jennifer Moreton, "Robert Grosseteste and the Calendar," in *Robert Grosseteste*, ed. James McEvoy (Turnhout: Brepols, 1995), 77–88; Moreton, "On Not Editing Grosseteste"; Matthew F. Dowd, "Astronomy and Computus at Oxford University in the Early Thirteenth Century: The Works of Robert Grosseteste" (PhD Diss., University of Notre Dame, 2003), 200–304.

maius and the supplementing *Opus tertium* (1266/68).[3] In his writings, Bacon referred to the present state of the calendar as "unbearable," "horrible," and "ridiculous." Not unlike Reinher of Padborn, he connected these charges with the complaint that Muslims, Jews, and other unbelievers were laughing at the Church's expense about its inability to calculate the date of Easter with sound astronomical means. Addressing the Pope directly, he proclaimed that despite the fact that the Council of Nicaea had once prohibited any changes to its Easter computus, it was now time for the Holy See to finally step up and "relieve the church from this *monstrum*."[4]

---

[3] See Roger Bacon, *Opus majus*, ed. John Henry Bridges, 3 vols. (Oxford: Clarendon Press, 1897–1900), 1:269–85. This section is largely identical with Roger Bacon, *Opus tertium* (chap. 67–71), edited in his *Opera quaedam hactenus inedita*, vol. 1, ed. J. S. Brewer (London: Longman, Green, Longman, and Roberts, 1859), 272–95. According to Bacon (ibid., 273–74), the chapters found in the *Opus tertium* were supposed to be a revised and corrected version of what had been included in the *Opus majus*. Since both versions are nowadays found to be identical, it seems that the original was replaced at a later point. See *Opus majus*, 1:269n2. There are numerous difficulties concerning the relationship between the *Opus majus, minus* and *tertium*. In particular, it is doubtful whether the latter was ever finished and sent. These questions have little or no bearing on the matters discussed here. For further detail, see Stewart C. Easton, *Roger Bacon and His Search for a Universal Science* (Oxford: Blackwell, 1952), 144–66, and Eugenio Massa, *Ruggero Bacone* (Rome: Edizioni di Storia e Letteratura, 1955), 7–80. On Roger Bacon's life and works, see Lynn Thorndike, *A History of Magic and Experimental Science*, 6 vols. (New York: Columbia University Press, 1923–58), 2:610–30; Theodore Crowley, *Roger Bacon* (Louvain: Éditions de l'Institut supérieur de philosophie, 1950); David C. Lindberg, *Roger Bacon's Philosophy of Nature* (Oxford: Clarendon Press, 1983), xv–xxvi; Jeremiah Hackett, ed., *Roger Bacon and the Sciences* (Leiden: Brill, 1997); Hackett, "The Published Works of Roger Bacon," *Vivarium* 35 (1997): 315–20; Robert Bartlett, *The Natural and the Supernatural in the Middle Ages* (Cambridge: Cambridge University Press, 2008), chap. 4.

[4] Bacon, *Opus tertium*, 272–73: "Secundum quod expono circa ecclesiastica est de corruptione kalendarii, quae est intolerabilis omni sapienti, et horribilis omni astronomo, et derisibilis ab omni computista. Unde omnes instructi in astronomia, et in computo, et in talibus, mirantur quod tam abominanda falstias sustinetur; sed impossibilie est quod sustineretur, nisi quia illi, qui habent auctoritatem super hac correctione, non sunt exercitati in astronomia, et computo, et in hujusmodi. Nullam enim percipiens talem abominationem susteneret eam. Et ideo quilibet sapiens Christianus, qui haec tractat, ostendit articulos istius corruptionis et docet remedia. Non tamen aliquis praesumit tradere kalendarium correctum propter hoc, quod concilium generale prohibet ni quis mutet kalendarium sine licentia sedis apostolicae speciali. Et hoc justum est. Sed illa sedes beatissima deberet hoc monstrum tollere de ecclesia." Ibid., 293: "Atque philosophi infideles, Arabes, Hebraei, et Graeci...abhorrent stultitiam quam conspiciunt in ordinatione temporum quibus utuntur Christiani in suis solemnitatibus." Bacon wrote this plea for calendar reform in 1267, as can be gathered from ibid., 277.

While Bacon's remarks on the calendar are well known, his closely related ventures into historical chronology, which take up considerable space in both works, have received only scant attention from modern scholars.[5] As a matter of fact, while John of Sacrobosco and Robert Grosseteste had remained silent on the chronological pathways connecting the reform of the calendar and the problem of the Passion date, Roger Bacon stated in no unclear terms that the error of the ecclesiastical calendar made it necessary to rethink traditional doctrines on the date of the crucifixion. Previous to him, Reinher of Paderborn and *Constabularius* had already shown that the approach of the 'critical computists', such as Gerland and Marianus, could not lead to a satisfactory solution, because they had equated the calendar of the Jews in first-century Jerusalem with the 19-year cycle of the Alexandrian Church. Not only was this an infelicitous anachronism, but the 19-year cycle did not even possess the kind of astronomical accuracy that would have been necessary to correctly display lunar data for periods at some remove from the time of its institution in the third or fourth century. As Bacon noted with some disdain, "every rustic" in his day could see that the calendar was behind the astronomical moon by two days. Since the ecclesiastical lunisolar calendar failed at correctly predicting the lunar phases in the present, there was no reason to suppose that it could make any more reliable predictions for the first century AD. Roger Bacon thus realized that an understanding that the calendar was in disarray could not leave the question of the crucifixion date untouched. In his *Compotus*, written between 1263 and 1265, he remarked:

> It is this error that renders us unable to ascertain by means of our lunar cycle for the year of the Lord's Passion that our Lord Jesus Christ suffered on a Friday and on the 15th day of the lunar month, but that instead we keep finding different dates, which either contradict the Gospels, which to impugn is not allowed to any Catholic, or go against some other known assumption on the duration of Christ's life or the certified opinion of all Saints on the month and day on which He is said to have suffered. And we have treated this more fully in the second book which we have composed *On the Times from Christ* and the one after that,

[5] A short and not fully accurate discussion of Bacon's dating of the Passion is included in Brigitte Englisch, "Artes und Weltsicht bei Roger Bacon," in *Artes im Mittelalter*, ed. Ursula Schaefer (Berlin: Akademie Verlag, 1999), 63–64.

> where we have pointed out this error and its causes and, to our best ability, explained the way to correct it.[6]

This passage is intriguing in more than one respect. To begin with, it contains a clear and unapologetic acknowledgement to the effect that the Passion date had to be newly investigated in light of the deficiencies of the calendar. While Bacon left it at a promissory note in the *Compotus*, he returned to this question a few years later in his works for Clement IV (*Opus maius* and *tertium*), which both contained lengthy expositions on the calculation of the Passion date (see below). At the same time, however, the passage above indicates that the same subject had by that time already been tackled in a separate work entitled 'On the Times from Christ' (*De temporibus a Christo*). Alongside another treatise called *De termino paschali*, the *De temporibus* is one of two lost Baconian works sporadically mentioned in the *Compotus* that seem to have contained some of the material on which he drew in this tractate.[7] While the *Compotus* approached the subject of lunisolar calendation in a more general astronomical way, *De termino paschali* may have been a treatise exclusively dedicated to the problem of finding the date of Easter. Quite possibly it already contained the seeds for Bacon's later complaints about the shifting equinoxes and full moons that caused the calculated Easter Sundays to differ by up to five weeks from what would have been the astronomically legitimate date. It is much more difficult to categorize the *De temporibus a Christo* with any conviction. Aside from the passage already cited, the title also appears, in curtailed form, in a short excursus "on the different years and epochs used among various nations" (*de diversis annis diversarum gentium et radicibus ipsorum*) that closes the first book of Bacon's *Compotus*. Towards the end of this chapter, Bacon remarks:

[6] Bacon, *Compotus*, 150: "Est autem hic error quod secundum racionem cicli et collocacionem aurei numeri in kalendario nequit inveniri in anno Dominice Passionis, Dominus noster Jesus Christus fuisse passus in feria 6a et luna 15a primi mensis lunaris, set invenitur aliud semper, seu contra veritatem ewangelicam, cui non licet alicui Catholico contraire, aut contra aliquam suppusicionem famosam de etate Christi in carne, aut contra omnem opinionem sanctorum autenticam de mense et de die mensis solaris in quo passus dicitur, et hec omnia lacius prosecuta sunt in secundo libro quem *De Temporibus a Christo* composuimus, et deinceps, ubi hunc errorem manifestavimus, et causas erroris et modum correcionis prout potuimus explanavimus, et ideo nunc dicta sufficiant de hiis que ad compotum naturalem et ecclesiasticum pertinent explanata."

[7] Ibid., 18, 86, 97, 150.

> These are the intervals (in years from the Incarnation) of the various eras, as they are found in the astronomical books that have come down to us. Further eras and intervals appear in historical works, on which we have treated more fully in the sixth chapter of the third part of our first book *On the Times*.[8]

This remark is followed by a table of ancient eras, which is fairly similar, in its choice of epochs, to the conversion tables customarily found in Arabic collections of astronomical tables, such as the aforementioned *zīj al-Sindhind* of al-Khwārizmī. We shall return to this table further below. For the moment, it will suffice to note that Bacon's table indicates each era's epoch or starting-date by giving its interval—in years and days—from the Dionysiac era of incarnation (with an epoch on 1 January, AD 1). The intervals are accordingly designated as *anni ante* and *anni post Incarnacionem*. Naturally, it would be tempting to connect this way of displaying eras to the title *De temporibus a Christo*. In any case, the reference to precise structures (chapter 6, part 3, book 1) hints at a work of considerable size, indicative of a serious and wide-ranging interest in historical chronology on Roger Bacon's part. This must be borne in mind when we later deal with the aforementioned chapter 21 of the *Compotus*, which appears to be no more than a précis of what had been recorded in much greater detail in *De temporibus a Christo*. Moreover, the fact that the latter work apparently also contained a treatment of the Passion date makes it possible to connect it to the chronological excursus in book four of the *Opus maius* and the corresponding passages in the *Opus tertium*, where the crucifixion likewise plays an important role. In the same excursus are included some allusions to calendars and eras, which bear a remote resemblance to what can be found in chapter 21 of his *Compotus*.[9] For these reasons, it seems legitimate to assume that the now lost tractate *De temporibus a Christo* was the original basis for the passages on time reckoning, found in the *Opera maius* and *tertium*, and that parts of its contents may be reconstructed on basis of the latter.[10] It is with this

[8] Ibid., 86: "Iste sunt differencie annorum qui reperiuntur in libris astronomorum qui ad nos pervenerunt ad annos Incarnationis sive annos post Incarnationem. Alie sunt annorum differencie que reperiuntur in hystorialibus, de quibus in primo libro *De Temporibus*, parte tercia, et capitulo sexta, plenius mencionem fecimus."

[9] Bacon, *Opus majus*, 1:187–210; Bacon, *Opus tertium*, 204–26.

[10] Steele, in his introduction to Bacon, *Compotus*, xxv–xxvi, arrives at similar conclusions.

assumption in mind that we shall now turn to Bacon's treatment of chronology in his works addressed to the pope.

2. *Astronomy to the Rescue: Biblical Chronology and Ptolemy's Eclipses*

In the *Opus majus*, the subject of chronology is introduced as an example for the "usefulness of mathematics in the study of divinity" (*mathematicae in divinis utilitas*), seven applications of which are discerned by Bacon: (1) astronomy, (2) geography, (3) *de temporibus*, (4) *accidentia et passiones temporum*, (5) geometry, (6) arithmetic and (7) music.[11] The close connection between categories (3) and (4) becomes immediately apparent from the peculiar choice of words, in which the 'substance' of time (or time reckoning) is discerned from its 'accidents and passions'. The latter are defined in the *Opus tertium* as "primations [i.e. new moons], lunations, embolisms and all that which is considered in the study of computus and calendars."[12] As we shall see further below, the section on the 'accidents and passions' is almost wholly dedicated to the Jewish calendar, whose knowledge Bacon deemed essential for a proper understanding of sacred scripture. This leaves us with the 'substance' of time, loosely definable as the problem of counting the years between important biblical events. For Bacon, this was by no means a task to be taken lightly:

> Of all the considerations of Holy Scripture, the course of history through all the ages and generations, from the beginning of the world to the time of Christ, is the most important one. The entire Old Testament is organized to this end and has the purpose of indicating His time, so that the Saviour of the world would not be expected before his time; and neither after, as the Jews do, who senselessly and foolishly continue to expect him until this day. And the Saracens believe that the time of salvation began with the advent of Muhammad. Hence, it is most necessary for

---

[11] Bacon, *Opus majus*, 1:180 (astronomy), 183 (geography), 187 (*de temporibus*), 195 (*accidentia et passiones temporum*), 210 (geometry), 219 (arithmetic), 237 (music).

[12] Bacon, *Opus tertium*, 212: "Nunc volo tangere ea, quae scripsi de passionibus temporum. Et voco passiones hic primationes, lunationes, et embolismos, et hujusmodi, quae in computo et kalendario considerantur." Bacon, *Opus majus*, 1:195: "Quarta vero radix mathematicae respectu theologiae est penes accidentia et passiones temporum, cujusmodi sunt primationes et caeterae aetates lunae et embolismi et hujusmodi.... Differt autem haec radix a praedicta, quod illa consistit penes substantiam temporum, haec vero penes proprietates et accidentia."

> us Christians...that we know how to ascertain [the number of years in] this history from the beginning to the time of Christ.[13]

The importance of a certain and precise count of years for the broader scheme of Christian existence was only rivalled by the difficulty of obtaining it. As Bacon pointed out with blunt honesty, the thread of biblical history broke down completely for the period between the book of Maccabees and the birth of Jesus. This was a serious problem, which could not be remedied by appeal to non-biblical sources, since these contained "contradiction without end" (*contrarietas infinita*). Worse than that, the Sacred Text itself was riddled with glaring chronological discordances between its various translations.[14] Eusebius, for instance, had counted 5198 years from Adam to Christ's birth, but based himself on a figure for the period between the Deluge and Abraham, which was diminished by 130 years compared to the 1072 years found in the Septuagint. A much lower sum of years awaited those who used the original Hebrew version of the Old Testament (a mere 3952 years from creation to Christ according to the Venerable Bede), whereas the world era of the Jews reckoned only 3760 years for the same period. Bacon went on to hint at the many further problems awaiting those who tried to ascertain the age of Terach when he fathered Abraham, the chronology of the Hebrew kings or the duration of the Babylonian Exile, making biblical chronology appear like the exegetical minefield it has remained until our day.[15]

---

[13] Bacon, *Opus tertium*, 204–5: "Nam maxima inter omnes considerationes Scripturae Sacrae est de cursu historiae a principio mundi usque ad Christum, per omnes aetates et saecula. Tota vero Scriptura Vetus ad ipsum, et ut tempus ejus habeatur, nec ante tempus ejus expectetur Salvator mundi; nec post, sicut Judaei adhuc inaniter et erronee expectant. Et Saraceni credunt quod tempus salutis incepit quando venit Machometus. Necessarium igitur est nobis Christianis, et maxime theologis, ut pro nostra utilitate, et intellectu Scripturae, et contra Judaeos, et Saracenos, et omnes sectas perversas, sciamus certificare historiam hanc a principio usque ad Christum." See also Bacon, *Opus majus*, 1:187–88. Bacon's disparaging remarks about Jewish messianic expectations may have been partly motivated by an encounter with existing eschatological predictions within thirteenth-century Judaism, which centered around the year AD 1240, the beginning of the sixth millennium according to the Jewish world era. See Yuval, *Two Nations*, chap. 6.

[14] Bacon, *Opus tertium*, 205: "Nec chronicae, nec historiae sanctorum et aliorum historiographorum certificant hic: nam apparet in eis contrarietas infinita, et translationes textus discordant crudeliter." See also Bacon, *Opus majus*, 1:189.

[15] Bacon, *Opus tertium*, 205–7. On the problems of biblical chronology, see Jeremy Hughes, *Secrets of the Times* (Sheffield: JSOT Press, 1990).

On the surface, none of this was entirely new: Augustine of Hippo had famously written on the problem of chronologically divergent Bible translations in *On the City of God*, whilst Bacon's contemporary Vincent of Beauvais noted the existence of conflicting world eras in his bulky *Speculum maius*.[16] Yet both the scope of Bacon's critique and the fearlessness with which he hammered it out seem remarkable. Having shattered the utility of standard historiographical method, he appealed to an altogether different discipline to bring light into the dark territory of the chronologer:

> But this diversity cannot be cleared away except through a sure source. No science can in this matter discover nor has it the ability to ponder on the means of settling such an important question except astronomy, whose function it is to consider the certain and unchangeable revolutions of eclipses and planetary conjunctions and other celestial events, the order of nature remaining fixed.[17]

With the luminaries moving through the skies like a celestial clockwork, astronomy was bound to provide the much needed check on a textual tradition riddled with contradictions or—as Bacon put it over and over again—to "achieve certainty about the times" (*certificare de temporibus*).[18] In the *Opus tertium*, he further elaborated on this insight by highlighting the chronological importance of eclipses of the sun and moon,

---

[16] Augustine, *De civitate dei* (15.10–13; 16.10), CCSL 48:466–72, 511–13; Vincent of Beauvais, *Speculum quadruplex*, 4 vols. (1624; repr. Graz: Akademische Druck- u. Verlagsanstalt, 1964–65), 4:84, 203–4.

[17] Bacon, *Opus majus*, 1:189: "Sed diversitas haec non potest certificari, nisi per aliquam radicem certam. Nulla vero scientia potest hic invenire nec habet unde cogitet de tanta certitudine nisi astronomia, cujus est considerare revolutiones certas et ratas eclipsium et conjunctionum planetarum et caeterarum revolutionum coelestium, stante ordine naturae." A slightly different translation of this passage can be found in Robert Belle Burke, trans., *The Opus Majus of Roger Bacon*, 2 vols. (1928; repr. New York: Russell & Russell, 1962), 1:209–10.

[18] Bacon, *Opus majus*, 1:188: "Sed nullus potest certificare de temporibus, nisi astronomus, nec aliqua scientia habet de his certificare nisi astronomia." Ibid., 1:191: "Haec autem alteratio est gravis valde, et ideo recurrunt homines periti ad scientiam cujus est certificare tempora, scilicet ad astronomiam." Bacon, *Opus tertium*, 204: "Nam nulla scientia potest tempora certificare nisi ista; nec aliqua se de hoc intromittit." Ibid., 208: "Sed, ut dixi, nulla scientia certificat de temporibus, nisi astronomia; quia ejus est considerare revolutiones motuum coelestium, qui fiunt in temporibus certis."

> which only occur at certain times, and in which there is no erring. And this is the reason why Ptolemy, in his *Almagest*, considered the times by using eclipses and thus ascertained the time of Nabonassar, and the whole period from him up to Alexander the Great and from him up to Octavian Augustus; in whose 42nd year our Lord was born according to all works of history and the testimony of the saints; and from there he ascertained how much time had flown until his own day, that is the time of the emperor Hadrian, under whom Ptolemy made his observations.[19]

The quote is telling in that it reveals the main source from which Bacon developed his remarkable program of applying astronomy to history. In the *Almagest* (3.7), which had become widely accessible thanks to Gerhard of Cremona's translation from Arabic (1175), Ptolemy specified the intervals between some of the ancient rulers, by whose reigns he dated eclipses and other celestial events, beginning with the Babylonian king Nabonassar:

From the first year of Nabonassar to the year of Alexander's death:	424 years
From Alexander's death to the first year of Augustus:	294 years
From the first year of Augustus to the 17th year of Hadrian:	161 years

All regnal years were synchronized to the Egyptian calendar via a system of ante-dating, which reckoned each reign from the beginning of the year (1 Thoth) in which a ruler succeeded to the throne, regardless of the actual accession date. A good example for this method is the era of Philippus Arrhidaeus, which is referred to by Ptolemy as the era of the death of Alexander. In Ptolemy's system, its epoch date corresponds to 12 November 324 BC even though Alexander is known to have died several months later, in June 323 BC. Another important feature of Ptolemy's regnal chronology is the use of the Egyptian

[19] Bacon, *Opus tertium*, 208: "Et similiter renovationes eclipsium, quae non cadunt nisi ad tempora certa, et in his non potest errare. Et ideo Ptolomaeus in Almagesti consideravit per eclipses tempora et certificavit tempus Nabuchadonozor, et totum tempus ab eo usque ad Alexandrum Magnum, et ab illo usque ad Octavianum Augustum; cujus quadragesimo secundo anno Dominus noster natus est secundum omnes historias, et secundum sanctos; et ab illo certificavit quantum fuit usque ad tempus suum, scilicet usque ad Adrianum principem, sub quo Ptolomaeus fecit considerationes suas." See also Bacon, *Opus majus*, 1:189. The corrupt rendering of the name of the Babylonian king Nabonassar (747–733 BC) in the Latin translation of the *Almagest* ensured that Bacon confused him with his successor Nabuchodonosor (or Nebuchadnezzar II, 605–562 BC), who is mentioned in the Old Testament, as most commentators did before the time of Copernicus. See Grafton, *Scaliger*, 2:124–26.

'wandering year', which assigned 365 days to each year without any intercalation. This was very convenient for astronomical calculations, but it also meant that 1 Thoth kept moving gradually in relation to the Julian calendar, coming full circle after $4 \times 365 = 1460$ years, the so-called 'Sothic period'.[20]

The dates thus found in the *Almagest* were derived from a list of Babylonian, Persian, Macedonian, and Roman kings and emperors, known as the *Astronomical Canon*, a version of which was later incorporated into Ptolemy's *Handy Tables*. Since its rediscovery in the West shortly after 1600, this list has been of central importance for the reconstruction of the chronology of the ancient world. Before that time, however, the much sparser account of the *Almagest* had to serve in its place as the backbone for an essential portion of sixteenth-century chronological scholarship.[21] Bacon anticipated this Renaissance appreciation for Ptolemy's epochs, adding to it the notion that they had been established with the help of eclipses. But whilst the *Almagest* did nothing to conceal the close affinity of astronomy and chronology, the idea that eclipses could be used to date history was far from explicit in Ptolemy's own words. All one could find, in the context of *Almagest* 3.7, was a passing remark on the era of Nabonassar to the effect that the most ancient astronomical observations on record dated from the reign of this king. Bacon may have searched the *Almagest* for traces of these ancient observations, finding them in the fourth book (4.6–8), where Ptolemy recorded a triple of lunar eclipses from the first year of the reign of Mardokempad, which was equivalent to the 27th year of Nabonassar. These eclipses in turn belonged to a set of fifteen lunar eclipses, spread over a period of 900 years, which served as the empirical foundation for Ptolemy's theory of lunar motion.[22] It is quite conceivable how Bacon might have grasped the utility of eclipses

---

[20] On time reckoning in the *Almagest*, see Olaf Pedersen, *A Survey of the Almagest* (Odense: Odense University Press, 1974), 124–28, and Ptolemy, *Almagest*, trans. G. J. Toomer (New York: Springer, 1984), 9–14.

[21] Grafton, *Joseph Scaliger*, 2:115–33, 720–29. See also Leo Depuydt, "'More Valuable than All Gold': Ptolemy's Royal Canon and Babylonian Chronology," *Journal of Cuneiform Studies* 47 (1995): 97–117; Otto Neugebauer, *A History of Ancient Mathematical Astronomy*, 3 vols. (Berlin: Springer, 1975), 3:1071–73; Bickerman, *Chronology*, 109–11; James Evans, *The History and Practice of Ancient Astronomy* (New York: Oxford University Press, 1998), 176–82.

[22] Pedersen, *A Survey*, 169–71. For a list of dated observations in the *Almagest*, see ibid., 408–22.

for historical dating by looking at this and other parts of the *Almagest*, with their steady flow of recorded observations, accompanied by the names and years of ancient rulers. His perceptive and creative reading of Ptolemy thus enabled him to formulate a foundational principle of technical chronology, as it is still practiced today.

If the temptation to put the English Franciscan into a direct line with Theodor Oppolzer—whose *Canon der Finsternisse* (1887) set the modern standard of making the possibilities of eclipse-dating available to the historian—is to be resisted, it is mainly because no concrete example of his connecting a dated eclipse and a historical event has survived. This does not mean, however, that Bacon was a mere theorizer on chronological matters. Some striking evidence to the contrary can be culled from the very condensed account of eras at the end of the first book of his *Compotus*, to which we now turn. The *Almagest* offered precise intervals for several key eras of ancient history, but from the perspective of a thirteenth-century reader, these had to remain virtually useless as long as there was no way of connecting them to the established framework of Christian chronography. A good indicator of both this framework and the standard of chronological knowledge that could be expected from an ordinary student of the Bible in Bacon's day is the *Historia Scholastica* of Peter Comestor. In his introduction to the Gospels, Peter established the birth year of Jesus through a series of synchronisms: the 42nd year of Augustus = the 752nd year since the founding of Rome = the third year of the 194th Olympiad = the 30th year of Herod. In combination, the data pointed strongly towards the year 2 BC. Peter Comestor, on the other hand, quite naturally tied them to the first year of the Christian era, assuming that Dionysius Exiguus had considered Jesus to have been born on 25 December, AD 1.[23]

Bacon reprimanded the *magister historiarum* rather heavily on several chronological accounts, but with regard to Christ's nativity he obviously agreed, dating it—as we shall see below—to the end of the

[23] Peter Comestor, *Historia Scholastica* (5), PL 198:1540. This passage is partly quoted in Vincent of Beauvais, *Speculum maius*, 4:203. Peter's text in Migné-edition (PL 198) has Olympiad 193.3, but it is clear that Eusebius's year Ol. 194.3 was meant. That Peter intended AD 1 may be seen from his claim that the year in question had the concurrent (weekday of 24 March) 5, meaning Thursday.

first year of the Christian era.[24] Yet although his above quote implied something to the contrary, he also knew that simply correlating this year and the 42nd year of Augustus was not the right key to unlock Ptolemy's treasures. The latter equated the 17th year of the reign of Hadrian (ca. 133/34) with the 162nd year of Augustus (*Almagest* 3.7), thereby indicating an accession date of Augustus considerably later than 42 BC. From the chronicle of Eusebius, Bacon could learn about the existence of an alternative count of Augustus' regnal years, which began 12 years later and marked his annexation of Egypt after the victory over Cleopatra and Anthony at Actium.[25] According to the framework outlined by Peter Comestor, this would have meant an epoch in 29 or 30 BC, which roughly matches the fact that Ptolemy is known to have counted the years of Augustus from 1 Thoth or 31 August 30 BC.

It is important to note, however, that Bacon, like any other thirteenth-century scholar with an interest in chronology, could have obtained the correct Ptolemaic result without brooding over ancient sources. All one needed to do was to consult the Toledan Tables or any similar *zīj*-style collection of astronomical tables, which had been imported from Muslim Spain during the previous century. These were typically accompanied by a list of eras, which gave intervals in days between important epochs, from which *inter alia* the precise beginning of Ptolemy's era Nabonassar (26 February, 747 BC) could be extracted.[26] Since the *Almagest* reckoned 424 + 294 = 718 Egyptian years between 1 Nabonassar and 1 Augustus, the latter date was a matter of simple calculation: (718 × 365) ÷ 365.25 = 717y 186d. This is precisely the difference between 26 February, 747 BC, and 31 August, 30 BC, Ptolemy's epoch for the reign of Augustus. Bacon's remarks in the *Compotus* as well as his specific choice of eras discussed leave no doubt that he was familiar with both these astronomical tables and their potential use as a chronological resource. At the same time, however, his list of epochs shows several points of divergence from the received standard dates:

---

[24] For Bacon's criticism of Peter Comestor, see his *Opus majus*, 1:200, 206, and *Opus tertium*, 215–17, 225.

[25] Peter Comestor, *Historia Scholastica* (5), PL 198:1540: "Annos enim duodecim qui a morte Julii duxerant, usque ad Actium bellum, regno Augusti connumeramus." See Eusebius, *Chronicle*, GCS 47:163; Bacon, *Opus majus*, 1:265.

[26] Pedersen, *The Toledan Tables*, 3:896–901; Neugebauer, *The Astronomical Tables*, 82–84, 143.

1) Nabonassar:	–746y 311d (24 Feb/26 Feb 747 BC)
2) Felix:	–329y 92d (1 Oct 330 BC)
3) Alexander's death:	–323y 92d (1 Oct/12 Nov 324 BC)
4) Seleucid era:	–311y 92d (1 Oct 312 BC)
5) Hispanic era:	–38y 92d (1 Oct/1 Jan 38 BC)
6) Incarnation era:	(1 Jan AD 1)
7) Era of Diocletian:	+289y (1 Jan/29 Aug AD 284)
8) Hijra:	+621y 195d (15 Jul AD 622)
9) Era of Yazdegerd III:	+631y 170d (20 Jun/16 Jun AD 632)[27]

For the date of Nabonassar, Bacon noted that he had encountered in the book of "a certain author" an interval of 742y 11m 5d from Christ, implying an epoch on 27 January, 743 BC. It is difficult to see how anyone could have arrived at such a figure other than through a corrupt text. In fact, Bacon himself rejected it in favour of another date, which he supposedly took from the *Almagest*: 24 February, 747 BC.[28] As this differs from the correct date by only two days, one might again suspect a scribal error, especially considering that Bacon tied this date to a Wednesday, which in 747 BC matched the 26th rather than the 24th of February. Some doubt, however, is cast on this solution by the fact that the interval between Nabonassar and the beginning of the Christian era (1 January, AD 1) is *twice* mentioned as 746y 311d, once supplemented by the expression *sexto kalendas Marcii*, which unequivocally supports 24 February, 747 BC. Unless the text was corrupted by an unusually attentive scribe, it seems tempting to suppose that Bacon arrived at the date independently. This assumption is backed by the

[27] See Bacon's list of eras in the *Computus* (chap. 21), as found in MS London, BL, Royal 7.F.VIII, 124v, and in Bacon, *Compotus*, 86. I have calculated the corresponding calendar dates on the assumption that the variable month lengths of the Julian calendar are presupposed and that all intervals are counted from 1 January, AD 1, with the latter date not being included in the interval. The interval for the Seleucid era has been emended to be consistent with Bacon's claims in the main text. Where Bacon's dates diverge from the standard numbers, found in the Toledan tables, I have given the 'canonical' date on the right hand side of the slash. See Pedersen, *Toledan Tables*, 3:877, 899–901. The dates found in the Toledan tables are essentially identical with those in the tables of al-Khwārizmī. See Neugebauer, *Astronomical Tables*, 82–83.

[28] Bacon, *Compotus*, 84: "Alia summa invenitur in libris philosophorum antique fuisse in terra Caldeorum a tempore Nabigonodosor, qui fuit ante Incarnacionem septingentis annis et 42 bus et 11 mensibus et 5 diebus, cujus inicium fuit feria 4a, sicut invenimus in quodam auctore. Set secundum quod ex 3. diccione *Almagesti* capitulo 8. apparet, fuerunt tantum 700 anni solares cum quarta accepti et 46 anni et 311 dies, et per revolucionem annorum et concurrencium fuit principium ipsorum, ut in kalendario collocatur, sexto kalendas Marcii in feria quarta." The reference to *Almagest* 3.8 rather than 3.7 accurately reflects the division in Gerhard of Cremona's Latin translation of the book. See Ptolemy, *Almagestum* (Venice, 1515), 33v.

formulation *sed secundum quod ex [Almagesto] apparet*, which indicates that Bacon had tried to extract the date from Ptolemy directly. Under these circumstances, it is perhaps informative that 24 February is precisely the result one would arrive at if one tried to calculate the epoch of Nabonassar from 29 August instead of 31 August, 30 BC. Since 29 August was the permanent seat of 1 Thoth (in common years) after the Julian leap day had been introduced in Egypt during the reign of Augustus, this may be more than just a coincidence, especially if it is taken into account that Bede had offered the very same equation of dates in *De temporum ratione*—a source Bacon may have exploited in his effort to establish the epoch of Nabonassar on his own.[29]

A less ambiguous example of Bacon's creatively using the *Almagest* to chronological ends involves the 'era of Felix', the only era on his list to have no counterpart in the astronomical tables:

> Next after the era of Nabonassar one encounters in the writings of the philosophers the years of a 76-year cycle, which was established by Felix or Philippus Arrhidaeus, about whom one can read in the *Almagest*; and from Dionysius's letter to Apollophanes we know that [this cycle] belonged to a famous set of astronomical canons. This era is used by Hipparchus [*Abrachis*] and began 6 years before the death of Alexander and 329 years before the Incarnation, on a Sunday.[30]

Bacon's note is once again based on the third book of the *Almagest*, where several observations of equinoxes and solstices, ascribed to Hipparchus, are dated according to the Callippic cycle, a 76-year expansion and modification of the 19-year lunisolar cycle instituted by the mathematician Meton of Athens in 432 BC. From the numerous instances where Ptolemy correlated this era to the years of Nabonassar, its epoch can be safely dated to 330 BC.[31] Similar to the name of Hipparchus, which is rendered *Abrachis* in Gerhard of Cremona's translation, the name of Callippus had been badly preserved in the

---

[29] Bede, *De temporum ratione* (11), CCSL 123B:318.

[30] Bacon, *Compotus*, 84: "Post annos Nabugodonosor inveniuntur in scriptis philosophorum anni secundum revoluciones 76 annorum, quas Felix sive Philippus Arideus, de quo in *Almagesti* legitur, instituit, et in epistola Dyionisii ad [Ap]ollophanem inveniuntur fuisse in canonibus astronomie famosi. Istis usus est Abrachis, et inceperunt 6 anni ante morten Alexandri et 329 annis ante Incarnacionem, et fuit inicium ejus in feria prima, id est, Dominica." The text of the cited edition has "ad Ptollophanem," but Steele submits the alternative reading in his annotations on p. 294.

[31] See, e.g., Ptolemy, *Almagest* (3,1; 6.5; 7.3), trans. Toomer, 138, 284, 334–38. For a full list, see Samuel, *Chronology*, 47.

process of translating the *Almagest* from Greek into Arabic and from Arabic into Latin. It eventually degenerated into *felis*,[32] which Bacon interpreted as the name of a certain Felix, whom he in turn identified with the Macedonian king Philippus Arrhidaeus. His ultimate source for this identification was an apocryphal letter, which Hilduin of St. Denis had inserted in his ninth-century biography of Dionysius the Areopagite. The letter has Dionysius recall the solar eclipse at the time Christ's crucifixion, which he and his pagan friend Apollophanes had supposedly witnessed in their youth, during a stay in the Egyptian city of Heliopolis. In order to ascertain themselves of the eclipse's miraculous nature, both men had recourse to a *regula Philippi Aridaei*.[33] In Albert the Great's commentary on the letters of Dionysius (ca. 1250), Bacon could find the word *regula* interpreted as the Latin equivalent for the Greek term *canon*, which was often used to refer to an astronomical table. He apparently also shared Albert's erroneous assumption that Philippus Arrhidaeus had been Alexander's father rather than half-brother, making him his predecessor on the Macedonian throne. Seeing that the 'era of Felix' was based on an astronomical cycle and began only several years before Alexander's death, he decided to connect the two testimonies, perhaps supposing that Philippus Arrhidaeus had been an active practitioner or sponsor of astronomy during his reign.[34]

As a result of these deliberations, Bacon could now prove that the Arabic astronomical tables were in error when they associated another era, which began 311 years and three months before the Incarnation, on 1 October, 312 BC, with the name of Alexander the Great. Seeing

---

[32] Ptolemy, *Almagestum*, 27r–28r. It is also worth noting that Gerhard's translation of Thabit ibn Qurra's *De anno solari* puts 'Philippi' instead of 'felis' in its quotations from the *Almagest*. See Carmody, *The Astronomical Works*, 67–68.

[33] Hilduin of St. Denis, *Areopagita sive Sancti Dionysii vita* (14), PL 106:33: "Obfusi namque orbe uniformiter tenebrarum caligine tabescente, ut purgatum rediit solis diametrum, regulam Philippi Aridei assumpsimus." On the background, see Max Buchner, "Die Areopagitika des Abtes Hilduin von St. Denis und ihr kirchenpolitischer Hintergrund," pt. 2.2, *Historisches Jahrbuch* 58 (1938): 61–71.

[34] Albertus Magnus, *Super Dionysii Epistulas*, ed. Paulus Simon (Münster: Aschendorff, 1978), 553: "*Assumpsimus* canones, idest tabulas ad aequandum planetas, *Philippi Aridaei*, qui fuit pater Alexandri, ut videremus, si posset esse eclipsis." Thoughts not altogether dissimilar to Bacon's have been entertained in more recent times by Franz Cumont, "Regula Philippi Arrhidaei," *Isis* 26 (1936): 8–12. A better explanation for the *regula Philippi Arrhidaei* would be that 'Dionysius' was talking about Ptolemy's *Handy Tables*, which used the reign of Philipp as its base date. See Otto Neugebauer, "Regula Philippi Arrhidaei," *Isis* 50 (1950): 477–78.

that Alexander had died just six years after the 'era of Felix', which began in 330 BC, this was obviously 12 years too late. Reading Eusebius and Bede, he could easily infer that the era mentioned in the tables was in reality the one that marked the beginning of Seleucus's reign over Asia, also known as the 'era of the Greeks' from the books of Maccabees.[35] The astronomical tables gave 1 October as its calendrical starting point, a date Bacon happily adopted. But for some reason, he also used it to mark the beginning of the two preceding eras of 'Felix' and 'Alexander's death' as well as the Hispanic era, which is correctly counted from 1 January, 38 BC. In the case of the era of Alexander's death, he was contradicted not only by the tables, but also by the *Almagest* itself, which gave a distance of 424 Egyptian years from Nabonassar, implying an epoch on 12 November, 324 BC (or 10 November, if based on Bacon's deviant date for Nabonassar). An explanation for this strange error may be found in Bacon's deference to the *auctores canonum* who had supposedly assigned an October epoch to Alexander. It would seem that Bacon was led astray by a statement found in the introductory canons to the Toledan Tables, sometimes ascribed to Arzachel (al-Zarqālī), which mention October as the first month of the Greeks in connection with the era of Alexander the Great.[36] As he should have known from his own deliberations, the name of Alexander in this case referred to the Seleucid era, not to the era beginning in 324 BC. Yet Bacon presumably inferred that, as a rule, all 'Greek' eras, including that of Alexander, were tied to a

[35] Bacon, *Compotus*, 85: "Post hos annos inveniuntur anni Grecorum qui, ut jam dictum est, inceperunt 311 annis et tribus mensibus ante Incarnacionem, et fuit inicium ipsorum a principio regni Seleuci regis Asie 27 [!] annis post morten Alexandri, teste Beda in libro suo *De Temporibus*. Et hiis annis utitur sacra scriptura in libris *Machabeorum*, et fuit inicium ipsorum feria secunda, et isti sunt anni qui in canonibus astronomie attribuntur Alexandro." The pertinent data are taken from Bede, *De temporum ratione* (66), CCSL 123B:489; Eusebius, *Chronicle*, GCS 47:126. Eusebius and Bede both date the beginning of the Seleucid era to the 13th year of Alexander's successor Ptolemy. This is in harmony with Bacon's epoch for the death of Alexander (324 BC). The above mention of 27 years is obviously the result of a scribal error.

[36] Bacon, *Compotus*, 85: "Usque ad Incarnacionem fuerunt anni 323 et 3 menses, et fuit inicium annorum horum in principio Octobris in feria secunda, ut putant auctores canonum." Note that *feria secunda* (Monday) matches the beginning of the Seleucid era instead of the era of Philipp/Alexander, which began on a Sunday. The canon is edited in Pedersen, *The Toledan Tables*, 1:218: "Cum hoc volueris, et quot anni transierint de annis Alexandri Magni.... Pone itaque eos menses, et accipe unicuique mensi numerum dierum eius, et incipe a Tesrin primo, qui est primus mensis anni Graecorum, et est October." See also ibid., 1:222, 2:390, 590, 592. Explicit references to the canons of Arzachel appear in Bacon, *Opus majus*, 1:192, 196.

calendar year that began on 1 October. The otherwise unknown epoch date of the 'era of Felix' was thereby obtained.

3. *Past and Future Connected: Bacon and Abū Maʿshar*

In his works for Clement IV, Bacon confidently presented Ptolemy as the key-holder to a precise reconstruction of ancient chronology from the time of Nabonassar onwards. As his list of epochs in the *Compotus* shows, this was a competent judgment, based on his own brave attempts at chronological research. Yet biblical chronology, whose imponderabilities had called for the application of astronomy in the first place, extended much further to the past than Ptolemy could possibly account for. A different model was hence needed for the first few millennia of world history. Bacon thought he had found such a model in the person of Abū Maʿshar, known in Latin as Albumazar (787–886), to whose chronological merits he gave only passing reference in the *Opus majus*, but became more precise on in the *Opus tertium*:

> From the creation of Adam until the Friday night, on which the flood took place, there were 2226 years and one month and 23 days and 4 hours. Hereunto, [Albumazar] speaks about the ways of verifying the entire time from the flood up to the destruction of the law of Mahomet, including all intervening [epochs], i.e. the times of Nabonassar, Alexander, Christ, and Mahomet.[37]

The specified interval of 2226y 1m 23d 4h along with the dating of the flood to a Friday famously appears in the first book of Abū Maʿshar's *Liber conjunctionum*, where it is attributed to Abtnūs or Anynūs (possibly Annianus?) in the Arabic manuscripts. The source reference is probably corrupt, but the interval is sufficiently close to the Septuagint, with its count of 2242 years for the same period, to

[37] Bacon, *Opus tertium*, 208: "Alii tamen non tacuerunt, et praecipue Albumazar in libro Conjunctionum. Nam ibi ponit principium mundi et primum hominem, scilicet Adam; et a creatione ejus notat tempus usque ad Diluvium, docens qua die et qua hora fuit Diluvium. Nam a plasmatione Adae usque ad noctem diei Veneris, in qua fuit Diluvium, fuerunt 2226, et mensis unus, et viginti tres dies, et quatuor horae. Deinde dicit modos verificandi totum tempus a Diluvio usque ad destructionem legis Machometi per omnia media, scilicet de tempora Nabuchadonozor, et Alexandri, et Christi, et Machometi."

warrant at least some cohesion with biblical chronology.[38] The same does not hold true for Abū Maʿshar's date of the flood, which corresponds to 17/18 February, 3102 BC. This was a startling contrast to the dates which could be arrived at by adding up the intervals found in the Hebrew Bible, where the flood would be found to have occurred in roughly the 2300s BC. The Septuagint, which reckoned 2242 years between the flood and a creation in roughly 5200 or 5300 BC, evidently came much closer, but the precise date of 17/18 February, 3102 BC, was nevertheless difficult to reconcile with any biblical account when precision was demanded. As it turns out, the original point of reference for Abū Maʿshar's flood date had not been the Bible, but a conjunction of all planets in 0° Aries, which marked the beginning of the current *kaliyuga* in Hindu astronomy (as found, for instance, in the *ārddharātrika*-system of Aryabhata I, ca. AD 500). Knowledge of this date was transmitted to the Islamic world and subsequently to the Latin West by Persian astrological sources, where similar periods of planetary motion (*yuga*s) were employed, and where the *kaliyuga* epoch had at one stage been conflated with the biblical deluge.[39] Bacon's silence about this date in the above quote is conspicuous, but it may be explained with his inability to extract it from Abū Maʿshar's work.[40] Its existence, however, cannot have been unknown to him, as

[38] See Abū Maʿshar, *On Historical Astrology* (1.1.26), ed. Keiji Yamamoto and Charles Burnett, 2 vols. (Leiden: Brill, 2000), 1:22 (Arabic), 2:105 (Latin). A nearly identical reference can be found in al-Bīrūnī, *Chronology*, ed. Sachau, 25. Sachau's identification of the source with Annianus of Alexandria (ibid., 374) has received fresh support from van Bladel, *Arabic Hermes*, 142–47. See also Mosshammer, *Easter Computus*, 359. Some doubt on this is shed by the fact that Annianus concurred with the Septuagint in dating the flood to AM 2242. See Elias of Nisibis, *Opus chronologicum*, CSCO 63**: 7–8; Adler and Tuffin, *Chronography*, 23.

[39] See David Pingree, "Astronomy and Astrology in India and Iran," *Isis* 54 (1963): 229–46 (243–44), and Pingree, *The Thousands*, 27–45. On the general background, see also Edward S. Kennedy, "The World-Year Concept in Islamic Astrology," in *Actes du Dixième Congrès International d'Histoire des Sciences*, 2 vols. (Paris: Hermann, 1964), 1:23–43; Edward S. Kennedy and B. L. van der Waerden, "The World-Year of the Persians," *Journal of the American Oriental Society* 83 (1963): 315–27; August Strobel, "Weltenjahr, große Konjunktion und Messiasstern: Ein themageschichtlicher Überblick," in *Aufstieg und Niedergang der Römischen Welt II*, vol. 20.2, ed. Wolfgang Haase (Berlin: de Gruyter, 1987), 988–1187.

[40] See Abū Maʿshar, *On Historical Astrology* (1.1.26), ed. Yamamoto and Burnett, 2:15, where the flood is separated by an interval of 3671 years from a planetary conjunction that indicated the rise of Islam. A correct understanding of this date, however, presupposes knowledge that this latter conjunction occurred in AD 571. Some manuscripts of *De conjunctionibus* contain a list of intervals between historical epochs, including the flood, but there is no evidence that it belonged to the version used

he could find it in many other sources, including some astronomical tables. In fact, he appears to mention it in the *Compotus*, albeit in the garbled form of 3104y 2m 28d before the Incarnation, adding that "others speak of fewer years and set its beginning to Thursday." This might be an allusion to 17 February, 3102 BC, (which was indeed a Thursday), the date found in the Toledan Tables.[41] In either case, Bacon did not include the *kaliyuga*-flood among his list of epochs, which only began with Nabonassar. As to his reasons for this omission he remarked:

> This sum is not fixed among philosophers, in so far as it is neither derived from any considerations of [the flood's] time, nor from any other point in time in the vicinity; but merely from the histories of Sacred Scripture or from other authors, and this is why we omit it.[42]

The passage is difficult to interpret. The "considerations" supposedly absent in the case of the flood apparently refer to astronomical calculation. While this might indicate that Bacon was unaware of the wholly astronomical nature of the *kaliyuga* date, his remark that the sum was "merely" (*tantum*) culled from scripture sounds rather daring. It seems to reflect his conviction that chronological dating not resting on astronomical considerations was of inferior value.

Be that as it may, Abū Maʿshar was evidently no match for Ptolemy when it came to providing usable epochs for world history. Nevertheless, he inspired Bacon to the idea that "the revolutions of the planets and their conjunctions" could play a role in chronology similar to that of eclipses.[43] In fact, whereas the latter were attached to historical events in a wholly contingent manner, the connection between planetary conjunctions and terrestrial fates seemed to be determined by certain laws, which man could hope to grasp. The paradigm for such

by Bacon. See ibid., 1:520–23, and David Pingree, "The Indian and Pseudo-Indian Passages in Greek and Latin Astronomical and Astrological Texts," *Viator* 7 (1976): 155–58.

[41] Bacon, *Compotus*, 84. See Pedersen, *Toledan Tables*, 3:899–900; Neugebauer, *Astronomical Tables*, 82–83.

[42] Bacon, *Compotus*, 84: "Hec autem summa non est fixa apud philosophos, eo quod non accipitur per consideraciones aliquas illius temporis, nec temporis propinqui; set tantum ex historiis Sacre Scripture seu aliorum auctorum, et ideo eam obmittimus."

[43] Bacon, *Opus majus*, 1:189: "Et si consideremus sententias Albumazar in libro Conjunctionum, videbimus quod ipse...ponit diem et horam quibus incepit diluvium, et per revolutiones planetarum et per eorum conjunctiones determinat sequentia secula, scilicet quando fuit Nabugodonosor, et quando Alexander, et quando Dominus Christus, et quando Mahometus, et sic de multis."

correlation was the doctrine of the 'great conjunctions', which had its roots in pre-Islamic Sassanian astrology. In the version divulged by Abū Maʿshar, terrestrial history was influenced by periodic conjunctions of Jupiter and Saturn. The lesser conjunctions recurred roughly every 20 years, but the most potent ones appeared only after the conjunctions had traversed through all signs of the zodiac to come back to their initial position after 960 years. Such great conjunctions were powerful enough to not only cause natural disasters, but to inaugurate new eras in history, precipitating the fall of empires, the rise of prophets, and the emergence of new religions.[44] As Bacon explained towards the end of the fourth book of his *Opus maius*:

> One can examine history at past periods, and study the effects of the heavens from the beginning of the world, as in the case of floods, earthquakes, pestilences, famines, comets, prodigies, and other things without number, which have happened both in human affairs and in nature. Having discovered these facts, one should consult the tables and canons of astronomy, and one will find that there are constellations corresponding in an appropriate way to the effects in each case. One should then study with the help of tables similar constellations in future time, either near or remote according to his wish; and one will then be able to express judgments on the effects, because they will be similar to those in the past, since if we assume a cause the effect is taken for granted.[45]

All this suggested that the contrast between chronological research and divination of the future was only a matter of degree. As Howard Hotson aptly puts in writing about an early modern context, the

[44] See Edward S. Kennedy and David Pingree, *The Astrological History of Māshā'allāh* (Cambridge, MA: Harvard University Press, 1971); Abū Maʿshar, *On Historical Astrology*, ed. Yamamoto and Burnett, 573–613. On the influence of this doctrine on the Latin Middle Ages, see Eugenio Garin, *Lo zodiaco della vita* (Rome: Laterza, 1976), 17–30; John North, "Astrology and the Fortunes of Churches," *Centaurus* 24 (1980): 181–211; Kocku von Stuckrad, *Locations of Knowledge in Medieval and Early Modern Europe* (Leiden: Brill, 2010), 115–34.

[45] Bacon, *Opus majus*, 1:389: "Et potest homo revolvere historiam ad tempora praeterita, et considerare effectus coelorum a principio mundi, ut sunt diluvia, terrae motus, pestilentiae, fames, cometae, monstra, et alia infinita, quae contigerunt tam in rebus humanis quam in naturalibus. Quibus comparatis, revolvat tabulas et canones astronomiae, et inveniet constellationies proprias singulis effectibus respondere. Deinde consideret per tabulas consimiles constellationes in futuro tempore propinquo vel remoto sicut vult; et poterit tunc pronuntiare in effectibus, qui consimiles erunt sicut fuerunt in praeterito, quia posita causa ponitur effectus." Translation according to Burke, *Opus majus*, 1:404.

> move from astronomically-based chronology to astrological history was in fact merely a matter of perspective, the result of reading an astronomical chronology, not as a work of historical scholarship, but as a natural history.... The very establishment of chronology on a scientific basis thus provided a goldmine of empirical evidence for the veracity of astrology and a clear research programme for those who wished to develop it further.[46]

What applied to the Protestant polymath Johann Heinrich Alsted in the seventeenth century was equally true for Bacon some 350 years earlier. Indeed, in his description of Abū Maʿshar's chronology, the method of 'certifying' time clearly went beyond past and present, extending to the destruction of "the law of Muhammad," that is the end of the Saracen threat to Christianity—an event which was evidently yet to occur, but which was just as certainly inscribed to the stars as the distant past.[47]

The idea according to which great changes in political and religious affairs could be predicted by astrological means was a patent source of fascination for medieval thinkers such as Bacon. Its impact is well-exemplified by the pseudepigraphic poem *De vetula*, written in the middle of the thirteenth century, but deceptively ascribed to Ovid, who is also the work's narrator. According to *De vetula*, Ovid was inspired by a conjunction of Jupiter and Saturn to envision the coming of a new religion, which was going to be founded by a prophet born of a virgin and dominated by the planet Mercury—a clear allusion to Christianity.[48] The text has been frequently attributed to Richard de Fournival, but recent considerations voiced by Charles Burnett make it seem worth speculating whether Roger Bacon may have been himself

---

[46] Howard Hotson, *Paradise Postponed* (Dordrecht: Kluwer, 2000), 41.

[47] Bacon, *Opus majus*, 1:266; Abū Maʿshar, *On Historical Astrology* (2.8.5), ed. Yamamoto and Burnett, 2:83. On the background, see John North, "Roger Bacon and the Saracens," in *Filosofia e scienza classica, arabo-latina medievale e l'età moderna*, ed. Graziella Federici Vescovini (Louvain-la-Neuve: Fédération Internationale des Instituts d'Études Médiévales, 1999), 129–60; Amanda Power, "'In the last days at the end of the world': Roger Bacon and the Reform of Christendom," *Canterbury Studies in Franciscan History* 1 (2008): 135–51; Bartlett, *The Natural*, 125–29; Brett Edward Whalen, *Dominion of God* (Cambridge, MA: Harvard University Press, 2009), 190–93.

[48] ps.-Ovid, *De vetula* (3.611–19), edited in Paul Klopsch, *Pseudo-Ovidius De vetula* (Leiden: Brill, 1967), 272: "Una quidem talis [coniunctio] felice tempore nuper/ Cesaris Augusti fuit anno bis duodeno/ a regni novitate sui. Que significavit/ post annum sextum nasci debere prophetam/ absque maris coitu de virgine. Cuius habetur/ typus, ubi plus Mercurii vis multiplicatur/ cuius erit concors complexio primo future/ secte, nam nusquam de signis sic dominatur/ Mercurius, sicut in signo virginis."

involved in the text's composition. In any case, he imported copious citations from the *De vetula* both into the *Opus maius* and into his *Metaphysica*.[49] In both works, he gave much attention to the idea that a conjunction of Jupiter and Saturn may have heralded the birth of Christ and even set out to identify the conjunction witnessed by Ovid by calculating a mean conjunction of Jupiter and Saturn in Aries on 26 December, 7 BC.[50] In Bacon's mind, the notion that important religious changes could be foreseen in the stars provided him with an important argument for the importance and legitimacy of astrology, which could be used to defend and further the Christian cause.[51] In attempting to approach the birth of Jesus Christ from an astrological vantage point, he also set a precedent for later scholars, who were likewise fascinated by the predictive possibilities of conjunction-theory. A number of them went on to draw up nativity charts for Jesus Christ (usually for 25 December, 1 BC)—a controversial practice, whose most famous representatives include Renaissance astrologers such as Girolamo Cardano, Luca Gaurico, and Tiberio Rosselli.[52]

---

[49] Bacon, *Opus majus*, 1:256, 263–64, 267; Bacon, *Metaphysica*, ed. Robert Steele (London: Moring, 1909), 9–10, 45–52. See Charles Burnett, "The Astrological Categorization of Religions in Abū Maʿshar, the *De vetula*, and Roger Bacon," in *Language of Religion—Language of the People*, ed. Ernst Bremer et al. (Munich: Fink, 2006), 127–38.

[50] Bacon, *Opus majus*, 1:264: "Si enim revolvamus motus Saturni et Jovid ad tempus illud, inveniemus eos fuisse conjunctos per medios cursos suos ante nativitatem Christi per sex annos, quinque dies, et tres horas; et erat medius cursus utriusque in Ariete decem gradus, lvi minuta, lii secunda." See also Bacon, *Metaphysica*, 51–52. In converting the date, I presume that Bacon reckoned from a standard epoch on 1 January, AD 1. If he instead referred to the 'true' nativity date on 25 December, AD 1, the interval would imply 19/20 December, 6 BC. Modern calculations show a near conjunction of Jupiter and Saturn in Pisces between 5 and 15 December, 7 BC. See Bryant Tuckerman, *Planetary, Lunar, and Solar Positions 601 B.C. to A.D. 1 at Five-Day and Ten-Day Intervals* (Philadelphia: American Philosophical Society, 1962), 330.

[51] Bacon, *Opus majus*, 1:253–54: "Magnum enim solatium fidei nostrae possumus habere, postquam philosophi qui ducti sunt solo motu rationis nobis consentiunt, et sectam seu professionem fidei Christianae confirmant et nobiscum concordant in stabilitate hujus sectae; non quia quaeramus rationem ante fidem, sed post fidem, ut duplici confirmatione certificati laudemus Deum de nostra salute quam indubitanter tenemus. Et per hanc viam mathematicae non solum certificamur de professione nostra, sed praemunimur contra sectam Antichristi, de qua simul cum secta Christi fit consideratio in mathematica."

[52] Luca Gaurico, *Calendarium ecclesiasticum novum* (Venice, 1552), 25r; David Renaker, "The Horoscope of Christ," *Milton Studies* 12 (1978): 213–33; John North, *Horoscopes and History* (London: The Warburg Institute, 1986), 163–73; Ornella Pompeo Faracovi, *Gli oroscopi di Cristo* (Venice: Marsilio Editori, 1999). An English translation cum commentary of Cardano's horoscope is found in Wayne Shumaker,

More importantly, the *Opus maius* presents us with the embryonic form of a peculiar strand of historical thought which Friedrich von Bezold dubbed *astrologische Geschichtsconstruction*. Inspired by Bacon's works, the learned cardinal Pierre d'Ailly (1350/51–1420) later went on to make the ambitious attempt of synchronizing world history from creation with significant occurrences of planetary conjunctions.[53] Belief in the feasibility of 'historical astrology' was still upheld at the beginning of the seventeenth century by the Protestant encyclopedist and theorist of millenarianism Johann Heinrich Alsted, but also by Johannes Kepler, who structured world history along 800-year periods of alignments of Jupiter and Saturn, which roughly coincided with the creation of the world, Noah's flood, the Exodus from Egypt, the birth of Christ, and the coronation of Charles the Great as Roman emperor.[54]

Kepler's marked interest in astrology also enabled him to devise a fresh chronological approach to the problem of Christ's nativity, which has remained influential until this day. In October 1604, a galactic supernova caused the appearance of what Kepler and his contemporaries interpreted as a 'new star' in the constellation of Ophiuchus. Since the same region of the zodiac had previously been host to a rare conjunction of Jupiter and Saturn, Kepler surmised a connection between both phenomena. Furthermore, he extended this theory to the astronomical events surrounding the birth of Jesus, as related in the Gospel of Matthew. Calculating backwards in time, he was able to connect the 'star of Bethlehem' to a similar conjunction, which had occurred in 7 BC, not long before Christ's birth; the latter accordingly had to be shifted some five years away from the beginning of the Dionysiac era.[55] Kepler's interest in historical astrology also enabled him

---

Renaissance Curiosa (Binghamton: Center for Medieval and Early Renaissance Studies, 1982), 69–89. See further Germana Ernst, *Religione, ragione e natura* (Milan: Angeli, 1991), 207–12; Anthony Grafton, *Cardano's Cosmos* (Cambridge, MA: Harvard University Press, 1999), 151–55.

[53] Friedrich von Bezold, "Astrologische Geschichtsconstruction im Mittelalter," *Deutsche Zeitschrift für Geschichtswissenschaft* 8 (1892): 29–72; Smoller, *History*, 43–84; Haeusler, *Das Ende*, 142–55; Steven Vanden Broecke, *The Limits of Influence* (Leiden: Brill, 2003), 47–48.

[54] See Johannes Kepler, *De stella nova in pede Serpentarii* (Prague: Sessius, 1606), 29 = KGW 1:183. See also KGW 7:487–88. On Alsted, see Hotson, *Paradise Postponed*, 41–84.

[55] Johannes Kepler, *De Iesu Christi Servatoris Nostri vero anno natalitio* (Frankfurt: Richter, 1606) = KGW 1:360–90. See also KGW 4:87–91; 5:7–126, 129–201, 205–17,

to suggest an explanation for the journey of the Magi, who had come from the East to see the new king:

> The Magi...were Chaldaeans both according to their nationality and their profession [sc. Astrologers], in whose regions astrology had first been born; and this discipline holds that great conjunctions of the higher planets in cardinal points, especially in the equinoctial points of Aries and Libra, signify a universal change of affairs; and a cometary star appearing at the same time announces the rise of a monarch.[56]

Although the point cannot be further pursued here, these examples suggests that astrology had an essential, yet still largely uncharted, role to play in the long history leading up to scientific chronology—one to which Bacon stands as an important medieval witness.

4. *Wrestling with Tradition: The Jewish Calendar and the Date of the Passion*

Just like the chronologer of today, Bacon considered the reconstruction of ancient year counts and epochs to be only one aspect of the business he termed *certificare de temporibus*. Equally important, especially from the viewpoint of biblical exegesis, was the interpretation of

---

221–370, 397–422; W. Burke-Gaffney, "Kepler and the Star of Bethlehem," *Journal of the Royal Astronomical Society of Canada* 31 (1937): 417–25; A. J. Sachs and C. B. F. Walker, "Kepler's View of the Star of Bethlehem and the Babylonian Almanac for 7/6 B.C.," *Iraq* 46 (1984): 43–55; Max Caspar, *Johannes Kepler*, 4th ed. (Stuttgart: Verlag für Gesch. der Naturwiss. und der Technik, 1995), 175–79; Anthony Grafton, "Chronology, Controversy, and Community in the Republic of Letters: The Case of Kepler," in *Worlds Made by Words* (Cambridge, MA: Harvard University Press, 2009), 124–27.

[56] Johannes Kepler, *De vero anno quo aeternus Dei filius humanam naturam in utero benedictae Virginis Mariae assumpsit* (Frankfurt: Bringer, 1614), 135 = KGW 5:96: "Erant Magi et nationis et professionis vocabulo Chaldaei, penes quos primum nata fuit Astrologia, cuius disciplinae pronunciatum hoc est; Coniunctiones maximas Planetarum superiorum in punctis Cardinalibus praecipue aequinoctialibus Arietis et Librae mutationem rerum universalem, et stellam Cometam sub illa tempora coniunctionis apparentem ortum alicuius Monarchae significare." That the Magi could have learned of Christ's birth from a conjunction is already hinted at in d'Ailly's *Sermo primus de adventu*. See Pierre d'Ailly, *Tractatus et sermones* (1490; repr. Frankfurt: Minerva, 1971), s4r; Smoller, *History*, 51. On the background, see further M. A. Screech, "The Magi and the Star (Matthew 2)," in *Histoire de l'exégèse au XVIe siècle*, ed. Olivier Fatio and Pierre Fraenkel (Genva: Droz, 1978), 385–409; Stephen M. Buhler, "Marsilio Ficino's *De stella magorum* and Renaissance Views of the Magi," *Renaissance Quarterly* 43 (1990): 348–71.

dates in different calendars and the ability to reduce them to a given standard of time reckoning. Luckily, collections of astronomical tables usually came with instructions and conversion tables, which aimed at making their contents adaptable to different calendrical standards (Muslim, Persian, Julian, etc.). In Bacon's view, they thus represented another striking example for the usefulness of astronomy in chronological matters:

> Since, therefore, there are contained in Scripture years lunar and solar as well as those of the Greeks [sc. Seleucid era] and Latins and the like, and we wish to reduce all the calendars to solar years and to the years of the Latins, which are the years of Christ, it is necessary for us in sacred history to know the differences between these calendars.... But this cannot be done, as Scripture requires, except by means of canons, tables, and other astronomical means.[57]

Naturally, the standard lunar calendar found in the 'canons and tables' was usually Islamo-Arabic and not Hebrew, as the biblical exegete might have hoped.[58] Bacon gladly made up for this deficit by devoting most of his section on the 'accidents and passions of time' to the Jewish lunisolar calendar. As he explained to his readers, the Jews divided the hour into 1080 'parts' (or *chalakim* = ch) and measured the length of the mean synodic month as 29d 12h 793ch (or 29d 12; 44, 3, 20h, in sexagesimal notation). They thus exceeded the value of the Arabic astronomers, whose calendar was based on a lunation of 29d 12h 44m or 29 12h 792ch by exactly 1ch. In doing so, the Jewish lunation came remarkably close to 29d 12; 44, 3, 15, 44h, which Bacon himself regarded as the most exact figure available, missing the target by only 0; 0, 0, 4, 16h. He suggested that this remaining inaccuracy was no more than the result of the Jewish astronomers' desire to express their value in whole numbers of 'parts', a judgment that reflected his admiration

---

[57] Bacon, *Opus majus*, 1:188–89: "Cum igitur in scriptura contineantur anni lunares et solares et Graecorum et Latinorum et hujusmodi, et volumus omnia tempora reducere ad annos solares, et ad annos Latinorum, qui sunt anni Christi, necesse est nobis in sacra historia scire horum temporum diversitatem, et ut sciamus quid est proprium cuilibet et quomodo aequantur ad invicem, et quomodo possumus extrahere majus de minori, et contrario, et quodlibet de quolibet. Sed impossibile est hoc fieri, ut exigit scriptura, nisi per canones et tabulas et caeteras considerationes astronomiae." Translation according to Burke, *Opus Majus*, 1:209. See Toomer, "A Survey," 14–26, and Pedersen, *Toledan Tables*, 1:216–32, 2:382–402, 588–602; 3:901–29, 940–43, for a variety of examples.

[58] Information on the Jewish calendar is only rarely found in medieval astronomical tables. A few exceptions are noted in Pedersen, *Toledan Tables*, 3:928–29, 943.

for a calendar whose rapport with astronomical reality was far superior to anything Christianity could lay claim to.[59] As it turns out, both the Jewish value and the 'optimal value' cited by Bacon came from the same source, namely the fourth book of Ptolemy's *Almagest* (4:2). The first value is there expressly given as 29; 31, 50, 8, 20d, whilst being connected to an estimate according to which 4267 lunar months equal 126007d 1h. Ptolemy's statements, however, are somewhat imprecise, since 126007 1/24d ÷ 4267 should yield an average lunation of 29; 31, 50, 8, 9, 20…d, which is the value used by Bacon. The problem was noticed by later Arabic redactors, who interpolated the more precise value into their versions of the *Almagest*, one of which was eventually translated into Latin by Gerhard of Cremona. If it had not been for this change to the original text, Bacon would have had no reason to assume any difference between the Jewish mean lunation and the best available estimate.[60]

In any case, Bacon was obviously impressed with the precision achieved by the Jewish calendar, whose 19-year cycle of intercalation the Christian church had apparently adapted.[61] Taking certain cues from al-Farghānī, he went on to reject the claim, made by some unnamed theologians, that the Hebrew lunar months had once been based purely on observation of the new moon crescent. In his opinion, such a method was barely practicable, because the interval between the last visibility of the old moon and the first appearance of the new one was never constant, but vacillated between one and three days.[62]

---

[59] Bacon, *Opus majus*, 1:197: "Et ideo Hebraei astronomi, volentes compere lunationem, apposuerunt unam partem, quia minus non potuerunt ponere secundum hanc divisionem qua usi sunt.... Et longe certior est eorum consideratio quam astronomorum utentium tabulis et canonibus apud alias nations, quanquam et plus aliquantulum computant quam praecise exigat lunatio." See also ibid., 1:280, 282, and Bacon, *Opus tertium*, 288–289, 291.

[60] Ptolemy, *Almagestum*, 36r. On the background, see Bernard R. Goldstein, "Ancient and Medieval Values for the Mean Synodic Month," *Journal for the History of Astronomy* 34 (2003): 65–74.

[61] Bacon, *Opus majus*, 1:197: "Nam sicut nos utimur cyclo decemnovennali, sic ipsi cyclo lunari, cujus primus annus incipit in quarto anno nostro. Et ideo falsum dicunt qui non posuerunt Hebraeos uti cyclo; habent enim embolismos in suo cyclo, sicut nos in nostro; immo nos habuimos ab eis." See also Bacon, *Opus tertium*, 218–19; Bacon, *Compotus*, 147; Jacob Guttmann, *Die Scholastik des dreizehnten Jahrhunderts in ihren Beziehungen zum Judenthum und zur jüdischen Literatur* (Breslau: Marcus, 1902), 147–48.

[62] Bacon, *Opus majus*, 1:195–96; Bacon, *Opus tertium*, 213–214. See also al-Farghānī, *Il "Libro dell'aggregazione delle stelle"* (1; 25), ed. Romeo Campiani (Città di Castello: Lapi, 1910), 56–57, 148; Bacon, *Compotus*, 59, 64.

In order to give their calendar a more stable framework, the ancient Hebrews instead allegedly used astronomical calculations to determine the time of mean conjunction. Since this point would not usually coincide with the moon's actual sighting, the Jews of antiquity

> lit torches on a high mountain at Jerusalem, that it might be known that at that moment the lunation began, so that men might be ready to perform the solemn acts and keep the festivals which they had to observe.[63]

Bacon's attempt at going into calendar history was not entirely successful. Although the Mishnah (tractate *Rosh Hashanah*, ch. 1–3) does indeed record the existence of a system of fire-signals of the kind alluded to by Bacon in Palestine during the first century AD, the new months indicated in this way were defined by actual observation of the new moon crescent, not by theoretical calculations. More striking than his errors, however, is the fact that he knew about these signals in the first place, the most readily available source for which was the Babylonian Talmud (tractate *Rosh Hashanah*, 22b–23b).[64] This knowledge points to the existence of personal contacts to learned Jews, some of whom he may have met during his stays in Paris in the 1240s and 50s. In his *Compendium studii philosophiae*, written in 1272, Bacon boasted the claim that Jews, as potential teachers of Hebrew, could be found "everywhere" (*ubique*), especially mentioning Paris and France as places where "men who know what is necessary in these things" lived.[65]

There is indeed some slim but valuable evidence that Bacon knew Hebrew and had access to source texts written in this language. Examples include his attempt to write a Hebrew grammar, a fragment of which has been preserved in a Cambridge manuscript.[66] Thirteenth-century

---

[63] Bacon, *Opus majus*, 1:196: "Et ideo Hebraei antiquitus per astronomiam certificaverunt primationem lunae, et cum non fuerat in visione novae lunae, nec potuit per visum cognosci, accenderunt faces in Jerusalem in monte alto, ut sciretur quod tunc fuit tempus primationis, quatenus homines essent parati facere solemnitates et festa quae habebant expedire." Translation according to Burke, *Opus Majus*, 1:217.

[64] See Stern, *Calendar*, 158.

[65] Roger Bacon, *Compendium studii philosophiae*, in *Opera quaedam hactenus inedita*, 434: "Doctores autem non desunt; quia ubique sunt Hebraei, et eorum lingua est eadem in substantia cum Arabica et Chaldaea, licet in modo different. Suntque homines parisius, et in Francia, et ulterius in omnibus regionibus, qui de his sciunt quantum necesse fuerit in hac parte."

[66] See Edmond Nolan and Samuel A. Hirsch, *The Greek Grammar of Roger Bacon and a Fragment of his Hebrew Grammar* (Cambridge: Cambridge University Press, 1902), 199–208. On Bacon's knowledge of Hebrew, see also Samuel A. Hirsch, "Early

materials on Hebrew grammar can also be found in manuscripts from Toulouse and Florence, which in addition include some letters by an unnamed scholar, who is probably to be identified with Roger Bacon.[67] In one of these letters, the anonymous scholar replies to a correspondent, who had asked him if he had seen a book *De canonibus Hebreorum*, which apparently dealt with calendrical matters. The writer of the letter answers that he possesses books in Hebrew that treat on the determination the new moon, "which has been investigated by the Hebrews with much greater certainty than by either Arabs or Latins" (*potissime de primatione lune, que certius longe excogitata est ab Hebreis quam ab Arabibus vel a Latinis*). Moreover, he also mentions that a Jewish correspondent has sent him Hebrew books from Germany, written by a certain Abraham [possibly bar Ḥiyya or Ibn Ezra?], which contained a wealth of useful information on astronomy and astrology, including sets of astronomical tables. Another of his Jewish correspondents, he claims, was unable to send him a complete exemplar, even though he lived in Toledo.[68] If the author of the letter was indeed Bacon—as has been proposed by Berger, Hirsch, and more

English Hebraists: Roger Bacon and His Predecessors," *Jewish Quarterly Review* 12 (1899–1900): 50–88; Hirsch, "Roger Bacon and Philology," in *Roger Bacon*, ed. Andrew G. Little (Oxford: Clarendon Press, 1914), 101–51; Horst Weinstock, "Roger Bacon und das 'hebräische' Alphabet," *Aschkenas* 2 (1992): 15–48.

[67] Samuel Berger, *Quam notitiam linguae Hebraicae habuerint Christiani medii aevi in Gallia* (Paris: Hachette & Soc., 1893), 37–45; Étienne Anheim, Benoît Grévin, and Martin Morard, "Exégèse judéo-chrétienne, magie et linguistique: Un recueil de *Notes* inédites attribuées à Roger Bacon," *Archives d'Histoire Doctrinale et Littéraire du Moyen Age* 68 (2001): 95–154. See also Hirsch, "Roger Bacon," 139–42; Benoît Grévin, "L'hébreu des franciscains: Nouveaux éléments sur la conaissance de l'hébreu en milieu chrétien au XIIIe siècle," *Médiévales* 41 (2001): 65–82.

[68] See the excerpt edited in Anheim, Grévin, and Morard, "Exégèse," 118–19n58: "Queritis a me utrum viderim librum qui intitulatur *De canonibus Hebreorum*, in quo certificatur quantitas anni. Respondeo: Habeo libros hebraicos de hac materia, potissime de primatione lune, que certius longe excogitata est ab Hebreis quam ab Arabibus vel a Latinis. Et sciatis quod missi sunt michi quidam libri hebraici de Alemannia a quodam judeo ingeniosissimo qui me novit ex fama tantum et jam aliquotiens scripsit michi in hebreo et ego sibi. Illos autem libros composuit Abraham, et est in eis plus de littera quam in *Prisciano maiori*, exceptis tabulis multis que site sunt in diversis partibus libri, sicut apud nos videmus factum in *Almagesti* Ptholemei. Et sunt illi libri astronomici subtilissimi et pulcherrimi et utiliores quam alias viderim, et locuntur de theorica astronomie et de iudiciis astronomicis, et sunt ibi multa mirabilia. Et diu laboraveram ad habendum aliquid de libris illis, quia per alia scripta judeorum noveram eos esse editos, et pluries scripseram cuidem judeo noto meo qui moratur in civitate tholetana in Hyspania, ut quereret michi libros illos, et iam semel rescripserat quod non inveniebantur Tholeti nisi pauca capitula ex eis. Modo habeo eos perfecte, benedictus Deus, et intendo eos tranferre cum tempus habuero. Valete."

recently by Anheim, Grévin, and Morard—this would also go some way towards explaining how he was able to get hold of the Hebrew calendar manuscript, which he sent to the papal curia as a companion piece to the chronological sections in the *Opus majus*. The table itself has been sadly lost, but some of its specifics can still be grasped from Bacon's commentary in the *Opus tertium*, which also indicates that he expected his student John, whom he had commissioned to deliver his books to Clement IV, to be able to explain its contents to the pontiff if necessary:

> I have included a complete [table] in Hebrew letters with the *Opus maius*, containing all canons, i.e. rules of operation, which teach the art of finding the beginning of the Hebrew year for any of our own years. I have also instructed John in the use of this table and its canons.[69]

It seems that the table displayed calendar data for thirteen consecutive 19-year cycles or 247 years, complete with accompanying commentaries. This probably means that it contained the calendrical 'circle' (*'iggul*) attributed to R. Naḥshon, the Gaon of Sura (died 889).[70] Bacon explicitly likened its function to the 532-year Easter table of the Christians and went on to characterize it as

> a wonderful work of astronomical art and highly useful for the understanding of the [Mosaic] Law and the feasts prescribed by it. For a person

[69] Bacon, *Opus tertium*, 220–21: "Et ideo totam posui Hebraicis literis in Opere Majori, cum omnibus canonibus, id est regulis suis ex expositione, docens artem inveniendi singulis annis nostris quando debet esse principium anni secundum Hebraeos; et docui Johannem hanc tabulam cum suis canonibus." Ibid., 215: "Et hanc tabulam literis Hebraicis misi in Opere Majori, cum ejus expositione et canonibus suis, secundum quod pertinet ad computum eorum." According to Bacon, John was capable of copying entire texts from memory. See his *Opus tertium*, 270: "Et si vultis copiosus videre, jubeatis Johanni ut faciat scribi de bona litera tractatum pleniorem." See also Andrew G. Little, ed., *Part of the Opus tertium of Roger Bacon* (Aberdeen: The University Press, 1912), 61: "Et ideo nunc mitto exemplar, ut Johannes cum suis sociis corrigat ea que remanserant incorrecta." Ibid., 82: "3m autem scriptum misi de manu per Johannes, ut Vestre Glorie transcriberetur." On John, see further ibid., 63; Easton, *Roger Bacon*, 150–51; Duhem, *Le système*, 3:500–02; Thomas A. Orlando, "Roger Bacon and the *Testimonia gentilium de secta christiana*," *Recherches de théologie ancienne et médiévale* 43 (1976): 215–18.

[70] Bacon, *Opus tertium*, 214–15: "Et posuerunt unam tabulam ex tredecim cyclis talibus, qua revoluta complentur omnes, et omnia redeunt ad idem temporis principium. Et hic cyclus cum canonibus suis et expositionibus est apud eos loci computi et kalendarii apus nod quantum ad multa." See also Bacon, *Opus majus*, 1:198; Steele's introduction to Bacon, *Computus*, xxvi; Salo W. Baron, *A Social and Religious History of the Jews*, 2nd ed., 18 vols. (New York: Columbia University Press, 1952–83), 8:192; Stern, *Calendar*, 193.

> ignorant of these [feasts] can never hope to understand the Law, neither can he confer with the Jews about such things, let alone convince them in any useful way.[71]

While these words obviously aimed at raising the pope's appreciation for an exotic Hebrew document, in whose acquisition Bacon had invested "much labour and expense,"[72] his own treatment of the Jewish calendar in the *Opus maius* was concerned with its usefulness in a different area: biblical chronology. Bacon's message was simple: anyone hoping to understand the time reckoning employed by Scripture had to become acquainted with the basics of the Hebrew/Jewish calendar, "for our Lord and the Apostles were Hebrews, as were the Patriarchs and Prophets."[73] As he went on to show, expanding on some topics that had been earlier discussed by the Venerable Bede, even a superficial understanding of the Hebrew lunar months and the way they were distributed over a year of 354 days could yield important insights about the duration of the flood and the date of the law giving on Sinai.[74] Yet these encounters with Old Testament history only served as a prelude to a far more exquisite problem, namely the date of Christ's Passion. By saving this theme for the grand finale of his chronological excursus, Bacon paid due respect both to its theological significance and technical difficulty. As we have seen in the previous chapters, medieval discussions had left the subject in such an unsatisfactory state that a fresh solution could be rightly considered overdue. Yet any new treatment of the problem meant to venture on sensitive terrain, where time-honoured traditions of the Latin Church could seem dangerously at stake.

Previous writers, such as Marianus Scottus, had decided to step beyond Bede's complaint about existing chronological contradictions and search for a definite solution. At the same time, however, it was difficult to overlook the awkward chronological problems that had, for instance, resulted from Marianus's new Christian era *secundum*

---

[71] Bacon, *Opus tertium*, 220: "Et in hac tabula est mirum artificium astronomiae, et summa legis intelligendae utilitas, et omnium festorum legalium, quam qui nescit numquam potest scire intellectum legis, ut oportet, nec cum Judaeis conferre de talibus, nec eius persuadere utiliter."

[72] Ibid., 221: "Et ideo hoc capitulum est unum de majoribus notabilibus in omnibus quae scribo, et magnis laboribus et expensis acquisitum."

[73] Ibid., 213: "Nam Dominus noster et apostoli fuerunt Hebraei, sicut patriarchae et prophetae."

[74] Bacon, *Opus majus*, 1:199–201; Bacon, *Opus tertium*, 215–218. See Bede, *De temporum ratione* (11), CCSL 123B:313–15.

veritatem evangelii, which began in 22 BC. Luke (3:1) firmly posited the life of Jesus in the context of Roman imperial history by tying the appearance of John the Baptist to the 15th year of the reign of Tiberius (AD 28/29) and since Jesus "began to be about thirty years of age" (3:23) around the same time, Marianus's dates had a lot to answer for. As we have seen, the Irish monk had not only been aware of the problem, but decided to dissolve it by postulating that the sum of Roman imperial regnal years since the birth of Christ had really been 22 years greater than commonly assumed. Bacon showed little sympathy for these chronographical manipulations. "However freely you may reckon the years of the emperors," he dryly noted, "[Marianus] still accepts 22, 21 or at least 20 years too many and should not be imitated."[75] Another authority, whose theories met with Bacon's outspoken disapproval was Gerland, "a famous man whom all computists and astronomers follow in computistical matters."[76] Gerland had tried his hands on a slightly different version of the same basic argument by dating the crucifixion two days earlier, to 23 March. In combination with Friday and *luna* 15, the date pointed towards AD 42 as the true Passion year, making the discrepancy from received tradition less extreme than in Marianus's case, but still unacceptable in Bacon's eyes.

Seeing that the 'critical computists' had failed at their task, Bacon had every reason to assume that the quest for the full moon at the time of Christ's Passion called for new astronomical tools of improved accuracy.[77] He presented such a tool in form of a table for the years AD 1–38, which listed mean opposition times for March and April as well as corresponding weekday letters. As he explained, the table had been constructed at Paris, but its actual values were calculated for the longitude of Novara.[78] It is not altogether clear whether this meant

[75] Bacon, *Opus majus*, 205: "Quantumcumque enim largius computentur anni imperatorum, inveniemus quod ipse superfluit in xxii annis, vel in xxi, vel ad minus in xx, et ideo non est imitandus."

[76] Ibid.: "Gerlandus autem famosus, quem omnes computistae sequuntur et astronomi in tractando ea quae ad computum pertinent..." Bacon, *Opus tertium*, 223: "...et ideo Gerlandus, omnium computistarum doctissimus, ei adhaeret." Gerland is also mentioned numerous times in Bacon's *Compotus*, 96–97, 102, 111–12, 121–22, 190.

[77] Bacon, *Opus tertium*, 224: "Qua proprium est astronomiae certificare haec tempora; et nunquam invenimus falsitatem secundum tabulas illas, sicut probamus per eclipses et caetera quae in coelo renovantur."

[78] Englisch, "Artes," 64, commits an obvious mistake when she claims that Bacon used the "Alfonsine Tables." The latter did not even exist by the time the *Opus majus* was written, neither did they circulate outside Spain before 1320.

to imply that the required calculations had been made by himself or someone else. The latter case would seem likelier considering that Novaran adaptations of astronomical tables appear rather frequently in medieval manuscripts. Moreover, at least some of these tables can be ascribed to Campanus of Novara (died 1296), whom Bacon praised as one of the best mathematicians of his time.[79] What seems to make Bacon's own table unique, however, is the underlying parameter for the length of the mean synodic month. Standard lunar tables of the twelfth and thirteenth century, including Novaran ones, were usually based on either the 'Muslim' month length of 29d 12h 44m = 29; 31, 50d, or its more precise 'Jewish' variant of 29; 31, 50, 8, 20d.[80] Bacon's table, by contrast, relied on the aforementioned 'optimal value' found in the Latin version of the *Almagest*, i.e. 29; 31, 50, 8, 9, 20d. Its epoch or *radix* corresponded to 14:31:02 on 28 March, AD 1 (reckoned, in astronomical fashion, from noon on the preceding day), from which the following dates of mean opposition in the spring of AD 29 to AD 34 resulted:[81]

AD	Opposition 1	Opposition 2
29	19 Mar 04:34 (Sat)	17 Apr 17:18 (Sun)
30	08 Mar 13:22 (Wed)	07 Apr 02:07 (Fri)
31	27 Mar 10:55 (Tue)	25 Apr 23:39 (Wed)
32	15 Mar 19:44 (Sat)	14 Apr 08:28 (Mon)
33	05 Mar 04:33 (Thu)	03 Apr 17:17 (Fri)
34	24 Mar 02:05 (Wed)	22 Apr 14:49 (Thu)

[79] Bacon, *Opus tertium*, 35. Novaran tables are sometimes accompanied by a canon with the incipit "Composui hanc tabulam ad inveniendum diem," probably authored by Campanus. See, e.g., the thirteenth-century MS Munich, Clm 17703, 27v–31r. See also Lynn Thorndike and Pearl Kibre, *A Catalogue of Incipits of Mediaeval Scientific Writings in Latin*, rev. ed. (London: The Medieval Academy of America, 1963), 241, 486, 1234; Francis S. Benjamin Jr. and G. J. Toomer, *Campanus of Novara and Medieval Planetary Theory* (Madison: The University of Wisconsin Press, 1971), 3–5, 15–16.

[80] Pedersen, *Toledan Tables*, 4:1327–50. The Novaran tables in MS Munich, Clm 17703, 30r, are based on M = 29d 12h 44m.

[81] See the fold-out table in Bacon, *Opus majus*, 1:209 (facing page), which is based on MS London, British Library, Cotton, Tiberius C.V, 76r. For the above table, I have reduced Bacon's time intervals to Julian calendar dates, with all values rounded to hours and minutes. Note that most of the numbers in Bridges's edition are obviously corrupt and have been silently emended.

To judge from Bacon's own words, these data suggested a fairly clear-cut solution for the historical date of the crucifixion:

> According then to this table the Lord suffered three days before the Nones of April [3 April] on a Friday at the mean opposition of the sun and moon... in the 33rd year from the incarnation according to the cycle of Dionysius, and that is in the 32rd year, according to the true age of the Lord... and so the most expert in these calculations have reckoned, who have labored hard to prove this.[82]

Notwithstanding such general declarations, he did little to elucidate the precise criteria by which he had selected his date. In a calendar based on the *molad* or mean conjunction, as it was used by the Jews in Bacon's own time, the date of mean opposition would have fluctuated between 14 and 15 Nisan, since both syzygies are roughly 14d 18h apart.[83] Contrary to what he had earlier proclaimed about the ancient Hebrew habit of calculating the time of mean conjunction, however, Bacon now apparently assumed that the Jews of ancient Palestine at the time of Jesus had in fact begun their months on the evening of the moon's first visibility (its 'kindling' or *accensio*). In such a scenario, the day of mean opposition would have mostly corresponded to 14 Nisan, but in some cases it could precede this date by a day or more. Bacon reacted to this fact in two not fully consistent ways. On the one hand, he denied out of hand that his calculations (despite the Thursday dates) allowed for a crucifixion on 15 Nisan,[84] but on the other hand he admitted that his calculated syzygies could not yield a definitive answer to the question whether Friday in a particular year had fallen on 14/15 Nisan or not. In fact, he apparently realized that

[82] Ibid., 1:209–10: "Secundum hanc tabulam passus fuit Dominus iii nonas Aprilis die Veneris in oppositione solis et lunae mediae xv anno cycli decemnovennalis et xiv cycli solaris, anno xxxiii ab incarnatione secundum cyclum Dionysii, et hoc est, xxxii secundum veram aetatem Domini. Et istud in tabula secunda accidit super b literam in directo xxxiii anni post duos dies de Aprili et xvii horas, et xvi minuta, et xxxiii secunda, l tertia, xxiv quarta, et sic sapientissimi in his considerationibus aestimaverunt, qui multum laboraverunt ad hoc probandum." Translation according to Burke, *Opus Majus*, 1:231–32. See also Bacon, *Opus tertium*, 222–23.

[83] To use an easy rule of thumb for the above table, only oppositions between 00:00 and 06:00h will still fall on 14 Nisan, whereas Bacon's date (3 April AD 33) would correspond to 15 Nisan. Note that the Jewish day begins at sunset, whereas Bacon's oppositions are all calculated from the preceding noon. See Bacon's comments on the Jewish evening epoch in his *Opus majus*, 1:195, and *Opus tertium*, 211.

[84] Bacon, *Opus majus*, 1:206: "Sed astronomi solliciti in hac parte non possunt invenire xv lunam nec viii calendarum in tempore passionis."

it was necessary to collate his own data with the evening of first lunar visibility, if all doubts were to be removed:

> Thus, if the mean opposition and the 14th day since the moon's kindling both coincide on the day of the Passion, then things are obvious both according to this table and according to the assessment of the wise. Yet if the opposition precedes the 14th day of the moon by a full day in case of the Passion, it would be necessary to have recourse to a table for the visibility of the new moon, made according to similar principles as the present one, and then doubts could be better excluded.[85]

Needless to say, a visibility table of the kind envisioned by Bacon was hardly part of the standard equipment of a thirteenth-century Christian astronomer and its composition would have required a daunting degree of sophistication, both in terms of astronomical theory and applied mathematics. For these reasons, Bacon apparently chose to contend with his more simplistic opposition table, which could still serve as a useful means of illustration to the pope and an approximation to a technical solution, whose principles he may have grasped, but which he was unable to execute in practice.[86] Accepting Bacon's principles, there should have been two further possible candidates for the crucifixion: 5 March, AD 33, and 22 April, AD 34. Both dates fell on a Thursday and thus implied a combination of 15 Nisan and Friday. As we shall see, however, Bacon was critical of the synoptic chronology, which explains his rejection of both dates, apart from the fact that they happened to fall extremely early (in the first case) or late (in the latter case) in the calendar year. From the little he said about the year of the crucifixion (AD 33), it would seem that he considered it a relatively unproblematic element of his proposed solution. As a matter of fact, it neatly corresponded to the chronicle of Eusebius, where the Passion of Christ was dated to the 33rd year from his nativity.[87] However, since Jesus was thought to have been born on 25 December,

---

[85] Ibid., 1:210: "Si igitur oppositio media et xiv luna ab accensione concurrant in unum in passione, res manifesta est secundum hanc tabulam et secundum aestimationem sapientum [!]. Si vero oppositio praecessit in passione xiv lunam per diem integrum, oporteret recurrere ad tabulam accensionis novae lunae factam consimiliter huic, et tunc magis excluderetur dubitatio."

[86] The difficulties involved in calculating lunar visibility are elucidated in Bernard D. Yallop, *A Method for Predicting the First Sighting of the New Crescent Moon* (Cambridge: HM Nautical Almanac Office, 1997), http://www.icoproject.org/pdf/yallop_1997.pdf.

[87] Eusebius, *Chronicle*, GCS 47:169 (birth: 42 Augustus/Ol. 194.3), 174 (death: 15 Tiberius/Ol. 202.3).

AD 1, whereas his Passion took place in springtime, Bacon rightly inferred that this was really only the 32rd year if one counted the actual years of his life (*secundum veram aetatem domini*), which meant that Jesus merely lived to the age of 31 years and three months. This was a notable departure from the Venerable Bede's teachings, who had assigned to Jesus a life of 33½ years and a Passion in AD 34.[88]

The real ground for controversy, however, lay elsewhere. Compelled by his astronomical evidence, Bacon decided to shift the crucifixion from the 15th to the 14th day of the moon, a notion which Bede had once prohibited as being a "great danger to the Catholic faith."[89] Although this statement would seem slightly exaggerated to the modern observer, from the medieval scholastic point of view the lunar age at the time of the Passion was doubtlessly a sensitive issue. Aside from its obvious bearing on the theological interpretation of events, it could also be employed to legitimize different ways of celebrating the Eucharist. As we have seen in the previous chapter, Latin-Greek debates over the form of bread used during the altar sacrament had been raging since the Great Schism of 1054 and they would continue to be a major point of contention between both churches up until the council of Ferrara-Florence (1438/39).

Before Bacon's day, the underlying chronological problem had been taken under fresh scrutiny by Pope Innocent III (1198–1216), who dealt with two further Gospel passages (besides John 13:1, 18:28, and 19:14, 31), which could lend ammunition to the Greek defence of a Last Supper on 13/14 Nisan: Luke 23:54–56 and Matthew 26:5. According to the first passage, the women prepared spices and ointments for Christ's body on the evening of his burial and rested throughout the following Sabbath. This implied that the Friday of the crucifixion was still an ordinary working day and not a day of rest, akin to the Sabbath, as 15 Nisan should have been according to Jewish law.

---

[88] Bacon, *Opus majus*, 1:209. A similar distinction between *anni aetatis* and *anni nativitatis* had previously been made by Marianus Scottus, who may have influenced Bacon. See ibid., 1:204, and Verbist, "Reconstructing," 304. Englisch, "Artes," 64, misunderstands Bacon when she claims that he corrected the epoch of the Christian era to AD 2. A shift of this kind was only suggested in the sixteenth century by Paul of Middelburg, on whom see below, chapter seven (section 3).

[89] Bede, *De temporum ratione* (61), CCSL 123B:452: "Tantum diligentissime cavendum ne hanc sexta decima luna, ut quidam, patratam confirmando, non solum inevitabile nostrae calculationis dispendium sed et gravissimum catholicae fidei incurramus periculum."

No less problematic was Matthew 26:5 (alongside Mark 14:2), which related how the Jewish chief priests plotted against Jesus, deciding to kill him "not on the feast day...lest there be an uproar among the people." Even the synoptic Gospels, on which Roman Christians based their understanding of the Last Supper and their use of azymes, thus seemed to contain passages favouring the 'Greek' crucifixion date on 14 Nisan. Innocent nonchalantly defused both objections with a few strokes of the quill. According to him, the women and Jesus's disciples had been allowed to break the rest on Friday, because they no longer stood under the Mosaic Law (*sub lege*), but under the age of Christian grace (*sub gratia*). Moreover, Mark 16:1 indicated that the women only bought the ointments "when the Sabbath was past," i.e. on the morning they returned to the burial site. Innocent tied this to the scene before the Last Supper, when a woman, identified by Roman Catholic tradition as Mary Magdalene, anointed Jesus's head or feet with expensive perfume (John 12:3–7; Matthew 26:7–12), thus assuming that she had already begun preparing the ointment back then, but did not finish her work until after the Sabbath. Concerning the priestly conspiracy, he simply stated that their plan to kill Jesus on a different day had been thwarted by divine providence, which was why Matthew 26:5 could not be used as a witness to the Johannine chronology.[90]

In 1204, only a few years after Innocent composed his treatise, Constantinople was brutally sacked by crusaders and large parts of the Byzantine Empire fell under the control of Latin rulers, putting much additional strain on the relationship between the Roman and Greek churches. Some futile attempts towards reconciliation were made at Nicaea and Nymphaion (1234), where a group of Dominican and Franciscan apocrisiaries, sent to Asia Minor by Pope Gregory IX, debated with Emperor John III Vatatzes, the patriarch Germanos, and other representatives of the Greek Church.[91] While in Nymphaion, the friars were handed a hostile Greek document, in which the Latin use

---

[90] Innocent III, *De sacro altaris mysterio* (4.4), PL 217:855–56. For further details, see Avvakumov, *Entstehung*, 142–44, who argues that Innocent's arguments must have been copied from somewhere else.

[91] For an official Latin report, see "Disputatio Latinorum et Graecorum seu Relatio Apocrisariorum Gregorii IX de gestis Nicaeae in Bithynia et Nymphaeae in Lydia 1234," edited in Hieronymus Golubovich, *Archivum Franciscanum Historicum* 12 (1919): 418–70, previously edited in Giovanni Domenico Mansi, *Sacrorum conciliorum nova et amplissima collectio*, 53 vols. (Paris: Welter, 1901–27), 23:279–307. See also the account in Chadwick, *East and West*, 238–43.

of azymes was once again openly condemned. They responded with a written text of their own, which they spiced with a number of citations from the works of Greek fathers, John Chrysostom and Epiphanius, in support of the view that Jesus had celebrated the legal Passover and eaten unleavened bread with his disciples.[92] Confronted with the traditional rebuttal that John 18:28 indicated that the Passover had only begun on the evening *after* the crucifixion, they answered that the *pascha* mentioned in the Gospel text was merely a reference to the 'paschal foods' (*cibos Paschales*), which were consumed throughout the week of unleavened bread. In making this exegetical move, they were apparently relying on some of the same passages in the Mosaic books that had been employed for similar arguments by Reinher of Paderborn in the twelfth and by the Humbertian *Dialogues* in the eleventh century. The *cibos Paschales* were hence either a reference to the additional sacrifices, made during the feast of unleavened bread (Numbers 28:19; Deuteronomy 16:2–3), or to the matsot, which were eaten for a full week, from 15 to 21 Nisan (Exodus 12:15–19; Leviticus 23:5–8; Numbers 28:16–18).[93]

While the negotations of 1234 ended in dissent and mutual accusations, some of the arguments advanced by the papal envoys resurfaced two decades later in a treatise entitled *Contra errores Graecorum* (1252), whose author was a member of the Dominican convent in Constantinople.[94] In his attempts to support the synoptic chronology of the Passion and, by consequence, the Roman Eucharistic liturgy, the anonymous author pointed to John Chrysostom's old charge that the Jews had deliberately postponed Passover in the year of the crucifixion, so that they could get hold of Jesus and kill him. According to this theory, Jesus had celebrated the Last Supper as a regular Passover meal on the legitimate date, whereas the Jews had waited until the following

[92] Golubovich, "Disputatio," 454–55, 459–61; John Chrysostom, *Homiliae in Matthaeum* (81.1), PG 58:729–30; Epiphanius, *Panarion* (30.22; 42.11.17 refut. 61), GCS 25:362–65; 31:148–49.

[93] Golubovich, "Disputatio," 461: "Ut comederent Pascha, id est cibos Paschales: sic enim legimus in veteri Testamento Pascha fuisse vocatum, et hoc erat dictum xv luna."

[94] On this text, see Raymond J. Loenertz, "Autour du traité de Fr. Barthélemy de Constantinople Contre les Grecs," *Archivum Fratrum Praedicatorum* 6 (1936): 361–71; Antoine Dondaine, "'Contra Graecos': Premiers écrits polémiques des Dominicains d'Orient," *Archivum Fratrum Praedicatorum* 21 (1951): 326–28, 338–50.

evening.[95] In addition, however, the Dominican author realized that there was a more elegant way of reconciling John with the synoptic Gospels, by claiming that 'eating the *pascha*' could sometimes simply stand for the unleavened bread (*panes azymi*) consumed during this whole period. This he alleged to have been the case in John 18:28, where the Jews refrained from entering the praetorum, in order to 'eat the *pascha*'. As a result, the scene in John could be understood to have taken place on 15 Nisan, the first day of unleavened bread.[96]

Another treatise of the same title (*Contra errores Graecorum*) was written in ca. 1264 by Roger Bacon's famous contemporary Thomas Aquinas (1225–74), who at the time was working as a lector in theology at the Priory of San Domenico in Orvieto, where the papal curia under Urban IV had its seat. Aquinas had been approached by the pope to provide his theological expertise on a *Libellus ss. Trinitatis*, which contained a wealth of (often spurious) citations from Greek fathers. This collection had been composed by Nicholas of Durazzo, bishop of Cotrone in Calabria, who had been sent to Italy by Emperor Michael VIII Paleologos in an effort to re-open negotiations with the Roman Church. Six of the quotations contained in Nicholas's florilegium directly concerned the contested consecration of unleavened bread. Thomas Aquinas later used them to demonstrate Jesus's adherence to the Jewish law, which had compelled him to eat unleavened bread on the evening before his death. As he pointed out both in his *Contra errores Graecorum* and in his influential *Summa contra gentiles*, the claims of the Greeks, which were based on the Johannine chronology, were easily answered by dissociating the eating of the *pascha* in John 18:28 from the Passover meal on 14 Nisan.[97]

---

[95] *Contra errores Graecorum*, PG 140:519. See John Chrysostom, *Homiliae in Matthaeum* (84.2), PG 58:754.

[96] *Contra errores Graecorum*, PG 140:521–22.

[97] Thomas Aquinas, *Contra errores graecorum* (2.39), in *Opera Omnia Iussu Impensaque Leonis XIII*, 50 vols. (Rome: Typographia Polyglotta S. C. de Propaganda Fide, 1882–), 40A:103; Thomas Aquinas, *Summa contra gentiles* (4.69), in *Opera omnia*, 15:218–19; Thomas Aquinas, *Summa Theologiae* (III.46.9; 74.4), in *Opera omnia*, 11:447–49; 12:149–50; Thomas Aquinas, *Scriptum super libros Sententiarum* (IV, d. 11, q. 2, a. 2, qc. 3), ed. P. Mandonnet and M. F. Moos, 4 vols. (Paris: Lethielleux, 1929–47), 4:465–71; Thomas Aquinas, *Super Evangelium S. Matthaei lectura* (26.2), ed. R. Cai, 5th ed. (Turin: Marietti, 1951), 332–32; *Super Evangelium S. Ioannis lectura* (13.1; 18.5), ed. R. Cai, 5th ed. (Turin: Marietti, 1952), 323–24, 436–37. Dondaine, "'Contra Graecos'," 387–93, shows that Thomas knew the Dominican *Contra errores Graecorum* of 1252. On the background, see Hyacinthe Dondaine, "Le Contra Errores Graecorum et le IV[e] livre du Contra Gentiles," *Les sciences philosophiques et*

The exegetical methods leveled by Aquinas and other scholastic writers not only amounted to an elegant defence of Latin rites and customs, but also worked to underline the conviction that the evangelists always agreed, even where they seemed to disagree. In keeping with this view, Aquinas condemned as "heretical" the judgment of those "who say that something wrong can be found, not just regarding the Gospels, but also in any other canonical Scripture." It was "hence necessary to say that all evangelists say the same and disagree in no point."[98] Yet while Reinher of Paderborn had been clearly aware of this problem, which is one reason why he decided to shift the crucifixion to 16 Nisan, lest the unanimity of the Gospels be questioned, Roger Bacon remained adamant in his reluctance to accept the synoptic chronology. Comparing the Passion accounts with what he knew about the Jewish calendar and the ritual aspects of the feast of unleavened bread, he sided with John and the Greeks and thus stood against his own Church in dating the Passion to the 14th day of the moon. Aside from the fact that this was the lunar date mentioned in the *Quaestiones Veteris et Novi Testamenti*, a patristic text which Bacon and his contemporaries falsely ascribed to St. Augustine (it is now attributed to Ambrosiaster), the main source for his contention were the Gospel texts themselves. In his support of a crucifixion on 14 Nisan, Bacon

théologiques 1 (1941–42): 156–62, and Palémon Glorieux, "Autour du Contra Errores Graecorum: Suggestions chronologiques," in *Autour d'Aristote* (Louvain: Publications Universitaires de Louvain, 1955), 497–512, who differ on the question whether the argument in *Contra errorum Graecorum* preceded that of *Summa contra Genteils* (Dondaine) or vice versa (Glorieux). See further Antoine Dondaine, "Nicolaus de Cotrone et les sources du Contra errores Graecorum de Saint Thomas," *Divus Thomas*, 3rd ser., 28 (1950): 313–40; Mark D. Jordan, "Theological Exegesis and Aquinas's Treatise 'against the Greeks'", *Church History* 56 (1987): 445–56.

[98] Thomas Aquinas, *Super Evangelium S. Ioannis lectura* (13.1), 324: "Sed haereticum est dicere, quod aliquid falsum, non solum in Evangeliis, sed etiam in quacumque canonica Scriptura inveniatur; et ideo necessarium est dicere, quod omnes Evangelistae dicunt idem, et in nullo discordant." For further thirteenth-century scholastic treatments of the azymes-question and the chronology of the Last Supper, see William of Auxerre, *Summa Aurea* (IV, tr. 7, cap. 8), ed. Jean Ribaillier, 5 vols. (Paris: CNRS, 1980–87), 4:185–93; Alexander of Hales, *Quaestiones disputatae "antequam esset frater"* (qu. 51, disp. 2, mem. 1–2), 3 vols. (Quaracchi: Collegium S. Bonaventurae, 1960), 2:905–12; William of Militona, *Quaestiones de sacramentis* (tr. 4, p. 3, qu. 9), ed. Caelestinus Piana and Gedeon Gál, 2 vols. (Quaracchi: Collegium S. Bonaventurae, 1961), 2:544–56; Albertus Magnus, *Commentarius in IV Sententiarum* (IV, dist. 12, C, art. 8), in *Opera omnia*, ed. Auguste Borgnet, 38 vols. (Paris: Vivès, 1890–99), 29:305–8; Bonaventure, *Commentaria in quattuor libros Sententiarum Magistri Petri Lombardi* (IV, dist. 11, p. 2, art. 2, qu. 1), in *Opera omnia*, 10 vols. (Quaracchi: Collegium S. Bonaventurae, 1882–1901), 4:260–62, and Avvakumov, *Entstehung*, 144–46.

also enlisted the synoptic passages that had previously been discussed by Innocent III, which included the conversation between the chief priests (Matthew 26:5, Mark 14:2) and the preparation of spices and ointments for Jesus's burial (Luke 23:54–56). As Bacon explained, Jewish law prohibited such activities on a holy day like 15 Nisan, on which no funeral was allowed to take place and people were commanded to abstain from work. The astronomical result that 3 April, AD 33—the only plausible date yielded by astronomical considerations—was the 14th day of the Jewish lunar month served to underscore Bacon's exegetical case.[99]

His crucifixion date, however, did not only contradict Thomas Aquinas and the synoptic Gospels, but all those patristic authorities which could be cited in defence of a Passion in March. Previous to Bacon, the date of the Passion had already been astronomically investigated by Albert the Great in his aforementioned commentary on the epistles of Dionysius the Areopagite (ca. 1250). In Dionysius's letter to Polycarp of Smyrna, the Areopagite gave an account of the solar eclipse at the time of the crucifixion, similar to the one in the spurious letter to Apollophanes, which underlined its miraculous nature. Although it was clear that solar eclipses could only occur during conjunctions of sun and moon, which foreclosed any natural explanation of the event, Albert used the opportunity provided by the passage to insert an elaborate astronomical excursus into his commentary. Armed with an array of different tables, he calculated the exact position of the sun and the moon at the time of Christ's crucifixion, finding that both luminaries had been in conjunction on 11 March, AD 34, before reaching opposition on 25 March at 15:55 (reckoned from the previous noon). A few hours later, on the afternoon of 25 March, Jesus died on the cross, as claimed by ecclesiastical tradition. This result naturally also confirmed the synoptic account of the Last Supper and Passion, according to which Jesus was crucified on *luna* 15.[100]

---

[99] Bacon, *Opus majus*, 1:206–8; Bacon, *Opus tertium*, 221–22; ps.-Augustine (Ambrosiaster), *Quaestiones Veteris et Novi Testamenti* (84; 94), CSEL 50:144, 166.

[100] Albertus Magnus, *Super Dionysii Epistulas*, 509: "Et hoc concordat Evangelio et dictis sanctorum, quia decima die mensis lunaris agnus ferebatur in castra, et illa fuit dies palmarum, quando praesentavit se dominus Ierusalem, et ita luna fuit undecima die luane post palmas, duodecima vero feria tertia et tertia decima feria quarta et quarta decima feria quinta, quando captus fuit." For further details, see B. B. Price, "The Use of Astronomical Tables by Albertus Magnus," *Journal for the History of Astronomy* 22 (1991): 221–40. On medieval perceptions of eclipses, see Robert

Yet while there was no reason to argue with Albert about the impossibility of a natural solar eclipse on that day, any astronomer equipped with an uncorrupted set of tables was in principle able to convict the learned Dominican of a serious blunder. One of the first scholars to take notice of this fact was the Flemish philosopher Peter de Rivo (ca. 1420–1500), who had asked an astronomer friend to double-check Albert's data. As it turned out, the true conjunction in AD 34 had taken place on 9 March in the morning, making the opposition fall on 23 March, two days earlier than Albert had supposed. Worse even, he had completely ignored the fact that 25 March was a Thursday and not a Friday in that year, making nonsense of the assignment of the crucifixion to that date.[101] While Albert's mishap seems to have been mainly caused by the corrupt parameters found in his own tables, it cannot be excluded that his results—especially with regard to the weekday—also reflected some sort of intentional manipulation in order to salvage ecclesiastical tradition with regard to the crucifixion date. The most prestigious representative of this tradition was St. Augustine, who, in *On the City of God*, had dated the crucifixion to 25 March in the consulate of the two Gemini, which is equivalent to AD 29.[102] Apart from the obvious problem of the opposition times, Bacon knew enough about the sequence of consulates in the first century to realize that this was entirely too early for the historical year of the crucifixion, especially if Jesus had been born at the end of AD 1. To him, the example merely revealed that Augustine, in speaking about the chronology of Jesus's life, could not be treated as a binding authority. On the contrary, if the bishop of Hippo mentioned such an absurd date, he obviously did so

Bartlett, *The Natural*, 51–70. On astrological interpretations relating to the crucifixion-eclipse, see John North, "Chaucer, Holbein, Libra, and the Crucifixion," in *Sic Itur ad Astra*, ed. Menso Folkerts and Richard Lorch (Wiesbaden: Harrassowitz, 2000), 461–73; North, "The Crucifixion and the Heavens," in *Seventh Centenary of the Teaching of Astronomy in Bologna, 1297–1997*, ed. Pierluigi Battistini et al. (Bologna: CLUEB, 2001), 17–35.

[101] Peter de Rivo, *Tercius tractatus de anno, die et feria dominice passionis atque resurrectionis* (Louvain, 1492), d8r: "Qui respondit quod calculans ex eisdem tabulis invenit anno quo christus passus creditur ad longitudinem iherosolimitanam veram coniunctionem luminarium fuisse octavo die martii completo, horis xix post meridiem, et minutis xl. Que coniunctio differt ab ea (quam invenit Albertus) uno die horis xix et minutis lvi." For true syzygy-times, see Hermann H. Goldstine, *New And Full Moons, 1001 B.C. to A.D. 1651* (Philadelphia: American Philosophical Society, 1973), 87. For more on Peter de Rivo, see chapter seven below.

[102] Augustine, *De civitate dei* (18.54), CCSL 48:655: "Mortuus est ergo Christus duobus Geminis consulibus ocatuum Kalendas Aprilis."

"by following the vulgar opinion" and not because he had investigated the problem himself at any length:

> For the Saints say a lot things in which they simply refer to common opinion, because they either had no basis on their own to decide upon the truth or they refrained from doing so because of the inherent difficulty or they simply were too occupied with other things; and so it sufficed for them at the time to speak with the majority, even though it was their task to think like the minority.[103]

Interestingly, he only offered this critique of patristic tradition in the *Opus tertium*, where he declared that he would proceed "with greater certainty" than he had done in the *Opus majus*, "wherefore the present text is to be preferred."[104] As a matter of fact, by the time Bacon composed the *Opus tertium*, the circumspection of his earlier remarks had paved way to an outspoken effort to convince Clement IV of the soundness of his chronological conclusions. Under these circumstances, it was only fitting that he now linked his re-dating of the Passion to the reform of the calendar. In both cases, new astronomical considerations stood against time-honoured ecclesiastical traditions. In both cases, too, Bacon hoped that the pope would answer to his requests and initiate a change.[105]

5. *Coda: Bacon's Successors*

It is not known if and how Clement IV, who passed away in 1268, reacted to this or any other of the suggestions Bacon brought before

---

[103] Bacon, *Opus tertium*, 223: "Et ideo hoc dixit secundum opinionem vulgi, non ex definitiva sententia. Multa enim dicunt sancti et referunt secundum vulgum, ubi aut non habuerunt posse determinandi veritatem, aut propter difficultatem reliquerunt, aut in aliis satis occupati, suffecit illud ad tempus loqui ut plures, quamvis fuit eis sentiendum ut pauci."

[104] Ibid., 221: "Et hic certius proceda quam in Opere Majori, propter quod haec scriptura magis tenenda est." Ibid., 226: "Multa plura scripsi in Opere Majori, quae hic non tango, sed certius scribo hic, et ideo magis est huic scripturae adhaerendum."

[105] Ibid., 212: "Quamvis enim astronomi sint, qui possent hujusmodi corruptionem tollere; tamen non sunt ausi, sine auctoritate summi pontificis, secundum quod concilium generale ordinavit. Sed de hoc inferius; nunc prae manibus sunt alia; inter quae una de maximis difficultatibus continetur, et similiter quod non terminabitur nisi per auxilium papae, quia nec certificabitur veritas, nec falsitas excludetur; et est de Passione Domini." Bacon, *Opus majus*, 1:202: "Sed nullus tantae auctoritatis est in ecclesia praeter summum pontificem, qui ausus esset dare sententiam contra sententias vulgatas in hac parte quamvis essent falsae."

him in his *Opus maius* and its supplements. The only indicator that these writings reached the papal curia safely is a treatise on optics (*Perspectiva*), written around the same time by the Polish mathematician Witelo. Since the work displays its author's familiarity with the *Opus maius* and since Witelo is known to have spent some time at the curia in Viterbo, it is likely that he encountered Bacon's works during his stay there.[106]

Although the details of the circulation of the *Opus majus* in the two centuries after its composition are not well-documented, there are some signs that it was known and read more widely than has oftentimes been supposed.[107] As has already been hinted at above, Bacon had an influential late medieval reader in the cardinal Pierre d'Ailly, who, as Laura Ackerman Smoller has shown in her study of d'Aillys astrological works, drew extensively from the English Franciscan's astrological ideas. In addition, it is well-established that Bacon was the source behind some geographical remarks in d'Ailly's *Tractatus de imagine mundi*, which later inspired Christopher Columbus in his attempt to find a westward passage to Asia.[108] Yet even previous to Pierre d'Ailly works we find references to the chronological portions of Bacon's *Liber ad clementem papam*, which is none other than the *Opus majus*, in the *Summa astrologiae iudicialis de accidentibus mundi*, written by the English astrologer John of Ashenden in Oxford in 1347/48.[109] Furthermore, some undisclosed lines of influence may have existed between the *Opus maius* and the work of another English Franciscan, named Robert of Leicester, who served as regent master at Oxford in 1321/22. His treatise *De ratione temporum, sive de compoto Hebraeorum, aptato ad kalendarium Latinorum*, written in

[106] David C. Lindberg, "Lines of Influence in Thirteenth Century Optics: Bacon, Witelo, and Pecham," *Speculum* 46 (1971): 66–83; Lindberg, *Roger Bacon's Philosophy of Nature*, xxv; Lindberg, *Roger Bacon and the Origins of* Perspectiva *in the Middle Ages* (Oxford: Clarendon Press, 1996), xix. On science at the papal curia, see also Klaus Bergdolt, *Das Auge und die Theologie* (Paderborn: Schöningh, 2007).

[107] In addition to the following observations, see Andrew G. Little, "Annual Lecture on a Master Mind: Roger Bacon," *Proceedings of the British Academy* 14 (1928): 265–66; Edward Lutz, *Roger Bacon's Contribution to Knowledge* (New York: Wagner, 1936), 66–72; Amanda Power, "A Mirror for Every Age: The Reputation of Roger Bacon," *English Historical Review* 121 (2006): 658–59.

[108] Pauline M. Watts, "Prophecy and Discovery: On the Spiritual Origins of Christopher Columbus's 'Enterprise of the Indies'," *American Historical Review* 90 (1985): 73–102.

[109] John of Ashenden [Eschuid], *Summa astrologiae iudicialis de accidentibus mundi* (Venice, 1489), 7v. Bacon's name is here misspelt as "bathon."

ca. 1294 and dedicated to Richard Swinfield, the bishop of Hereford (died 1317), contained an astonishingly comprehensive and accurate depiction of the Jewish conjunction-based calendar, whose intricacies were demonstrated in no fewer than ten lavish tables.[110] Like Bacon and his twelfth-century predecessors, Robert of Leicester applied his extensive knowledge of the Jewish calendar to fix the historical date of the Passion. His unwillingness to let go of patristic tradition eventually led him to opt for 23 March, AD 42, which was 14 Nisan according to the medieval Jewish calendar.[111]

The Baconian date of 3 April, AD 33, eventually resurfaced in the writings of Jean des Murs, who worked during the first half of the fourteenth century and belonged to the foremost astronomers of his time.[112] Among his earliest known works is a treatise carrying the incipit *Autores kalendarii*, which has only been preserved in two fifteenth-century manuscripts, found in the Austrian National Library. The first four chapters of this work happen to be an almost word-for-word paraphrase of various passages in Roger Bacon's *Compotus*, who is never cited by name.[113] Exactly at the point where Bacon had connected the errors found in the calendar to the problem of the Passion chronology, Jean des Murs inserted an elaborate astronomical discussion of the dates of Christ's life, coming to the conclusion that the correct date of the crucifixion was marked by a true opposition

[110] The only two MSS known to me are Oxford, Bodleian Library, Digby 212, 2r–10r; Erfurt, UB, Amploniana, Quart. 361, 80rb–85rb. The Oxford MS contains two additional tracts by the same author. On Robert of Leicester, see Andrew G. Little, *The Grey Friars in Oxford* (Oxford: Oxford Historical Society, 1892), 168–69; Josiah C. Russell, *Dictionary of Writers of Thirteenth Century England* (London: Longmans, Green, 1936), 139; Conrad Walmsley, "Two Long Lost Works of William Woodford and Robert of Leicester," *Archivum Franciscanum Historicum* 46 (1953): 458–70; Alfred B. Emden, *A Biographical Register of the University of Oxford to A.D. 1500*, 3 vols. (Oxford: Clarendon Press, 1957–59), 2:1142; John North, "Astronomy and Mathematics," in *The History of the University of Oxford*, ed. Trevor H. Aston, 8 vols. (Oxford: Clarendon Press, 1984–2000), 2:132–33.

[111] MS Erfurt, UB, Amploniana, Quart. 361, 85rb.

[112] See in general Duhem, *Le système*, 4:30–39, 41–60; Thorndike, *A History*, 3:268–70, 294–324; Max Lejbowicz, "Présentation de Jean de Murs, 'observateur et calculateur sagace et laborieux'," in *Méthodes et statut des sciences à la fin du Moyen Âge*, ed. Christophe Gerrard (Villeneuve d'Ascq: Presses Universitaires du Septentrion, 2004), 159–80; Joël Plassard, "Projets de réforme du calendrier à Paris au début du XIVe siècle: Textes édités et commentés," *École Nationale des Chartes: Positions de thèses* (1975): 175–81.

[113] Vienna, ÖNB, CVP 5273, 91r–93r = CVP 5292, 199r–200v, citing passages from Bacon, *Compotus*, 13–15, 55–58, 146–50.

of sun and moon on 3 April, 02:06 pm, in AD 33.[114] Twenty years later, he revisited the subject in his widely copied *Sermo de regulis computistarum* (1337), a short treatise in which the errors of the traditional ecclesiastical calendar were exposed in a remarkably scornful tone. As Jean pointed out, there was an intrinsic connection between the accuracy of the calendar and the chronology of the Passion, "for if the computists cannot find the [astronomically correct] date of the first Easter with their rules, how can they ever arrive at the truth?"[115] Only exact astronomical computations could yield the true date of the crucifixion, which Jean again asserted to have occurred on 3 April, AD 33. Just like Bacon, he inferred from this a lifespan of 31 years and 100 days, assuming that Jesus had been born on 25 December, AD 1. Contrary to Bacon, however, he treated the day of conjunction, which in AD 33 fell on 19 March, as 1 Nisan. Accordingly, the crucifixion took place on 15 Nisan, thereby validating the synoptic account.[116]

---

[114] Vienna, ÖNB, CVP 5273, 94v: "Et fuit media oppositio lunarium sicut examinavi super tertiam diem aprilis elapsis annis 32bus, primo mense solari, duobus diebus, 16 horis, 48 minutis. Vera autem oppositio quantum ad locum fuit 0 in signis, 10 in gradibus, 37 minuta, 7 secunda, 3 tertia; luna autem erat 0/10/37/7/30. Et erat verum tempus et equale annis Christi 32 bus, primo mense, 3 diebus, 2 horis, 6 minutis, 54 secundis transactis post meridiem." John's calculations were based on Toledan-style tables for the meridian of Toulouse, which he would later discard in favour of the Alfonsine Tables.

[115] MS Brussels, Bibliothèque Royale, 1022–1047, 203r: "Quero quoque quid inveniant per suas regulas quoto anno sue etatis passus est christus, mensem insuper et diem in qua a mortuis resurrexit, quoniam tunc fuit primum pascha, hoc autem non debet ignorari. Si igitur compotiste primum pascha nesciunt per eorum regulas invenire quomodo revera reperirent?" A list of MSS can be found in Thorndike and Kibre, *Catalogue*, 389. Besides the MS cited, which contains versions of the same text on both fol. 40, and fols. 203r–204v, I have also studied MS London, British Library, Royal 12.C.XVII, 213ra–vb.

[116] MS Brussels, Bibliothèque Royale, 1022–1047, 203v: "Adhuc dico quod anno passionis domini ex veritate scripture sacre passus est christus die veneris et luna 15. Sed haec duo convenerunt anno christi trecesimotertio currente ab eius incarnacione non alibi nisi longe invenires. Nam secundum tabulas hebraicas et latinas anno predicto scilicet 33° fuit media coniunctio solis et lune 19. die marcii completa a meridie sumens ortum et in 20. die marcii fuit luna unius diei perfecte et dicta prima completa. In qua secundum eos deberet 15 pro primacionis numerus collocari et in 3. nonis aprilis fuit 15...in qua passus est Christus.... Fuerunt ergo dies vite christi a nativitate 31 anni solares 100 dies ut estimo per predicta." See Goldstine, *New And Full Moons*, 87, who notes a conjunction on 19 March, 13:41, and an opposition on 3 April 18:02 in AD 33, for the meridian of Babylon.

Jean's concern over the want of accuracy of the ecclesiastical calendar and its failure to provide the correct date of Easter would continue to occupy him in later years. In 1344, he was invited by Pope Clement VI alongside the astronomer Firmin de Bellavalle to come to Avignon to work out a reform of the lunisolar cycle. The fruit of these labours was the *Epistola super reformatione antiqui kalendarii*, written in 1345, but no factual improvement of the calendar was achieved.[117] Following this fourteenth-century papal initiative, the calendar continued to be an object of concern in the context of the great church councils of the early fifteenth century. At the Council of Constance (1414–18), the subject was principally advertised by Pierre d'Ailly, who also penned a reform proposal entitled *Exhortatio super correctione calendarii* (1411). As Kaltenbrunner was able to show, many ideas of the *Exhortatio* are taken straight from Bacon's *Opus majus*, without the famous cardinal's marking his debt.[118] After d'Ailly's intervention at Constance had remained without much success, some abortive steps towards an actual reform were again taken at the Council of Basel (1431–49), where a special commission for questions pertaining to the calendar was established in 1434. One of the driving forces behind the commission's work was Cardinal Nicholas of Cusa, whose treatise *De reparatione kalendarii* (1436) again shows traces of Baconian influence.[119] This influence emerges even more clearly in the case of Nicholas's contemporary, the Cistercian monk Hermann Zoestius, who proposed his own plan of a calendar reform in a work entitled *Phaselexis* (1437). Commenting on the date of the Passion, Zoestius was able to cite both Roger Bacon's *Epistola ad clementem papam* and Jean des Murs's *Sermo de regulis*

[117] Edited in Gack-Scheiding, *Johannes de Muris*, and Chris Schabel, "Jean des Murs and Firmin of Beauval's Letter and Treatise on Calendar Reform for Clement VI," *Cahiers de l'Institut du Moyen-Âge Grec et Latin* 66 (1996): 187–215. See also Schabel, "*Ad correctionem calendarii*... The Background to Clement VI's Initiative?" *Cahiers de l'Institut du Moyen-Âge Grec et Latin* 68 (1998): 13–34. On the background, see Kaltenbrunner, "Die Vorgeschichte," 315–22; Eugène Déprez, "Une tentative de reforme du calendrier sous Clement VI: Jean des Murs et la Chronique de Jean de Venette," *Mélanges d'archéologie et d'histoire* 19 (1899): 131–43.

[118] Pierre d'Ailly, *Exhortatio super correctione calendarii*, edited in Mansi, *Sacrorum conciliorum*, 28:370–81. See also Kaltenbrunner, "Die Vorgeschichte," 326–36.

[119] Nicholas of Cusa, *Die Kalenderverbesserung (De correctione kalendarii)*, ed. Viktor Stegemann (Heidelberg: Kerle, 1955), lix, 107–8nn14–17. On the background, see Kaltenbrunner, "Die Vorgeschichte," 342–50; Martin Honecker, "Die Entstehung der Kalenderreformschrift des Nikolaus von Cues," *Historisches Jahrbuch* 60 (1940): 581–92; Tom Müller, *"Ut reiecto"*, 141–49.

computistarum as modern sources who had convincingly argued in favour of a Passion on 3 April, AD 33. Subsequent chronologers would frequently repeat this ascription and thus preserve the memory of Bacon's seminal role in the long history of attempts to establish the date on which Jesus died on the cross.[120]

[120] Hermann Zoestius, *Phaselexis*, MS Munich, BSB, Clm 3564, 139rb–va. For more on Hermann Zoestius, see chapter seven (section 2) below. For further examples, see Paul of Middelburg, *Paulina de recta Paschae celebratione* (Fossombrone, 1513), G5r–v; Johannes Lucidus Samotheus, *Opusculum de emendationibus temporum* (Venice, 1537), 181r; Pietro Pitati, *Verae Solaris atque lunaris anni quantatis...explicatio* (Basel, 1568), 50r, 72v, 74v; Jacob Christmann, *Muhamedis Alfragani Arabis chronologica et astronomica elementa* (Frankfurt, 1590), 372. See also Johannes Kepler, *Eclogae Chronicae* (Frankfurt, 1615), 85 = KGW 5:284, who critically remarks: "Primo Rogerio placuit hic annus [AD 33], quem ille hodiernae Chronologorum genti, Astronomiam ridicule iactanti, in haereditatem reliquit."

CHAPTER SEVEN

TIME FOR CONTROVERSY: CATHOLIC CHRONOLOGERS AND THE DATE OF THE PASSION IN THE FIFTEENTH AND SIXTEENTH CENTURIES

1. *A "Presumptuous and Arrogant" Idea: The* 'Defensorium' *of Alfonso Tostado (1443)*

On 21 June, 1443, the papal curia in Siena became witness to a scholastic disputation, conducted by the illustrious Castilian scholar Alfonso Fernández de Madrigal, better known as 'el Tostado' or 'Tostatus', who publically defended 21 'conclusions' or propositions over the course of two days. What seems to have been planned as a scholarly showcase, designed to display Tostado's skill as a theologian and scholastic debater, soon ended in a whirl of controversy and calumny. Having been alarmed by charges of impiety spread by some of his detractors, the disputant saw himself compelled to write an obsequious letter to Pope Eugene IV, in which he assured his orthodoxy, claiming that he never intended "to establish or reject anything by my own, except for that which the Holy Roman Church and Your Holiness wish to establish or reject," while also admitting that "some of my conclusions have seemed to certain people not to be in sufficient harmony with common doctrine."[1]

The incriminated Alfonso Tostado, who had been dispatched to Siena as an envoy by the Castilian king Juan II, hailed from the town

[1] Alfonso Tostado, *Defensorium trium conclusionum*, in *Opera omnia*, 27 vols. (Venice, 1728), 25:90a: "Beatissime pater, pridie exercitandi ingenii causa, sicut ceteris scholasticis viris solitum est in hac sacra curia, Sanctitati vestrae quasdam conclusiones scholastice, & disputative tenui, nihil ex meipso determinare, aut reprobare intendens, nisi quod Sacrosancta Romana Ecclesia, & Sanctitas vestra determinant, & reprobant. Hoc enim semper mihi propositum fuit, & est: & ego nunquam intendo recedere a veritate doctrinae Sanctae Rom. Ecclesiae, & Sanctitatis vestrae, & omnia mea dicta semper illi, & Sanctitati vestrae submisi, & semper submissa esse volo, qualitercumque contingat me loqui. Quaedam tamen conclusionum mearum visae sunt aliquibus non satis consonare doctrinae communi doctorum." Most of the letter is also reproduced in Cesare Baronio, Odorico Rinaldi, and Jacopo Laderchi, *Annales ecclesiastici*, ed. Augustin Theiner, 37 vols. (Bar-le-Duc: Guérin, 1864–83), 28:411–12.

of Madrigal de las Altas Torres in the province of Ávila, where he had been born in ca. 1410. As a young man, Tostado enrolled at the University of Salamanca, where he acquired a vast body of learning, which secured him the status of one of the foremost humanistic scholars to come out of Castile during the fifteenth century. Having attained master's degrees in the arts and in theology, he went on to teach poetry, moral philosophy, and theology, while also serving as rector at the College of San Bartolomé. He died on 3 September, 1455, and was buried in the cathedral of Ávila, to whose episcopal see he had only ascended in the previous year.[2] Tostado's written work is said to have encompassed more than 60,000 pages, an output so prolific that it has become proverbial to say of a writer that he *escribe más que el Tostado*. Multi-volume editions of his collected writings in Latin appeared no less than six times from the sixteenth to the eighteenth century, the last one of which, printed in Venice in 1728, encompasses 27 volumes in folio. Within these writings, the majority of which were biblical commentaries, there are occasional hints at an independent and critical mind. His commentary on Deuteronomy contained remarks on the presence of non-Mosaic material in the Pentateuch, which may have been an influence on later commentators and critics, including free-thinkers such as Isaac La Peyrère, Thomas Hobbes and Baruch Spinoza.[3]

Some controversial views on Church government were set forth by Tostado in his *Defensorium trium conclusionum* (1443), in which he

---

[2] For biographical details, see Joaquín Blázquez Hernández, "El Tostado alumno graduado y profesor de la Universidad de Salamanca," in *XV Semana española de Teologia (19–24 Sept. 1955)* (Madrid: CSIC, 1956), 411–47; Blázquez Hernández, "Madrigal, Alonso o Alfonso Fernández de," in *DHEE* 2:1390–91; Vicente Beltrán de Heredia, *Cartulario de la Universidad de Salamanca (1218–1600)*, 6 vols. (Salamanca: Universidad, 1970–73), 1:474–99; Roxana Recio and Antonio Cortijo Ocaña, "Alfonso de Madrigal 'El Tostado': un portavoz único de la intelectualidad castellana del siglo XV," *La Corónica* 33, no. 1 (2004): 7–15; Nuria Belloso Martín, "Sobre la guerra y la paz en Alfonso de Madrigal, el Tostado," *La Corónica* 33, no. 1 (2004): 17–38 (with bibliography of Tostado's works); Carmen Parilla, "*Qui scit, docere debet:* Acerca de Alfonso Tostado, el Tostado," *Archivum: Revista de la Facultad de Filología* 54–55 (2004–5): 367–90.

[3] See Noel Malcolm, *Aspects of Hobbes* (Oxford: Clarendon Press, 2002), 404–14; Malcolm, "*Leviathan*, the Pentateuch, and the Origins of Modern Biblical Criticism," in *Leviathan after 350 Years*, ed. Tom Sorell and Luc Foisneau (Oxford: Clarendon Press, 2004), 243–48. On Tostado's exegesis, see also Alastair J. Minnis, "Fifteenth-Century Versions of Thomistic Literalism: Girolamo Savonarola and Alfonso de Madrigal," in *Neue Richtungen in der hoch- und spätmittelalterlichen Bibelexegese*, ed. Robert E. Lerner (Munich: Oldenbourg, 1996), 163–80.

denied papal infallibility and argued that the general council should stand above the pope in questions of authority.[4] As the title suggests, the *Defensorium* was mainly a response to the accusations levelled against him in the wake of the public disputation of June 1443. As he makes clear in his letter to Pope Eugene IV (who was not himself present at the event), Tostado had agreed to defend 21 propositions, covering diverse subjects including the remission of sin and the date of Christ's Passion, three of which were regarded as scandalous by some of the high-ranking prelates in attendance. Although his letter to the pontiff apparently ensured that he did not have to face any judicial consequences, the pope nevertheless assigned a commission of three cardinals with the mandate of evaluating Tostado's claims. The commission proceeded to find fault with five of his propositions, two of which directly concerned historical chronology: that (i) Christ had died in the 33rd rather than the 34th year of his life and (ii) on 3 April rather than on 25 March. In addition, the cardinal Juan de Torquemada (1388–1468), uncle to the inquisitor Tomás de Torquemada, further explicated the reasons for this condemnation in a *Tractatus* against Tostado's propositions.[5] It seems to have been partly in reaction to Torquemada's *Tractatus* that Tostado went on to pen his *Defensorium*, in whose second part he responded at length to all the criticisms that had been raised against his chronological claims. As Tostado explained in the preface with some satisfaction, his opponents had tried hard to expose his arguments on the Passion date as erroneous or even heretical, but only ever succeeded in denouncing them

---

[4] Tostado, *Defensorium* (chap. 30–38; 68–75), 120a–24a, 138a–41b.

[5] Juan de Torquemada, *Tractatus in quo ponuntur impugnationes quarumdam propositionum Alphonsi de Madrigal*, found in MSS Vat. lat. 976, 118r–31v; Vat. lat. 2580, 81va–87va; Vat. lat. 5606, 213r–39v; Ott. lat. 718, 119v–23v. See Jacinto M. Garrastachu, "Los manuscritos del Cardenal Torquemada en la Biblioteca Vaticana," *La Ciencia Tomista* 22 (1930): 197, 215, 296, 303, 319. See also Blázquez Hernández, "El Tostado," 428–29; Beltrán de Heredia, *Cartulario*, 1:484; Belloso Martín, "Sobre la guerra," 38, with reference to MS Madrid, BN, 13250. For a summary of the contents of both Torquemada's *Tractatus* and Tostado's *Defensorium*, see Lazzarato, *Chronologia*, 465–73. On Torquemada, see also Stephan Lederer, *Der spanische Cardinal Johann von Torquemada* (Freiburg i. Br.: Herder, 1879). Tostado's views on the date of Christ's death were later also rejected by Pedro Martinez de Osma, a theologian at the University of Salamanca, who penned a *Disputatio de anno in quo possimus dicere Dominum fuisse passum et de quibusdam erratis in kalendario* (1468). See Friedrich Stegmüller, "Pedro de Osma: Ein Beitrag zur spanischen Universitäts-, Konzils- und Ketzergeschichte," *Römische Quartalschrift* 43 (1935): 224.

as "presumptuous and arrogant, because they in no way dared to call them false."[6]

Notwithstanding the controversy surrounding his *Defensorium*, Tostado's profuse explanations in the first ten chapters of his defence were far from novel or subversive. Just like previous medieval theologians, he pledged allegiance to the Latin tradition, which held that Jesus had celebrated a regular Passover meal on the evening before his death, eating unleavened bread with his disciples. Accordingly, he sided with the synoptic chronology, which dated the Passion to 15 Nisan, although he had to admit that the objections voiced by the Greeks, who based themselves on the Gospel of John, were not without merit.[7] In making his case for an unleavened Last Supper, Tostado was in perfect harmony with his critic Juan de Torquemada, who had previously written an *Apparatus* to accompany the decree of union between the Eastern and Western churches, which had been signed at the council of Ferrara-Florence (1438/39). On 6 July 1439, the Bull *Laetentur caeli* was read out after a mass in the Florentine Duomo, solemnly declaring that

> the body of Christ is truly effectuated in both unleavened and leavened wheaten bread, and priests should effectuate the body of Christ in one of these, according to the custom of his own Church, be it Western or Eastern.

As Edward Gibbon would later remark with characteristic sarcasm, one may "bestow some praise on the progress of human reason" by observing that the use of *azyma* "was now treated as an immaterial rite, which might innocently vary with the fashion of the age and country."[8] Yet while this was in reality a hard-won compromise, designed to lay

[6] Tostado, *Defensorium*, 91: "Cum ergo multorum dierum super eis laboriosas perdidissent, quod satis sibiipsis ad ignominiam cedere arbitrabantur, ne nihil omnino fecisse putarentur, eas praesumptuosas, vel superbas vocaverunt: falsas tamen dicere eas nequaquam ausi sunt: cum tamen per multos dies ante plurimi aemulorum ipsorum eas erroneas, & haereticas apertissime testarentur."

[7] Ibid. (2–10), 92a–103a.

[8] See Edward Gibbon, *The History of the Decline and Fall of the Roman Empire*, ed. J. B. Bury, 7 vols. (London: Methuen & Co., 1909–14), 7:114, and the "Bulla unionis Graecorum," in *Epistolae Pontificae de rebus in Concilio Florentino annis 1438–1439 gestis*, ed. Georg Hofmann (Rome: Pontificium Institutum Orientalium Studiorum, 1944), 71–72: "Item, in azimo, sive fermentato pane triticeo, corpus Christi veraciter confici, sacerdotesque in altero ipsum domini corpus conficere debere, unumquenque scilicet iuxta sue ecclesie sive Occidentalis sive Orientalis consuetudinem." See also Joseph Gill, *The Council of Florence* (Cambridge: Cambridge University Press, 1959),

to rest an ancient dispute between Roman Catholics and Greeks, it did not necessarily imply that all readings of the Passion chronology were equally valid. As Juan de Torquemada made plain in his *Apparatus* to the decree, it was by no means correct to assume that Christ had eaten fermented bread at the Last Supper. Making use of the same arguments that had already been advanced by Thomas Aquinas, the cardinal firmly tied the Last Supper to the regular evening of the Passover meal on 14/15 Nisan.[9] While Tostado's position on the Jewish date of the Passion thus conformed to established doctrine, his views on the Julian calendar date of the crucifixion and the age of Jesus at the time of his death were somewhat more risqué. As the Castilian scholar explained to his readers in Baconian fashion, only astronomy (*astrologia*) could provide the means to answer these latter questions with any certainty:

> And this proof is wholly necessary, in the sense that sophistry has no place there, whether by means of omission, anticipation or recapitulation, unlike in the previous argument, which was based purely on the Gospels.[10]

Naturally, however, all astronomical knowledge was of little help if the calendar presupposed by the four evangelists was badly understood. Where Roger Bacon had demanded that Christian exegetes should be acquainted with the different types of calendars in use among ancient nations, Tostado presented his readers with a long excursus on calendar reckoning, which included some insightful comments on the Jewish calendar. As he pointed out, the Jewish year began in autumn, a custom which he suspected to have formed under the influence of the Egyptian calendar during the period of servitude:

> The knowledgeable among the Jews, however, instead say that this was because they calculated their years from the beginning of time and because the world was created in September, although the year itself

414, for the text, as well as ibid., 270–304, and Chadwick, *East and West*, 258–73, on the background.

[9] Juan de Torquemada, *Apparatus super Decretum Florentinum Unionis Graecorum*, ed. Emmanuel Candal (Rome: Pontificium Institutum Orientalium Studiorum, 1942), 66–68.

[10] Tostado, *Defensorium* (13), 105b: "Astrologia est via probativa ad id, quod intendimus. Et ista probatio est omnino necessaria, ita quod non habeat locum aliqua cavillatio de omissione, anticipatione, vel recapitulatione, sicut haberet locum contra praecedentem probationem, quae erat pure Evangelica."

> begins in October. For they say that the world was created on 25 September, and that the sixth day, on which Adam was created, was the last day of September, and the following day, on which God abstained from all work, which is the Sabbath, was the first day of October, and this was the first day of the year, while the six preceding days, on which creation took place, belong to another year, because the Jews treat these six days like a full year.[11]

Once we replace Tostado's use of the Julian month names September and October with the Hebrew month names Elul and Tishri, we can see that his comments were essentially correct: the first day of creation, which precedes the *molad vayad* on Friday, 26 September, 3760 BC, is the 25th day of Elul. However, since September is 30 days in length (as opposed to the 29 days of Elul), Tostado, in his account, was forced to shift the first day of the next month (October) to the Sabbath of God's rest, whereas Jewish calendar tradition ties 1 Tishri to the Friday of Adam's creation. Yet although Tostado thus displayed some impressive knowledge of exotic details of the Jewish calendar, this kind of information did little to settle the important question of which segment of the lunation was actually equivalent to the 15th day of Nisan, on which Jesus was crucified. Similar to Jean des Murs in the fourteenth century, Tostado simply stayed content with the assumption that the date of Christ's crucifixion could be found by calculating the time of mean opposition in early spring, a task for which he recommended the Alfonsine Tables. "The method of this calculation," he explained

> cannot be described here in any detail, but he who wishes may inquire through these selfsame tables, or through others, the opposition of sun and moon, and he will find that in the aforementioned 31st year of Christ, 27 March was the 15th day of the moon, and that this lunation was the first month of the Jews.[12]

---

[11] Ibid. (14), 106b: "Periti tamen inter Iudaeos dicunt, quod hoc erat, quia ipsi computabant annos suos a principio seculi, & quia mundus creatus est in Septembri, incipit annus in Octobre. Nam dicunt, quod die vigesimaquinta Septembris creatus est mundus, & dies sexta, in qua creatus fuit Adam, fuit dies ultima Septembris, & sequenti die, in quo Deus cessavit ab omni opere, scilicet dies Sabbathi, fuit prima dies Octobris, & illa fuit prima dies anni, & sex dies praecedentes, in quibus fuit creatio, pertinent ad alium annum, & illos sex ponunt Judaei tamquam unum annum." Elsewhere, Tostado agreed with the Jewish dating of the creation to autumn as opposed to spring. See ibid. (89), 156. On his knowledge of Hebrew, see Santiago García, "La competencia hebraica de Alfonso de Madrigal," *La Corónica* 33, no. 1 (2004): 85–98.

[12] Tostado, *Defensorium* (15), 107b–08a: "Modus autem calculationis istius non potest describi, sed qui voluerit, inquirat per ipsas tabulas, vel per alias oppositionem

Tostado went on to examine the dates of all oppositions of 'Nisan' between AD 30 and 50, taking these years to be the lower and upper boundaries for Christ's death, as they could be deduced on the basis of Luke 3:23 and John 8:57. As it turned out, the opposition fell on Friday in only a handful of these years, namely in AD 33, 37, 38, and 47. Since it was commonly agreed that Jesus did not die later than in his 34th year, AD 33 was the only truly convincing candidate for the historical Passion date. In this year, the calculated mean opposition occurred on 3 April, 4:30, on the meridian of Jerusalem. Since he interpreted the day of opposition as the 15th day of the lunar month, Tostado's findings thereby also confirmed the synoptic chronology. In fact, the only remotely relevant combination of Friday with the 14th day of the month could be found on 30 March, AD 36, which on his showing was too late for the historical crucifixion. The logical conclusion, as Tostado put it, was that Jesus had been crucified on 3 April, AD 33, at an age of 32 years and 101 days, which shows that he—contrary to Roger Bacon and Jean des Murs—assumed the nativity to have taken place on 25 December, 1 BC, rather than in AD 1.[13]

As he went on to admit in his *Defensorium*, these astronomically derived conclusions had been attacked by his detractors precisely on the grounds that they contradicted the *communis opinio*, according to which Christ's Passion had taken place in the 34th year of the Dionysiac era on 25 March, the same calendar date which also marked the Annunciation of the Virgin in medieval calendars. The latter assumption was backed by the full weight of patristic opinion, in particular by St. Augustine's works *On the City of God* and *On the Trinity*. To believe otherwise could be regarded as an act of impiety.[14] His opponent Juan de Torquemada, in attacking Tostado for his audacity, had in addition also referred to the calculations of Albert the Great and of

Solis, & Lunae, & inveniet, quod in illo anno trigesimoprimo Christi, die vigesimaseptima Martii erat luna quintadecima, & illa lunatio erat primus mensis Iudaeorum." According to Goldstine, *New and Full Moons*, 86, there was indeed an opposition on that day.

[13] Tostado, *Defensorium* (15), 108b: "Dicendum ergo, quod anno 32. aetatis Christi finito, & diebus centum, & uno de anno 33. die tertia Aprilis Christus mortus est, & erat eadem die luna quintadecima, & fuit oppositio verissima Solis, & Lunae decem horis, & dimidia ante meridiem eiusdem diei, secundum meridianum urbis Romae octo horis, & aliqua particula parva, & secundum meridianum Jerusalem septem horis, & dimidia ante meridiem…quae probavi certissime ex radicibus tabularum regis Alphonsi. Ex quo patet, quod Christus non anticipavit festum Paschale, sicut Graeci dixerunt, sed comedit Luna quartadecima cum caeteris Iudaeis."

[14] Ibid. (17), 111b–12a.

"other most knowledgeable astrologers," who had supposedly proven to everyone's satisfaction that the 15th day of the moon and the sixth day of the week had fallen on 25 March in AD 34—a claim which, as we have seen, does not reflect positively on his astronomical competence.[15] As Torquemada went on to complain:

> Moreover, how can it possibly be believed that the blessed Augustine, Jerome, Isidore, and Bede, whose authority has always prevailed in the Church and who, as can be read, have calculated the years of the Lord so very diligently, could have erred in this matter? Surely, it seems that they deserve more faith than some astrological calculations. Furthermore, it is to be taken into consideration that the tables that astrologers use for their calculations, are not infallible rules, but deficient ones, which is why they are often in need of correction. And it certainly seems like an act of great temerity to claim just because of the calculation of one astronomer that that which is maintained by the doctrine of the holy Fathers and the common faith of the Church cannot be true.[16]

Tostado took these assertions of patristic authority very seriously, but could not help concluding that they failed to square with the astronomical facts. In allowing for the possibility that Jesus died on the day of his conception, Tostado even went out of his way, calculating opposition times for all years of the Christian era up to AD 175, which was the first year in which Friday and 25 March were found to coincide on a full moon.[17] In the end, the chronological facts carried too much weight to allow any retreat into pious trust in authority. Like Bacon, Tostado defended his position by pointing to the fallibility of Augustine and his contemporaries. The mere fact that Augustine had composed a collection of *Retractions* towards the end of his life

---

[15] Juan de Torquemada, *Tractatus*, cited after Lazzarato, *Chronologia*, 468: "Unde Magister Historiarum, Albertus Magnus…demonstrat, per viam astrologicam, quod anno aetatis Christi XXXIV, in quo anno juxta sanctorum Patrum sententiam dici Christum fuisse mortuum, quod luna fuerit XV, et feria VI, die XXV Martii. Item, alii peritissimi astrologi, de hac materia loquentes, reliquerunt, hoc idem probatum in libris suis. Sunt praeterea viri docti, qui hoc demonstrare possunt."

[16] Ibid.: "Amplius, quomodo credendum est, quod beati Augustinus, Hieronymus, Isidorus, et Beda, quorum auctoritas in ecclesia semper viguit, qui tam diligentissime annos Domini calculasse leguntur, errasse credantur in hac re, quibus certe magis credendum videtur, quam quibuscumque aliis calculationibus astrologis. Ceterum considerandum est, quod tabulae, quibus calculatores astrologi utuntur, non sunt regulae infallibiles, sed deficientes. Unde saepe correctionibus indiguerunt.… Magna certe temeritas est, videtur, ut, propter calculationem unius astronomi, praesumat quis dicere, maxime in facie Ecclesiae Romanae—quae magistra est fidei, quaeque solita est omnes haereses destruere—quod id, quod tenet sanctorum Patrum doctrina et fides communis ecclesiae non possit esse verum."

[17] Tostado, *Defensorium* (16), 109a–11b.

exposed the bishop of Hippo's human imperfection. While the canonical books of the Bible had been dictated by the Holy Spirit, Augustine was a mere man and sometimes had to rely on hearsay in order to collect his information. Since 25 March was astronomically impossible, there was every reason to suppose that Augustine had been simply mistaken when he mentioned this date in his writings.[18] If he had paid closer attention to the arguments of the astronomers, Tostado claimed, Augustine would himself have realized his error. Yet even though Augustine was undoubtedly "a man of the highest intelligence and above average erudition about all these things," he had been occupied with more important theological questions and hence

> never came to set his mind on this type of speculation and—as a result—did not find out the truth, which is no surprise, seeing how the same happens every day in so many other cases, where the minds of the moderns make new discoveries.[19]

In the end, it had to be concluded that Augustine and other learned fathers of the church had merely followed an already existing opinion when they dated the crucifixion on 25 March. But how could this manifestly erroneous tradition have arisen in the first place? As Tostado went on to point out, early Christians had been generally uninterested in chronology, because they faced graver problems such as violent persecution by the hands of the Roman authorities and the emergence of various heresies. Moreover, since Easter was a movable feast day based on the Jewish calendar, there had initially been relatively little incentive to settle for a single and accurate Julian date of either the crucifixion or resurrection of Christ. Indeed, the foremost concern of early Christians had been

> how they could remain in the Christian faith either in hiding, or by dying a violent death in public, wherefore it was enough for them to care about their bare lives and how they could enrich restless souls through Christ; and they had no time for the curiosities of learned disciplines nor for the investigation of truths that to them seemed more subtle than useful.

---

[18] Ibid. (18–19; 85–86), 112a–15a, 150a–51a.

[19] Ibid. (18), 114a: "Beatissimus Augustinus, licet altissimi ingenii, & eruditus, plus quam mediocriter circa omnia haec, cum esset occupatus circa alias considerationes altiores, & utiliores, non accommodavit animum huic speculationi, & ita non invenit hanc veritatem. Quod non admirandum est, quia ita quotidie fit de multis aliis, quas inveniunt moderna ingenia."

By the time the Church had consolidated during the reign of Constantine the Great, the true date of the Passion had consequently already fallen into oblivion, which made it possible for a false date, brought about by the carelessness of an unknown scholar, to firmly establish itself in Christian literature and ecclesiastical tradition.[20] As Tostado's *Defensorium* made clear, however, it was perfectly legitimate to reverse this error and unearth the truth, which could be demonstrated by astronomical means. As the subsequent two centuries of chronological scholarship were to show, his approach was there to stay.

2. *Jesus and the* Deḥiyyot: *Paul of Burgos's* 'Additio' *to Matthew 26*

The scholarly altercation between Alfonso Tostado and Juan de Torquemada in many ways set a standard for how the debate on the Passion date was to continue within Catholic scholarship up until the second half of the sixteenth century, a period during which positions taken by individual authors conspicuously oscillated between the astronomically calculated 3 April and the traditional date of 25 March. At the same time, there is evidence that Tostado's interest in chronology was more far-ranging than only the life of Jesus Christ. For instance, he concluded his *Defensorium* with a lengthy complaint about the lack of astronomical precision in the Julian calendar, as it was displayed in medieval manuscripts, which gave wrong dates for the sun's momentary positions in the zodiac as well as for the equinoxes and solstices. This awareness of the problems of the calendar apparently also led Tostado to write a separate treatise in his native Castilian language, dealing with the *Error del calendario*, which is listed among his unedited works but has been lost to posterity.[21]

---

[20] Ibid. (20), 115b: "Item in principio Ecclesiae, cum urgeret Christianos persecutio Gentilium, solum erant soliciti, quomodo possent in fide Christi manere in latibulis, vel in publico fortiter mori, & cum satis esset illis de vita curam agere, & quomodo errantes animas Christo lucrifacerent, non vacabant curiositatibus disciplinarum, & inquisitioni quarundam veritatum, quae magis subtiles, quam utiles esse videntur.... Et si non est mirandum, si talis veritas, immo aliae magis necessariae ad Christianum religionem illis temporibus incognitae fuerunt."

[21] Tostado, *Defensorium* (87–97), 151a–64b. For a listing of his works, see Luisa Cuesta, "La edición de las obras del Tostado, empresa de la Corona Española," *Revista de archivos, bibliotecas y museos* 56 (1950): 323.

Previous to Tostado's controversial appearance in Siena, the problem of the calendar had already played a prominent role at the Council of Basel (1431–49), where a special commission was set up to negotiate a reform of the ecclesiastical lunisolar calendar. One of the most outspoken calendar reformers at the council was Hermann Zoestius (ca. 1380/90–1445), a Cistercian monk from the monastery of Marienfeld, near Münster in Westphalia. In 1432, he wrote a *Tractatulum exhortationum*, which is probably identical to a *cedula* that was read out at a council session on 18 June, 1434, and whose exhortations were instrumental in establishing the aforementioned calendar commission. Once the latter, led by Nicholas of Cusa, had presented its official (and ultimately ineffective) decree in March 1437, Hermann went on to write another reform treatise of his own, entitled *Phaselexis*, in which he cited from a broad range of sources, including John of Sacrobosco, Robert Grossteste, Roger Bacon, and Pierre d'Ailly, but also Jean des Murs and even his obscure Westphalian compatriot Reinher of Paderborn.[22] In addition to the *Phaselexis*, Hermann also composed a *Calendarium Hebraicum novum* (1436) and a treatise *De fermento et azymo* (1436), in which he commented on the time-honoured conflict between Latin and Greek Christianity concerning the bread used druing the altar sacrament. In both the *Phaselexis* and the *De fermento et azymo*, Hermann endorsed the astronomical approach of dating the Passion to 3 April, AD 33, which he correctly attributed to the pioneering work of Roger Bacon and Jean des Murs. The same view is present in the *Calendarium Hebraicum novum*, which took the

[22] Preserved manuscripts of the *Phaselexis* include Munich, BSB, Clm 3564, 138ra–43brb; Clm 18470, 4r–19r; Clm 24868, 1r–23v. See also Kurt Heydeck, *Die mittelalterlichen Handschriften der Universitätsbibliothek Rostock* (Wiesbaden: Harrassowitz, 2001), 80–81, for reference to MS Rostock, UB, Mss. math.-phys. 1, 15r–24r. For further manuscript references and biographical data, see Friedrich Zurbonsen, *Hermannus Zoestius und seine historisch-politischen Schriften* (Warendorf, 1884); Zurbonsen, "Hermann Zoestius von Marienfeld und seine Schriften," *Westdeutsche Zeitschrift* 18 (1899): 146–73; Kaltenbrunner, "Die Vorgeschichte," 336–42, 350–54; Wilhelm Wattenbach, "Über Hermann von Marienfeld aus Münster," *Sitzungsberichte der königlich preussischen Akademie der Wissenschaften zu Berlin*, phil.-hist. Kl., 9 (1884): 93–109; Josef Tönsmeyer, "Hermann Zoestius von Marienfeld: Ein Vertreter der konziliaren Theorie am Konzil von Basel," *Westfälische Zeitschrift* 87 (1930): 114–91; Wilhelm Kohl, *Die Zisterzienserabtei Marienfeld* (Berlin: de Gruyter, 2010), 442–43; Müller, "*Ut reiecto*", 265–66. See further Olivier de Solan, "Les propositions de réforme du calendrier au XVe siècle," *École Nationale des Chartes: Positions des thèses* (1998): 275–81; de Solan, "La réforme du calendrier dans une question quodlibétique d'Henri de Runen (1444)," *Bibliothèque de l'École des Chartes* 157 (1999): 171–220.

form of a historical calendar for the thirteen months of Jewish calendar (including the intercalary Adar II), with important events from the Old and New Testament assigned to the individual days. Unsurprisingly, the crucifixion appears on 15 Nisan, augmented by the note that this corresponded to 3 April, AD 33, in the Julian calendar.[23]

While the dating problem was thus seemingly solved, the question of whether Jesus and his disciples had anticipated the ordinary Jewish Passover, when they convened for a meal on Thursday evening, remained controversial as ever. Back in the twelfth century, Rupert of Deutz and Reinher of Paderborn had noted how the Jewish postponement rules or *deḥiyyot*, which prohibited the celebration of Passover on a Friday, implicitly backed the Johannne account of the Passion, which set the Last Supper on the evening of 13/14 Nisan. This version of events was later also corroborated by the findings of Roger Bacon, who realized that the events described in the Gospels were in conflict with the obligation of sabbatical rest on 15 Nisan. Yet even though he could have thus provided his arguments with additional leverage, Bacon never invoked the *deḥiyyot* in this context, despite the fact that he was well-acquainted with both these rules and the ritual justifications behind them.[24] One can only suspect that Bacon refrained from using the postponements as evidence, because he realized that the rules of the conjunction-based Jewish calendar of present times could not be projected back into the ancient past.

Whether this is true or not, subsequent discussions of the Passion chronology seem to have lost the *deḥiyyot* out of sight. They were never evoked in thirteenth-century disputes over the form of bread used during the altar sacrament, nor did Alfonso Tostado mention them in his *Defensorium*. That Hermann Zoestius found this state of discussion rather unsatisfactory is indicated by an intriguing addition to his *De fermento et azymo*, which precedes the text's prologue on a separate leaf in a manuscript from the Bavarian National Library, which was copied by the Augsburg humanist Sigismund Gossembrot (ca. 1417–after 1488). The short note, which seems to have been

[23] For the *Calendarium Hebraicum novum*, see Munich, BSB, Clm 18470, 19v–27v; Clm 24868, 23v–32r. For the *De fermento et azymo*, see Munich, BSB, Clm 3564, 145ra–55ra; Clm 18536, 201r–18v. On the latter text, see also Manuel Candal, "El 'De fermento et azymo' del Ottoboniano Latino 718," in *Mélanges Eugène Tisserant*, 7 vols. (Vatican City: Biblioteca Apostolica Vaticana, 1964), 6:289–311, who is unaware of Hermann's authorship.

[24] Bacon, *Opus majus*, 1:201; Bacon, *Opus tertium*, 219–20.

originally composed by Hermann himself, faults the Christian scholars for their ignorance of the rule *lo BaDU Pesaḥ* and states that—to the author's knowledge—only two people have ever given sufficient attention to the Jewish postponement rules in their treatment of the Passion chronology: Reinher of Paderborn and Paul, the bishop of Burgos, "who wrote an excellent commentary on Matthew 26 in his *additions* to Nicholas of Lyra's *postilla* on the Bible." These *additions*, however, "only reached my hands in the ninth month after compiling the present book," which is why the author of the note asks prospective copyists of *De fermento et azymo* to include all the addenda, which he provided in the margins as a result of his reading of Paul of Burgos, into future copies of the finished text.[25]

The work that had spurred all this scholarly excitement, the *Additiones ad postillam Nicolai Lyrani* (1429/31), had indeed only been completed a few years before Hermann Zoestius's activity at the Council of Basel. It had been composed by Paul or Pablo de Santa María (ca. 1351–1435), who had served as the bishop of Burgos from 1415 until his death. Paul's career had begun under unusual circumstances, having been born as Solomon Halevi into a wealthy Jewish family. In his youth, Solomon appears to have enjoyed a thorough education in both Jewish and Gentile learning, which eventually led him to choose the career of a rabbi. He had risen to considerable fame in this position, when, in 1390 or 1391, he abandoned the faith of his ancestors and accepted baptism along with a new, Christian name. During

[25] MS Munich, BSB, Clm 3564, 144v: "Et non vidi aliquem michi mentionem facientem duobus exceptis, scilicet Reinhero et Burgensi episcopo, qui quidem dominus Paulus Burgensis in addicionibus super postillas Biblie Nicolai de Lyra luculenter et optime scripsit super illo verbo Mathei 26, *primo autem die azymorum*. Cuius quidem addiciones nono mense post presentis libelli compilacionem ad manus meas pervenerunt. Extraxi michi quaedam et ad margines huius in quibusdam locis posui in testimonium et confirmationem. Patet igitur quod Christus prevenit pasca Iudeorum. Et an ipse in cena usus sit azimo pane vel fermentato pane asserere non tentabo, sed audacter dico quod iudei fermentato pane usi sunt quinta feria, quando christus cenavit, et sexta feria quando passus est usque ad vesperam. Precor igitur eos qui huius libelli copiam habere concupiscunt, ut etiam ea que in marginibus annotata sunt, diligenter ad suas margines scribere non omittant." In Gossembrot's copy of the text, the addenda are duly inserted into the main text on fols. 146va and 147va. The above note is partly transcribed in Erwin Rauner, *Katalog der lateinischen Handschriften der Bayerischen Staatsbibliothek München: Die Handschriften aus Augsburger Bibliotheken*, vol. 1, *Stadtbibliothek: Clm 3501–3661* (Wiesbaden: Harrassowitz, 2007), 268–69, whose cautiously expressed doubts that Gossembrot was the author of these lines seem quite justified.

the following years of his life, he studied theology, was ordained a priest and embarked on a steep ecclesiastical and political career. As a churchman and confidant to the Avignonese antipope Benedict XIII, he rose to the rank of bishop of Cartagena in 1403, before succeeding to the episcopal see of his home town Burgos. Furthermore, he played an important role at the Castilian royal court, where he became a councilor to Enrique III and a tutor to his son Juan II. Paul even temporarily served as chancellor of the kingdom and, upon the king's death in 1406, as one of the officials in charge of Enrique's testament. Shortly before his death in 1435, the former rabbi was named titular bishop of Philippi in Greece.[26]

Similar to Petrus Alfonsi, the critical, or rather polemical, engagement with his former faith played an important role in Paul of Burgos's theological writings. In 1432/34, he penned his *Scrutinium Scripturarum*, in which a Jew (Saulus) and a Christian (Paulus) argued over the question whether Jesus was the true Messiah. Before that, Paul had already composed his monumental *Additiones*, which comprised more than one thousand critical additions and remarks on Nicholas of Lyra's *Postilla litteralis super totam Bibliam* (1322/31) and in which he often chastised the famous commentator for relying too heavily on rabbinical literature.[27] His attacks were severe enough to motivate another fifteenth-century theologian, the German Franciscan Matthias Döring (ca. 1390–1469), to write a *Defensorium postillae* in an attempt to protect Nicholas against the criticism levelled in Paul's *Additiones*.

---

[26] Francisco Cantera Burgos, "La conversión del célebre talmudista Salomón Levi (Pablo de Burgos)," *Boletin de la biblioteca de Menéndez Pelayo* 15 (1933): 419–48; Yitzhak Baer, *A History of the Jews in Christian Spain*, 2 vols. (Philadelphia: Jewish Publication Society of America, 1961–66), 2:139–50; Nicolás López Martínez, "Santa María, Pablo de," in *DHEE* 4:2180–81; Norman Roth, *Conversos, Inquisition, and the Expulsion of the Jews from Spain* (Madison: The University of Wisconsin Press, 1995), 136–50.

[27] On Paul's exegesis, see Ch. Merchavia, "The Talmud in the *Additiones* of Paul of Burgos," *Journal of Jewish Studies* 16 (1965): 115–34; Nicolás López Martínez, "Pablo de Santa María y el sentido literal bíblico en las controversias con los judíos," in *Biblia, exégesis y cultura*, ed. Gonzalo Aranda Pérez, Claudio Basevi, and J. Chapa (Pamplona: Ediciones Universidad de Navarra, 1994), 475–83; Ryan Szpiech, "Scrutinizing History: Polemic and Exegesis in Pablo de Santa María's *Siete edades del mundo*," *Medieval Encounters* 16 (2010): 96–142. For manuscripts and printed editions of the *Additiones*, see Klaus Reinhardt and Horacio Santiago-Otero, *Biblioteca bíblica ibérica medieval* (Madrid: CSIC, 1986), 240–44; Horacio Santiago-Otero, *Manuscritos de autores medievales hispanos*, vol. 1 (Madrid: CSIC, 1987), 85–96; Edward A. Gosselin, "A Listing of the Printed Editions of Nicolaus de Lyra," *Traditio* 26 (1970): 399–426.

All three commentaries became highly influential during the early modern period, owing to their inclusion in numerous Bible editions of the fifteenth to eighteenth centuries.

In his commentary on Matthew 26, referred to by Hermann Zoestius, Paul reacted to Nicholas of Lyra's reading of the Passion chronology, which essentially followed Thomas Aquinas in upholding the notion that Jesus was crucified on the afternoon of 15 Nisan, the Johannine version being in harmony with the synoptic one.[28] As a former rabbi, well-acquainted with the Jewish calendar and the feast days attached to it, Paul was certain that the events surrounding the crucifixion could not possibly have taken place on the first day of the feast of unleavened bread, on which work was prohibited, not to speak of the various judicial activities described in the Gospels such as "accusations in court, floggings, the summoning and interrogation of witnesses, and the pronouncement of definitive verdicts."[29] The mere fact that Jesus had been hastily and quickly buried on the evening of his death in order to respect the start of the Sabbath was enough to demonstrate that the day of the Passion had by no means been a high feast day. Yet these important insights also led into a dilemma, insofar as the three synoptic Gospels still seemed steadfast in their implication that Jesus and his disciples had eaten the Passover meal on the evening before his trial and execution. Simply to assume that the fourth Gospel was correct in putting the crucifixion on 14 Nisan, as Roger Bacon had done back in the thirteenth century, was problematic in that it implied that the four evangelists were unable to agree on such a crucial detail.

In Paul's mind, the only route out of this exegetical predicament was opened by the *deḥiyyot*. In his commentary on Matthew 26, he drew attention to a particular element of the Jewish postponements, which prescribes that any *molad Tishri* falling later than noon (or 18:00 in the Jewish system of reckoning days from evening) has to result in 1 Tishri's being postponed to the next day. The precise rationale behind this rule is unknown, but it may have been partly designed with the

---

[28] See Nicholas's comments on Matthew 26 in *Textus biblie (V)* (Basel, 1507), 78vb–79ra.

[29] Ibid., 82va: "Sed assumptum probatur, nam multa fecerunt Hebraei in eadem sexta feria, quae nullo modo licebant fieri in sabbato, nec in tali solemnitate, videlicet accusare, flagellare, testes praesentare et examinare, sententiam definitivam ferre, quae omnia exercuerunt circa personam christi, ut evangelistae testantur. Quae quidem iudicalia etiam apud nos in solemnibus diebus festivis sunt illicita, multo magis apud eos qui nimis literalizant."

aim of increasing the likelihood that the thin new moon crescent could already be seen on the first day of the year. In Paul's own words, a postponement of the first day was necessary, because in such a case the conjunction "has already passed three quarters of this day and touches upon the last quarter," hence verging more strongly towards the following day than towards the one it actually belongs to. Accordingly, this type of postponement was also known as 'old molad' or *molad zaqen* (מולד זקן), a term which Paul correctly translated as *novilunium vetus seu antiquum*.[30] As he went on to explain, this postponement-type could be combined with the well-known rule *lo BaDU Pesaḥ*, according to which 15 Nisan could never fall on the 2nd, 4th, or 6th day of the week. Since the latter rule, by way of calendrical necessity, also applied to the first day of the month, i.e. to the day of the *molad Nisan*, a postponement of the *molad* due to rule *molad zaqen* could sometimes trigger rule *BaDU*, leading to a double postponement in certain years. According to the learned bishop of Burgos, this had also been the case in the year of the Passion, for which he reckoned the *molad Nisan* to have occurred on Thursday, shortly after 19:00 (reckoned from sunset), a result which has been "calculated by me with great certainty and which I have found out and collated most diligently with the learned men among the Jews."[31] Paul's reference to a *molad Nisan* on Thursday, shortly after 19:00, implies that he assumed the Passion to have occurred in the year 3793 of the Jewish world era, in which 14 Nisan fell on Friday, 3 April, AD 33, thereby revealing that he agreed with Roger Bacon and his contemporary Alfonso Tostado in dating the Passion to this day. The *molad Tishri* of 3793 JE can be retrospectively calculated to have fallen on Tuesday, 14h 737ch, which leads to a *molad Nisan* on Thursday, 19h 95ch (2.14.737 + 2.4.438 =

---

[30] Ibid., 82vb: "Qui quidem casus duo sunt. Primus: quando talis coniunctio transcendit xviii. horam eiusdem diei in qua est. Cuius repulsionis ratio est, quia tunc magis declinat talis coniunctio ad diem sequentem quam ad illam in qua est: cum iam transit tres quartas partes illius diei, attingit quartam ultimam, et talis coniunctio vocatur apud eos novilunium vetus seu antiquum." On rule *molad zaqen*, see Feldman, *Rabbinical Mathematics*, 191–93; Stern, *Calendar*, 195–96. As so often, the first Latin author to draw attention to this rule was Reinher of Paderborn. See van Wijk, *Le comput*, 30.

[31] *Textus biblie*, 82vb–83ra: "Septimum est quod anno quo christus passus fuit concurrit coniunctio luminarium prope aequinoctium vernale iuxta veram computationem Hebraeorum quae est secundum medium cursum solis et lunae, ut dictum est in feria quinta post xix. horam cum certis minutis. Hoc est certissime calculatum per me, et cum peritis Hebraeorum diligentissime repertum et concordatum."

4.19.95), a date which matches the information given by Paul of Burgos. Since the *molad* fell later than 18:00, the beginning of the month had to be shifted to Friday, which in turn necessitated another shift to Saturday. As a result, in the year of the Passion, both the beginning of the month (1 Nisan) and the first day of unleavened bread (15 Nisan) had fallen on a Sabbath, which conveniently explained the phrasing in John 19:31, where it was stated that the Sabbath was a "high day."[32]

In Paul's opinion, this double postponement of Passover at the time of Jesus's crucifixion was the key to a correct understanding of the, in his mind, merely apparent chronological contradiction between John and the synoptic Gospels. As he explained to his readers, one had to look at the date of Passover from the perspective of Jesus and his followers, who had to decide whether to postpone the beginning of Nisan in the year in question. In particular, it was important discern the rationale behind each of the two postponement-types involved. The rule *molad zaqen*, or so it seemed, was an out-and-out rational device, introduced to prevent the conjunction from falling into the last quarter of the day (in which case the new moon crescent could have hardly become visible in the same evening), and there was no reason why Jesus and his disciples should not have observed it in the year of the Passion. But what about rule *lo BaDU Pesaḥ*? As Paul explained to his readers, the reason for not celebrating 15 Nisan on a Friday had nothing to do with Passover itself, but was dependent on the date of *Hoshana Rabbah*, on 21 Tishri. The latter was prohibited in rabbinic law from falling on a Saturday, because the ritual beating of willow branches on said day was thought to conflict with the commandment of sabbatical rest. Naturally, such a postponement of 21 Tishri had to automatically affect other dates in the Jewish calendar. Since the interval between Nisan and Tishri was always 177 days, the postponement of *Hoshana Rabbah* from Sabbath to Sunday logically implied

---

[32] Ibid., 83ra: "Nam in anno passionis christi…habuit ergo repelli seu transferri neomenia ad diem sabbati, et per consequens xv. luna fuit eodem anno secundum hebraicam computationem legitimam et assuetam in die sabbati, in qua concurrunt duae festivitates solemnes apud eos, sabbatina scilicet et paschalis quae est prima dies solemnitatis azymorum, et sic erat magna dies illa sabbati secundum Ioh. et in feria sexta immediate praecedenti fuit secundum eos xiv. luna in qua legitime erat pascha immolandum." Although the rule *molad zaqen* technically only applies to the *molad* of Tishri, it had a knock-on effect on the other months of the calendar, including Nisan. One may premuse that Paul of Burgos intentionally simplified these matters, which do not in any way affect the accuracy of his calendrical argument, for his readers.

a postponement of 15 Nisan from Friday to Sabbath. Yet this in turn meant that the postponement of Passover in the year of the crucifixion depended on a Jewish feast day (*Hoshana Rabbah*), which at the time of the Last Supper still lay several months in the future. This feast day was part of Sukkot (or Tabernacles, on 15–21 Tishri) and hence part of the law of the Old Covenant, which was about to be dissolved through Christ's death and resurrection.

The calendrical rules attached to *Hoshana Rabbah* were thus bound to be null and void by the very next instance on which this feast was to be celebrated, in the autumn following the crucifixion. As such, it had no more legal authority to compel the postponement of the preceding Passover, especially from the point of view of Jesus, who was able to foresee his own Passion and the subsequent dissolution of the Old Covenant. Failing a binding reason to succumb to rule *BaDU*, Jesus was content with postponing the Passover date by one day instead of two days, as the majority of Jews in his time would have done. By invoking the consequences of Christ's resurrection on the validity of Jewish law, the former rabbi Paul of Burgos was hence able to explain why Jesus celebrated the Passover on the evening before his death, whereas the other Jews did the same only on the following evening. Owing to their different interpretations of the *deḥiyyot*, both parties were in principle observing different calendars in the year of the Passion.[33] Paul summarized his conclusions in the following manner:

> It follows that at the time of Christ's Passion there were two versions of the 14th day of the lunar month and of the beginning of the first month. The first of these was the true and legitimate version according to the truthful calculation of Christ, who knew all mysteries of the Law, both present and future. The other version, by contrast, received its justification only from the common and customary calculation of all Jews other than Christ.[34]

[33] Ibid.: "Christus autem in huiusmodi neomenia prima repulsam tantum tenuit et non secundum. Prima enim est rationalis in se et sine respectu ad festa futura, sed solum propter vetustatem coniunctionis, unde neomenia illius anni apud ipsum fuit translata ad sextam feriam, sed de sexta ad septimam nullatenus debuit transferri apud eum, nam ratio illius repulsae ex festis futuris in mensi septimo dependebat, ut patet ex iam dictis, quae festa, scilicet, futura cum christus scivisset per passionem suam de propinquo imminentem cassari, sicut et cetera legis cerimonialia, nullam vim habebant transferendi seu repellendi praesentem neomeniam, ut patet intuenti. Nam cessante causa praesertim finali, cessare debet effectus."

[34] Ibid.: "Ex quibus habetur quod tempore passionis christi duplex fuit xiiii. luna sicut et neomenia primi mensis. Una, scilicet, vera simul et legitima secundum veram

Michael Döring, whose *Defensorium postillae* is usually found alongside Paul's *Additiones* in early modern glossed bibles, was demonstrably unimpressed by the bishop of Burgos's "long and peculiar calculation, which gives rise to many doubts."[35] It should be noted, however, that Paul's interest in the Jewish calendar was not motivated by a fascination for its technical details, but by the question of how knowledge of these details could help solve the nagging exegetical and theological problems caused by the contradiction between the Gospels with regards to the date of the Passion. Hermann Zoestius, who hastily supplemented his *De fermento et azymo* with Paul's arguments, was by no means the only scholar who found this solution cogent and exciting. As we shall also partly see below, Paul's words had a lasting impact on Renaissance chronologers as diverse and eminent as Paul of Middelburg, Sebastian Münster, Johannes Lucidus Samotheus, Onofrio Panvinio, Gerhard Mercator, Heinrich Bünting, and the great Joseph Scaliger, all of whom were at one point wrongly convinced that the *deḥiyyot* were an authentic component of the first-century Jewish calendar.[36] By the seventeenth century, the works of the Castilian bishops Paul of Burgos and Alfonso Tostado had long become classical points of reference, in which the principal reasons for assuming a crucifixion on either 14 Nisan (Paul) or 15 Nisan (Alfonso) could be found. As Gerhard Johannes Vossius (1577–1649), himself the author of a *Dissertatio* on the Passion date, put it:

computationem christi cuncta legis mysteria tam praesentia quam futura scientis. Alio vero legitima tantum secundum communem et assuetam computationem aliorum omnium praeter christum, et in utraque xiiii. uterque agnus fuit immolatus verus et typicus."

[35] Ibid., 83rb: "Cum postillator solvisset omnes difficultates circa coenam domini et esum agni paschalis moveri consuetas praesertim a graecis, Burgensis facit unam longam singularem computationem in qua multa ponit multis dubia, cui placet videat."

[36] See, e.g., Johannes Lucidus Samotheus, *Opusculum de emendationibus temporum* (Venice, 1537), 162r, 183v–84v; Alexander Scultetus, *Chronographia* (Rome, 1546), *4r; Onofrio Panvinio, *Fastorum Libri V* (Venice, 1558), 311; Gerhard Mercator, *Chronologia* (Cologne, 1569), d6r–v; Scaliger, *De emendatione temporum* (1583), 265–66; Heinrich Bünting, *Chronologia* (Zerbst, 1590), 234v. On Sebastian Münster and Paul of Middelburg, see the following chapter. See further the accounts in Deborah Kuller Shuger, *The Renaissance Bible* (Berkeley: University of California Press, 1994), 35–37, and Anthony Grafton and Joanna Weinberg, *"I Have Always Loved the Holy Tongue"* (Cambridge, MA: Harvard University Press, 2011), 214–30. An interesting reference to the *deḥiyyot* in the polemical tradition of Petrus Alfonsi, but in rhymed form, can be found in Christoph Mandel, *Beweisung aus der Juden Gesatz* (Neuburg an der Donau: Kilian, 1557), B3r–Cr.

> From these two wells, all younger writers irrigate their gardens, in such way that they hardly add anything to the matter, which they have not read in [Paul of Burgos's and Tostado's] works; or in the works of those, who copied them.[37]

Alas, no garden remains green forever and as so often the old authorities gradually fell into oblivion. By the 1920s, the English Jesuit Matthew A. Power was able to sell the "ingenious, tricky, and occult rule Badhu" and its application to the Passion chronology as a minor sensation, not shying away from statements such as:

> Truly the wise men of Israel, *in camera* assembled, can afford to chuckle over the gullibility of the *goyim* who cling to the so-called purely lunar calendar of the Jews, and shutting their eyes to the foreign body, called Badhu, refuse to see how it works as a camouflage.[38]

Power's idiosyncratic claims, which contained more than just a hint of anti-Semitic paranoia, were justly rebuked in the same year by the eminent Jewish scholar Solomon Zeitlin, who reminded the Jesuit and his readers that there were no traces of rule *BaDU* in the calendar of Second Temple Judaism.[39] That their exchange was essentially the restaging of a late medieval debate will have hardly occurred to either of them.

3. *Night Visions: The Strange Case of Paul of Middelburg*

Despite the strenuous efforts of Hermann Zoestius and Nicholas of Cusa, the Council of Basel failed to produce an effective decision on

[37] Gerhard Johannes Vossius, "Dissertatio secunda; quae est de tempore dominicae passionis," in *Dissertatio Gemina* (Amsterdam, 1643), 54: "Diversum diem statuit Paulus Burgensis in Matth. cap. XXVI: eodem autem die celebrasse, contra eum operose contendit Tostatus Abulensis Defensorii sui Parte II. Atque ex horunce fontibus juniores fere hortos suos irrigarunt: ut qui vix adferant, quod non apud duos illos legerint; vel eos, qui hos excripserant." See also Dionysius Petavius, *De doctrina temporum*, 3 vols. (Antwerp, 1703), 2:438a: "Ex amborum scriptus, velut fontibus, omnium, qui deinceps in ea controversia versati sunt, disputationes, & commentarii profluxerunt."

[38] Matthew A. Power, "Nisan Fourteenth and Fifteenth in Gospel and Talmud: A Study in Jewish Camouflage," *American Journal of Theology* 24 (1920): 258, 262.

[39] Solomon Zeitlin, "'The Secret of *BADHU*': A Specimen of 'Jewish Camouflage'," *American Journal of Theology* 24 (1920): 502–11. See also Zeitlin, "The Judaean Calendar During the Second Commonwealth and the Scrolls," *Jewish Quarterly Review*, n.s., 57 (1966): 32–33.

how to cure the ecclesiastical calendar. Nevertheless, the subject of reform continued to occupy many a creative mind during the fifteenth century, including the great astronomer Johannes Regiomontanus (1436–76). In the Latin version of his popular *Calendarium*, which was published in several editions between 1474 and 1499, he clearly pointed out how the incorrect estimate of the lunar month and solar year in the ecclesiastical calculation caused Easter Sunday to differ from the astronomically legitimate date in no fewer than 30 instances between 1477 and 1531. The spread of printed astronomical calendars, as they were produced by Regiomontanus and others in the late fifteenth and early sixteenth century, had the side-effect of making the discrepancies between the Church's festive dates and astronomical reality more widely known, which in turn increased the pressure on ecclesiastical authorities to finally take care of the problem. Regiomontanus was eventually invited to Rome by Pope Sixtus IV, who was interested in winning the famous astronomer as a consultant in a future revision of the ecclesiastical calendar. Unfortunately, he died soon after his arrival, in July of 1476, before any tangible results could be produced.[40]

Aside from the computation of Easter, Regiomontanus had previously also dealt with the chronology of the life of Jesus Christ. During an earlier stay in Italy, where he lived and worked at Cardinal Bessarion's house in Rome, he had gotten into contact with his German compatriot Jacob of Speyer, who served as a personal astrologer to one of Bessarion's friends, Federico da Montefeltro, the duke of Urbino. In a short exchange of letters that took place in the spring of 1465, both men discussed a range of mathematical and astronomical problems, which also touched upon the celestial constellations that had influenced Christ at his birth and the date of his Passion. The latter subject was raised by Jacob of Speyer, who argued on the basis of Dionysius the Areopagite's famous letter to Polycarp of Smyrna that the crucifixion had taken place on a day on which sun and moon were in opposition, the simultaneous eclipse of the sun being an evident miracle. Since the Church traditionally taught that Jesus had died in his 34th year, these raw data made it possible to extract the corresponding calendar date and even the weekday of the crucifixion, which

[40] Johannes Regiomontanus, *Calendarium* (Venice, 1476); Kaltenbrunner, "Die Vorgeschichte," 367–74; Ernst Zinner, *Regiomontanus* (Amsterdam: North-Holland, 1990), 89, 112–15, 125–30. See also Karin Reich, "Problems of Calendar Reform from Regiomontanus to the Present," in Zinner, *Regiomontanus*, 352–55.

in Jacob's view was not automatically restricted to Friday.[41] Complying with his colleague's request to investigate the date, Regiomontanus found that in the 34th year of Jesus's life the first opposition of sun and moon after the vernal equinox had fallen on Monday, 11 April, a date that he seemed to be willing to accept despite the crass departure from everything theologians and astronomers had previously held plausible.[42] The only thing his calculation really showed, however, was that continued belief in the historicity of the Dionysiac era with regard to Jesus's birth date was bound to cause problems whenever one tried to combine it with the traditional view according to which the saviour had died in his 34th year. This valuable lesson had already been learned by the 'critical computists', but by 1465 their onslaughts on the Christian era had been largely forgotten. In Regiomontanus's case, the assumed nativity date had obviously been 25 December, AD 1, which led to a Passion in AD 35, a year in which—contrary to Bacon's AD 33—no spring full moon could be found on a Friday.

Meanwhile, the increasingly urgent problem of the calendar lingered on until it was officially addressed once again during the Fifth Lateran Council of 1512–17, when Pope Leo X ordered a new commission of experts to work out the proposal of a reform.[43] In preparation for the

[41] See the exchange of letters between Jacob of Speyer and Regiomontanus in Maximilian Curtze, "Der Briefwechsel Regiomontans mit Giovanni Bianchini, Jacob von Speier, und Christian Roder," *Abhandlungen zur Geschichte der Mathematischen Wissenschaften* 12 (1902): 292–309. Ibid., 301: "Quero, quota dies erat mensis sacre passionis sue, et que dies septimane esse potuit, Iovisne an Veneris vel sabbati, salvo loco Lune a beato Dyonisio invento et salvis nominibus dierum quibus quottidie utimur." Jacob's letter is dated "Urbino, 6 April 1465." On this correspondence, see Zinner, *Regiomontanus*, 79–82. On the crucifixion-eclipse in Renaissance art and scholarship, see also Eileen Reeves, *Painting the Heavens* (Princeton, NJ: Princeton University Press, 1997), 126–29.

[42] Curtze, "Der Briefwechsel," 307: "Diem passionis salutifere dico fuisse feriam secundam. Anno enim 34° etatis salvatoris nostri equinoctium vernale fuerat vigesima secunda die Martii, oppositio autem vera luminarium eodam anno proxime secta equinoctium fuit completis 10 diebus Aprilis horis 23 et minutis 21 fere, diebus non equatis, videlicet die undecima Aprilis currente, qui fuit feria secunda." Regiomontanus's letter is signed "Apud Balneas Viterbiensis," but lacks a date.

[43] See Demetrio Marzi, *La questione della riforma del calendario nel Quinto Concilio Lateranense (1512–1517)* (Florence: Carnesecchi, 1896); Marzi, "Nuovi studi e ricerche intorno alla questione del calendario durante i secoli XV e XVI," in *Atti del congresso internazionale di scienze storiche (Roma, 1–9 Aprile 1903)*, 12 vols. (Rome: Accademia dei Lincei, 1904–7), 3:637–50; Kaltenbrunner, "Die Vorgeschichte," 375–97; North, "The Western Calendar," 94–100; Frederic J. Baumgartner, "Popes, Astrologers, and Early Modern Calendar Reform," in *History Has Many Voices*, ed. Lee Palmer Wandel (Kirksville, MO: Truman State University Press, 2003), 41–56.

commission's task, briefs were sent to Emperor Maximilian I as well as to universities, to the Christian princes, and their theologians and astronomers throughout Europe, in which they were encouraged to submit their suggestions. Among the scholars who sent letters in reply was Nicholas Copernicus (1473–1543), who later, in the preface to his *De revolutionibus orbium coelestium* (1543), stated that he had been led onto his present road of inquiry "at the instigation of that most distinguished man, Paul, bishop of Fossombrone, who was then in charge of this matter."[44] The man whom Copernicus chose to honour in this way was Paul of Middelburg (1446–1534), a renowned scholar and churchman, who had been given presidency over the council's calendar commission. As his name suggests, Paul's birthplace was Middelburg in Zeeland. He received his early education at a public school in Bruges before studying theology and medicine at the University of Louvain, where he earned a doctorate in the arts. After an ill-fated return to his hometown, where he spent some time as a canon at the local cathedral before getting involved in private quarrels, Paul left his native Low Countries for Italy. From 1479 till 1481, he lectured in astrology at the University of Padua, where he was awarded a doctorate of medicine in 1480, before entering the services of Federico da Montefeltro (and later his son Guidobaldo) as an astrologer and physician—a position previously held by the aforementioned Jacob of Speyer. He was on friendly terms with scholars such as Marsilio Ficino and became the godfather of Giulio Cesare Bordon, better known as Julius Caesar Scaliger, Joseph Scaliger's father, whom he baptized in Padua in 1484. Finally, in 1494, he ascended to the episcopal see of Fossombrone, a position he held for four decades. He died on 13 December, 1534, at

[44] Nicholas Copernicus, *De revolutionibus orbium coelestium, libri VI* (Nuremberg, 1543), 4v: "Nam non ita multo ante sub Leone X cum in Concilio Lateranensi vertabatur quaestio de emendato Calendario Ecclesiastico, quae tum indecisa hanc solummodo ob causam mansit, quod annorum et mensium magnitudines, atque Solis et Lunae motus nondum satis dimensi haberentur. Ex quo equidem tempore, his accuratius observandis, animum intendi, admonitus a praeclarissimo viro Domino Paulo episcopo Semproniensis, qui tum isti negotio praeerat." See also Edward Rosen, "Galileo's Misstatements about Copernicus," *Isis* 49 (1958): 320–22; D. J. K. O'Connell, "Copernicus and Calendar Reform," *Studia Copernicana* 13 (1975): 189–202; Miguel A. Granada and Dario Tessicini, "Copernicus and Fracastoro: The Dedicatory Letters to Pope Paul III, the History of Astronomy, and the Quest for Patronage," *Studies in History and Philosophy of Science* 36 (2005): 464–70.

the age of 88, eight days after his arrival in Rome, where had been summoned by Paul III to receive the cardinalician purple.[45]

During his lifetime, Paul published a wide variety of books and pamphlets, mostly astrological prognostics, but also mathematical tracts. His most significant string of works dealt with questions of chronology, in particular with the ecclesiastical calendar and the calculation of Easter. In 1513, Paul published his *magnum opus*, the *Paulina de recta Paschae celebratione*, a monumental collection of tracts, divided into two parts, each of which encompassed roughly 200 folio-pages printed by Ottaviano Petrucci.[46] In the first part of his work, Paul dealt extensively with the error of the ecclesiastical calendar, railing against the suggestions of previous scholars such as Nicholas of Cusa, who wanted to omit a certain number of days from the present calendar in order to restore the vernal equinox to 21 March. In Paul's opinion, Easter Sunday was best determined according to the present vernal equinox on 10 March, while the solar calendar was to be adjusted by dropping a day in every 134 years. In the second part of the *Paulina*, he returned to the vexed question of the date of Christ's Passion, providing the longest and most evolved discussion of the subject that had ever been produced up to that time. While the book on calendar

---

[45] The most important source on Paul's life is the *vita* composed by Bernardino Baldi (1553–1617) between 1582 and 1600, which was first edited by Marzi, *La questione*, 233–50, and again in Bernardino Baldi, *Le vite de' matematici*, ed. Elio Nenci (Milan: Angeli, 1998), 356–95. See further Marzi, *La questione*, 39–53; Augusto Vernarecci, *Fossombrone dai tempi antichissimi ai nostri*, 2 vols. (Fossombrone: Monacelli, 1907–14), 2:552–73; Adolf de Ceuleneer, "Paulus van Middelburg en de Kalenderhervorming," in *Handelingen van het eerste Vlaamsch Taal- en Geschiedkundig Congres* (Antwerp: De Vos en Van der Groen, 1910), 276–89; Dirk Jan Struik, "Paolo di Middelburg e il suo posto nella storia delle scienze esatte," *Periodico di Matematiche*, 4th ser., 5 (1925): 337–47; Struik, "Paulus van Middelburg (1445–1533)," *Mededeelingen van het Nederlandsch Historisch Instituut te Rome* 5 (1925): 79–118; Struik, "Sull'opera matematica di Paolo di Middelburg," *Rendiconti della Reale Accademia Nazionale dei Lincei: Classe di scienze fisiche, matematiche e naturali* 1 (1925): 305–8; Antoine de Smet, "Savants humanistes et astrologie," in *Acta Conventus Neo-Latini Lovaniensis*, ed. J. I. Jsewijn and E. Keßler (Louvain: Leuven University Press, 1973), 191–97; Paolo Sambin, "Il dottorato padovano in medicina di Paolo da Middelburgo (1480)," *Quaderni per la storia dell'Università di Padova* 9–10 (1976–77): 252–56; Steven Vanden Broecke, "Paulus von Middelburg, *Paulina*," in *De geleerde wereld van Keizer Karel*, ed. Tineke Padmos and Geerd Vanpaemel (Louvain: Leuven University Press, 2000), 291–93; Vanden Broecke, *The Limits*, 61–65.

[46] See Lorenz Welker, "Ottaviano Petrucci and the Political-Cultural Elite of his Time: The 1513 Print of Paul of Middelburg's *Summa Paulina de Recta Paschae Celebratione*," in *Venezia 1501: Petrucci e la stampa musicale*, ed. Giulio Cattin and Patrizia Dalla Vecchia (Venice: Edizioni Fondazione Levi, 2005), 107–15.

reform comprised 14 chapters, symbolizing the 14 years of Christ's life under the reign of Augustus, the second book was divided into 19 parts, mirroring the notion that Jesus had been crucified in the 19th year of Tiberius's reign.[47]

In this second book, which was dedicated to Emperor Maximilian, Paul filled many pages with discussions of previous research on the Passion date, citing and classifying the opinions not only of numerous church fathers, but also of medieval scholars such as Marianus Scottus, Peter Comestor, Albert the Great, Roger Bacon, Nicholas of Lyra, Paul of Burgos, and Jean des Murs.[48] He was even aware of the dating-attempt made by Regiomontanus in his letter to Jacob of Speyer, according to which the crucifixion should have fallen on Monday, 11 April. As Paul of Middelburg disapprovingly pointed out, this matched the astronomical situation in AD 35 and completely failed to square with the chronological data found in the Gospels. The letter in question is now uniquely preserved in a Nuremberg manuscript containing Regiomontanus's correspondence, but since Paul and Jacob of Speyer had both been on Federico da Fontefeltro's payroll, it seems likely that Paul was shown the original or otherwise notified of its contents when both men met at the ducal palace in Urbino.[49]

One of the unifying themes of Paul's discussion of the crucifixion date was its anti-Judaic thrust, which led him to claim that his extensive search for the true date of the Passion was in part a defence against Jewish attacks. In his dedicatory epistle to the emperor, he complained that the Jews were in the habit of criticizing and ridiculing Christians for their views on the chronology of the Passion. According to Paul, the Jews held that their calendar prevented any coincidence of Friday and Passover. In addition, they mobilized astronomical calculations, which were intended to prove that Jesus could have been crucified neither on

[47] Paul of Middelburg, *Paulina*, a2v: "Binos ex fasciculis collectis acervos composui, quorum primus quatuordecim constans fasciculis iuxta numerorum annorum quibus Salvator sub Augusto vixit veram paschae rationem quatuordecim libris exponit; alter vero decem et novem fasciculis integratus pro numero annorum quibus Christus sub Tiberio vixit."

[48] Ibid., E7v–G7v.

[49] Ibid., G7v. On Paul's acquaintance with Jacob of Speyer and their Urbinese connection, see Baldi, *Le vite*, 358–59; Zinner, *Regiomontanus*, 154; Cecil H. Clough, "Federigo da Montefeltro's Patronage of the Arts, 1468–1482," *Journal of the Warburg and Courtauld Institutes* 36 (1973): 132–33; Heinz Hofmann, "Literary Culture at the Court of Urbino during the Reign of Federico da Montefeltro," *Humanistica Lovaniensia* 57 (2008): 5–59.

luna 14 nor on *luna* 15. Moreover, they complained that Christ's stay in the sepulchre was too short for him to count as a typological mirror of the prophet Jonah, who spent three full consecutive days and nights inside the stomach of a whale (Matthew 12:40). According to the Gospel narratives, Jesus was buried on Friday evening and rose from the dead on Sunday, which was barely two days.[50] In denying that Passover could have fallen on a Friday in the year of Christ's Passion, the Jews allegedly referred to a law from the time of the founding of the Second Temple, which prohibited such a coincidence of dates. Paul of Middelburg supported this claim by citing and translating a Hebrew 'legal document', in which the establishment of the Jewish calendar's postponement rules (*deḥiyyot*) was described:

> Thus did decree our sages the world renowned, the Sanhedrin, when the second Temple was completed. Whereupon for their sake there appeared a throne of fire, and upon it the Sovereign of the world, exalted, stood between the vestibule (and the altar). They received the crown (of the Torah) and held fast to the hidden seal. And they designated and decreed an everlasting decree. And He gave it into the hands of R. Eliezer, the greatest of them all. And he decreed that Purim must not ever fall on Monday, Wednesday, or Saturday; neither Passover on Monday, Wednesday, or Friday; neither the Eighth Day of Solemn Assembly on Tuesday, Thursday, or Saturday; neither the New Year on Sunday, Wednesday, or Friday; neither the Day of Atonement on Sunday, Tuesday, or Friday.[51]

---

[50] Paul of Middelburg, *Paulina*, Ar–v: "Solent enim perfidi & obstinati iudaei calumniari & reprehendere nos, quia dicimus Christum dominum ab eis crucifixum fuisse die veneris in festo paschae ipsorum, autumantes festum paschae nunquam die veneris posse contingere, per constitutionem in secunda templi instauratione multis annis ante Christi nativitatem aeditam." Ibid., A2r: "Etsi multa sunt argumenta quibus iudaei magnam nobis calumnia solent astruere, & fidem speratae a nobis resurrectionis stulta gurralitate deridere, in hac tamen lucubratiuncula nostra ea duntaxat confutare aggrediemur, quae dominicae passionis & resurrectionis materiam concernunt." On the problem of the *triduum* of Jesus's stay in the grave in the works of Paul of Middelburg, Erasmus, and Jacques Lefèvre d'Étaples, see Sheila M. Porrer, *Jacques Lefèvre d'Étaples and the Three Maries Debates* (Geneva: Droz, 2009), 51–61.

[51] כך גזרו רבותינו סנהדרין גאוני עולם /בבית מקדש שני כנבנה ונשלם. ואז נראה כסא אש בשלם /ועליו מלכו של עולם מסתלסל ועומד בין האולם/ וקבלו כתר ותפשו חותם נעלם. ומנו ותקנו תקנת עולם /ונתן ביד רבי אליעזר גדול שבכלם. והתקין לא בד״ז פור, ולא בד״ו פסח, ולא גה״ז עצרת, ולא אד״ו ראש השנה, ולא אג״ו יום כפור לעולם. Ibid., D7r. I follow the rendition of the text and the English translation in Leon J. Weinberger, "On the Provenance of Benjamin B. Samuel Quštani," *Jewish Quarterly Review*, n.s., 68 (1977): 52–53. On Paul's use of this and other Hebrew citations in the *Epistola apologetica*, see Adri K. Offenberg, "The First Use of Hebrew in a Book Printed in the Netherlands," *Quaerendo* 4 (1974): 44–54.

This alleged charter for the fixed calendar was in reality a *petiḥah* (a reading that precedes a *piyyut*), which can be found in manuscripts of the *Maḥzor Romania*, the prayer-book used by Jews in the Byzantine area of influence. According to Paul's own testimony, he had initially refused to accept the document's implications for the history of the calendar: "Since the Jews, in their perfidy, make a habit of arbitrary invention to enable them to slander the Christian religion," he used to think that they had "invented this institution after the Passion of Christ." He only changed his opinion after coming across yet another Hebrew text, whose title he translated as *Liber de solis et lunae motibus, annique ratione*. Its alleged author was Rabbi Gamaliel, "the pupil of Christ and teacher of Paul, at whose feet (it is said) the Apostle himself studied the Law and the Prophets."[52] Nothing is known about the source which Paul had in mind here, but it would seem that, similar to Petrus Alfonsi before him, Paul fell prey to a confusion of Gamaliel I, mentioned in the Acts of the Apostles (5:34–39; 22:3), with his grandson Gamaliel II, who was a well-known authority on the calendar. This reputation might explain why a book on astronomy and calendar lore, if it actually existed, was later falsely attributed to him.[53] Paul used both texts as proof that the *deḥiyyot* had been in existence since the earliest days of the Second Temple. This chronological point was underscored by the mention of the first-century R. Eliezer (ben Hyrcanus) in the *petiḥah*, whom Paul falsely identified with the biblical priest Eleazar,

---

[52] Paul of Middelburg, *Paulina*, D8r–v: "Verum quoniam iudaei perfidia pleni multa pro arbitrio solent confingere, ut christianam religionem calumniari valeant: arbitrabar hanc constitutionem post Christi passionem ab eis confictam fuisse, donec eandem apud sanctum Gamalielem Christi discipulum, & Pauli praeceptorem conscriptam invenirem, ad cuius pedes apostolus ipse se legem, & prophetas didicisse gloriatur, qui in libro quem de solis & lunae motibus, annique ratione conscripsit sic dicit. טעם דחיית המועדים שהם נדחיסמן הימים הללו מפני שני דברים האחד מפני יום הנפורים שלא יבא ביום א או ביום ו והשני כדי שלא יבא יום ערבה בשבת לפי שבזמן הזה אין ערבה דוחה שבת statuerunt doctores, & patres nostri neomenias, & festa in tempore conveniente ne sabbati honor minuatur, aut alia praecepta violentur. Statuerunt itaque anni neomeniam nunquam feria prima, feria quarta, vel sexta posse contingere. Festum expiationis, nunquam feria prima, feria tertia, neque feria sexta. Pascha vero nunquam feria secunda, feria quarta, neque feria sexta observari posse, sicque in caeteris iudaeorum festis." See further ibid., M4v.

[53] Offenberg, "The First Use," 51–53, identifies the Italian Jewish humanist scholar and Christian convert Flavius Mithridates as the source who may have furnished Paul of Middelburg with his Hebrew texts and their Latin translations, but cf. Grafton and Weinberg, *"I Have Always Loved"*, 219n184, who suspect an invention on Paul's part.

who was active after the time of the Babylonian Exile (Ezra 8:33).[54] As a result, Paul felt compelled to harmonize the Gospel chronologies in the same way as his namesake Paul of Burgos, namely by assuming that Jesus had celebrated the Passover one day earlier than the Jews, because he did not observe the postponement of 15 Nisan to Sabbath, caused by rule *lo BaDU Pesaḥ*.[55]

Paul's arguments and Hebrew citations were later re-used by Sebastian Münster (1488–1552), who was one of the foremost Christian Hebraists of his day. In 1527, Münster published a volume entitled *Kalendarium Hebraicum*, in which he assembled an impressive number of Hebrew chronological texts, supplemented by Latin translations and commentary sections, making it the very first stand-alone work on the Jewish calendar to appear in print.[56] In his dedicatory epistle to bishop Bernhard of Trent, Münster located the origins of the present-day Jewish calendar in the early days of the Second Temple and noted that the *deḥiyyot* could be used to harmonize the Passion chronology.[57] Later into his work, he illustrated the various postponement-types by citing the very same *petiḥah* already found in Paul's *Paulina*, although this time vowel points were added. As Münster saw it, the rule *lo ADU Rosh* contrasted with the astronomically justified rule *molad zaqen* in being an illegitimate form of postponement, which had been imported into the calendar against the original Mosaic Law (*ab eis contra legem conficta*). It was this illegitimacy of *ADU* and its derivates which explained why Jesus had refrained from employing it in the year of his death.[58] Münster later reiterated the same position, alongside yet

---

[54] Paul of Middelburg, *Paulina*, Or: "Hanc itaque veram paschalis solemnitatis rationem post transmigrationem babylonis in secunda templi instauratione per Eleazarum legisperitum iudaeis tunc praesidentem magna arte aeditam, atque omnium voto sancitam, hoc libro exponendam sumemus." See Weinberger, "On the Provenance," 53–54.

[55] Paul of Middelburg, *Paulina*, D8v–E6r, Xr–X3v.

[56] Sebastian Münster, *Kalendarium Hebraicum* (Basel, 1527). For a description, see Joseph Prijs, *Die Basler hebräischen Drucke (1492–1866)* (Olten: Graf, 1964), 45–48. See now also Carlebach, *Palaces of Time*, 49–50.

[57] Münster, *Kalendarium*, a3v: "Sunt qui hinc putant Evangelistas concordandos de agni Paschalis esu. Nam Christus, qui non venit solvere legem, celebravit Pascha cum discipulis suis iuxta traditionem legis divinae, scilicet in vespera feriae quintae: Iudaei vero seniorum traditiones mordicus retinentes, transtulerunt festum Paschale de feria sexta ad sabbatum, edentes scilicet agnum vespera feriae sextae sive parasceves."

[58] Ibid., 127–29.

another rendition of the *petiḥah*, in his edition of the Hebrew version of the Gospel of Matthew, printed in 1537.[59]

As Joanna Weinberg has shown, Münster's work on the calendar was known to the Jewish scholar Azariah de' Rossi (ca. 1511–77), whose famous *Me'or Enayim* ('The Light of the Eyes'), printed in Mantua between 1573 and 1575, contained extended discussions of chronological matters, which often challenged received wisdom and angered the rabbinic establishment into censuring the work. As de' Rossi's astute source criticism revealed, the present use of counting the years from a creation date in 3761 BC was not an ancient, let alone Sinaitic, institution, but had been adopted, at the earliest, "at the time of the close of the Babylonian Talmud, if not a long time subsequent to that date."[60] The same could be said about the conjunction-based system of calculating the beginning of the year, whose institution de' Rossi attributed to the mid-fourth century patriarch Hillel II. Before that time, the Jewish calendar had been based on the visibility of the new moon crescent. In keeping with these views, he later also dismissed Münster's attempt to impute old age to the postponement rules, referring to the latter's claims in his work *Matsref la-Kesef* ('Refinement of Silver'), written in 1576.[61]

But even back in the fifteenth century, the notion that the Jewish calendar, with its intricate postponement rules, had remained stable ever since Second Temple times had incurred criticism from within Christian scholarship. In fact, most of Paul of Middelburg's chronological views had been strenuously impugned two decades before the publication of the *Paulina* by the philosopher Peter de Rivo (ca. 1420–1500). Peter, who held the chair of rhetoric at the University of Louvain since 1460, is principally known for his involvement in the philosophical quarrel over future contingents with Henry of Zomeren (ca. 1417–72), which unfolded in 1465 and led to the condemnation of his position by Pope Sixtus IV in 1474.[62] The incident must have

[59] Sebastian Münster, תורת המשיח *Evangelium secundum Matthaeum in Lingua Hebraica, cum versione latina atque succinctis annotationibus* (Basel, 1537), 142.

[60] Azariah de' Rossi, *The Light of the Eyes*, trans. Joanna Weinberg (New Haven, CT: Yale University Press, 2001), 369.

[61] Ibid., 480–517; Joanna Weinberg, "Invention and Convention: Jewish and Christian Critique of the Jewish Fixed Calendar," *Jewish History* 14 (2000): 317–30.

[62] Chris Schabel, "Peter de Rivo and the Quarrel over Future Contingents at Louvain: New Evidence and New Perspectives," *Documenti e studi sulla tradizione filosofica medievale* 6 (1995): 363–473; 7 (1996): 369–435; Schabel, *Theology at Paris, 1316–1345*

been still fresh on the mind of his contemporaries when Peter began to lock horns with Paul of Middelburg, who had paid a visit to his *alma mater* Louvain in 1487. In that same year, Paul produced an *Epistola ad Universitatem Lovaniensem de Paschate recte observando*, which caused irritation owing to his unusual claim that the Passion of Christ had taken place on Monday, 22 March, and the Resurrection on the following Thursday, thereby extending the period during which Christ remained in the realm of the dead to a literal three days in parallel to the time Jonah had spent inside a whale. In his *Epistola apologetica*, dated 26 February, 1488, and printed by Johannes de Westfalia in Louvain, Paul responded to the criticism his views had incurred. The tone of his 'apology' was often violent and contained a number of *ad hominem* attacks against Peter de Rivo, whom he accused of spreading slanderous and harmful rumors at his old university. In addition, Paul closed his letter with a barrage of insults against his native Zeeland, which prompted a critical response by Thomas Basin (1412–90), bishop of Lisieux and professor of canon law at the Sorbonne, who is famous for his history of the reigns of Charles VII and Louis XI.[63]

While Basin's *Liber contra errores et blasphemias Pauli de Middelburgo* (1490/91) remained unpublished, Peter de Rivo composed an extensive *Opus responsivum*, consisting of two separate treatises on the date of Christ's Passion, which were printed together in 1488, the same year as Paul's *Epistola*. He eventually supplemented them with a *Tractatus tercius*, which was published in 1492 and added further nuances to his chronological arguments.[64] In Peter's view, the claim that the *deḥiyyot* had already existed in the first century AD was devoid of credibility and he accordingly refused to put any trust in the *Liber de solis et lunae motibus* by Rabbi Gamaliel that Paul of Middelburg had cited in support of his theory. Doubts regarding its authenticity were

(Aldershot: Ashgate, 2000), 315–36; Jonathan Evans, "Peter de Rivo and the Problem of Future Contingents," *Carmina Philosophiae* 10 (2001): 39–55. For biographical details, see also Marzi, *La questione*, 16–17; Vanden Broecke, *Limits*, 49–52.

[63] See Henri de Jongh, *L'Ancienne faculté de théologie de Louvain au premier siècle de son existence (1432–1540)* (Louvain: Revue d'histoire ecclésiastique, 1911), 83–87; Jules Quicherat, "Thomas Basin, sa vie et ses écrits," *Bibliothèque de l'École des Chartes* 3 (1841–42): 374–75; Quicherat, *Histoire des Règnes de Charles VII et de Louis XI par Thomas Basin*, 4 vols. (Paris: Lahure, 1855–59), 4:105–22, 136; Bernard Guenée, *Entre l'Église et l'État* (Paris: Gallimard, 1987), 430–34.

[64] Peter de Rivo, *Opus responsivum ad Epistolam apologeticam M. Pauli de Middelburgo de anno, die et feria dominicae passionis* (Louvain, 1488); Peter de Rivo, *Tercius tractatus de anno, die et feria dominice passionis atque resurrectionis* (Louvain, 1492).

aggravated by the fact that 'Rabbi Paulus', the erudite bishop of Burgos, who was obvously well versed in Hebrew literature, had omitted any mention of this supposedly early source in his discussion of the *deḥiyyot*.[65] At the same time, however, Peter's respect for Paul of Burgos's expertise did not go so far as to make him agree with the latter's harmonization of the Gospel chronologies. When it came to explaining the patent discrepancy between John and the synoptic evangelists, Peter instead preferred to resort to the old anti-Jewish interpretation of John Chrysostom, according to whom the Jews had decided an ad hoc postponement of Passover in order to be able to capture and kill Jesus without the feast day of 15 Nisan coming in their way. Unwilling to accept the existence of the *deḥiyyot* at the time of Jesus, Peter de Rivo instead proclaimed that he could not think of any other reason for their invention than the Jewish wish to bring Christ into disrepute by alleging that he had eaten the Passover against divine ordinance:

> And inquiring thereafter more deeply into the reasons why such a regulation was thought up by the Jews, I cannot imagine anything other than that they wanted to detract from Christ, who [according to them] ate the Passover against the commonly accepted regulation. And that his murderers and maybe also their fathers, who had feigned said regulation, may be excused for eating the Passover on the 15th day of the moon against the Law.... For it was not Christ, but these most audacious of all people, who postponed the time of the Passover, thereby transgressing a thousand commandments (because they were so enraged).... And the chief priests and all the others may have excused themselves before the common people for their transgression by claiming that they had unlawfully postponed the feast out of their zeal for the fulfilment of the Law, because they hunted to death the one whom they alleged to be the greatest breaker of the Law.[66]

---

[65] Peter de Rivo, *Opus responsivum*, c4v–5v.

[66] Ibid., c5r–v: "Dehinc profundius inquirens causam cur huiuscemodi constitutio a iudeis conficta est, nil aliud arbitrari possum, quam ut detrahant christo, qui contra constitutionem approbatam a deo pascha manducaverit. Et ut eius occisores ac patres forsan illorum qui dictam constitutionem confinxerant excusentur, qui pascha luna decima quinta contra legem manducaverunt...Non enim christus sed illi omnium audacissimi et mille transgredientes precepta (quoniam ita inflammati erant), tempus pascha transposuerunt.... Excusabant se forte apud plebem summi sacerdotes et ceteri de hac festorum translatione, dicentes quod ex zelo legis contra legem hec festa transposuernt, quoniam illum usque mortem persecuti sunt, quam legis subversorum maximum reputarunt." Cf. John Chrysostom, *Homiliae in Matthaeum* (84.2), PG 58:754, and Hugh of St. Cher's commentary on John 18:28 in his *Opera omnia in universum Vetus & Novum Testamentum*, 8 vols. (Lyons, 1669), 6:391ra.

In Peter's view, as explicated in the *Tractatus tercius*, the present-day Jewish calendar had in fact only come into existence at a very late stage, possibly not much more than two centuries ago. His most important source for this suggestion was Roger Bacon, who had compared the different values for the mean synodic month, noting that the Arabs used 29d 12; 44h, which was equivalent to 29d 12h 792ch, whereas the Jews had improved upon this estimate by using 29d 12h 793ch or 29d 12; 44, 3, 20h, a figure that Peter was able to confirm from a Hebrew calendar table, which had been sent to him by Paul of Middelburg. A comparison with further known values of the mean lunation seemed to show that this Jewish estimate was closer to the one found in the Alfonsine tables, which had been assembled under the auspices of King Alfonso X of Castile (1221–84) and whose mean month length Peter wrongly cited as 29d 12; 44, 3, 58, 51, 36h (the actual Alfonsine value is 29d 12; 44, 3, 2, 58, 51, 22h and hence less than the Jewish one). In Peter's mind, this affinity gave rise to the idea that the Jewish calendar might have been influenced by the Alfonsine data, which would have meant that the conjunction-based Jewish calendar had not been instituted earlier than the late thirteenth century.[67]

But what had been the shape of the Jewish calendar previous to this reform? Peter realized that the question of the Passion date could only be adequately solved if the precise shape of the Jewish calendar in first-century Jerusalem was correctly determined. He essentially concurred with the view of earlier medieval computists, according to

---

[67] Peter de Rivo, *Tercius tractatus*, e6r–v: "Opinor autem iudeos in errore predicto permansisse, donec ab astronomicis aliarum nationum didicerant qualiter per calculum astronomicum invenire possent neomenias suorum mensium lunarium. Primo quidem ab astronomis arabum dehinc ab his qui calcularunt ex tabulis regis Alphonsi. Cur sic opiner paucis edicam. Arabes enim (quorum anni inceperunt circiter annum domini secentesimussextus) dicunt (sicut scribitur in epistola Rogeri bachon) tempus lunationis equalis seu tempus medians inter duas coniunctiones medias luminarium esse dies xxix horas xii et minuta xliiii. Alphonsus autem rex (qui sicut michi dixit meus interpres posuit eram suam super anno domini Millesimo ducentesimoquiquagesimosecundo currente, ultima die maii completa) subtilius inquirens ponit tempus lunationis equalis esse dies xxix, horas xii, minuta quatuor [!], secunda tria, quarta lviii, quinta li, sexta xxxvi.... Hec attendens opinatus sum hebreos eodem ciclo usos non fuisse, nisi postquam defectum arabum (quo ad tempus lunationis equalis) Alphonsus vel alius quispiam astronomus (qui me latet) supplevisset. Liquet igitur ex premissis quod ciclus lunaris astronomicus hebreorum longe post christi nativitatem inventus est, immo post nycenam synodum, postquam scilicet vetus eorum ciclus lunaris in tantum aberravit, ut astronomos consulere cogerentur. Nec video cur ab hac sententia resilire debeam, nisi forte vetustiores cicli lunares hebreorum in medium proferantur."

whom the 19-year lunisolar cycle of the Alexandrian Church had been taken over from the Jews; but he also found evidence that this cycle had gone through several transformations since its inception in the remote past. Relying on the testimony of Josephus (*Antiquities* 1.167–68), he assumed that Abraham, who was credited with transmitting Chaldaean mathematics and astronomy to Egypt, had also brought the 19-year cycle to the Nile, where it was adopted by the Israelites during their time of servitude. In contrast to earlier scholars, Peter thus came to believe that the 19-year cycle was even older than the first celebration of Passover at the time of Moses.[68] Since he also assumed that the basic structure of this 19-year cycle had never changed from the time of Abraham up until the council of Nicaea, it followed that there had to have been periodic reforms of the epacts, seeing how the error of the 19-year cycle led to a calendrical 'drift' of the new and full moons at a rate of one day in every 304 years. The first of these reforms, Peter believed, had taken place during the reign of David, who, according to Ben Sira (47:10), "gave beauty to the festivals, and arranged their times throughout the year." The second reform was a purely Christian affair, having been decided at the Council of Nicaea, which, according to the chronicle of Eusebius, convened 1328 years after the reign of King David. Assuming that the year in which David had reformed the lunisolar calendar served as the starting point of the new 19-year cycle, Peter conjectured that the Nicene reform took place in the 17th year of this cycle ($1328 = 69 \times 19 + 17$).

In the intervening period, the error had accrued to roughly four days, which could be easily compensated by 'jumping over' three years in the cycle. The 17th year of the 19-year cycle, which in the Alexandrian version had an epact of 26 days, was a particular suitable year for such a reform. By counting it as the first year of a new cycle and thus resetting the epact from 26 to 30 (which, for computistical purposes, is equivalent to zero), it was possible to correct the error which had accrued since the time of King David. By postulating this three-year calendrical 'leap' at the time of the Nicene Council, Peter was also able to explain why the Alexandrian 19-year cycle, which resulted from this reform, and the 19-year intercalation cycle of the Jewish calendar had different starting points (the Jewish cycle beginning three years later than its Christian equivalent). Even better, the theory provided

[68] Peter de Rivo, *Opus responsivum*, b2r–v.

him with a striking support for the traditional crucifixion date on 25 March, which had been vouched for by so many patristic authorities. According to the post-Nicene Easter cycle, known from the Dionysian table, AD 34 was year 16/19, with an Easter full moon (*luna* 14) on 21 March. Yet in Peter's hypothetical pre-Nicene 'Jewish' reckoning it was equivalent to year 13/19, with 14 Nisan on 24 March. It was thus easily explainable how Jesus, who was crucified on 15 Nisan, could have died on 25 March, as ecclesiastical tradition claimed.[69]

Peter's basic position was in some ways akin to that of Tostado's opponent Juan de Torquemada. Not only did he assume that there were profound technical differences between the Jewish calendar of antiquity and the present one, but he also resolutely denied that the church fathers had erred in dating the crucifixion to the typologically attractive equinoctial date of 25 March. If the 28-year solar cycle and the 19-year lunisolar cycle, let alone more sophisticated astronomical methods, seemed to disprove this tradition, this only meant that the history of the calendar had taken a different route than commonly supposed. The fact that 25 March had not been anywhere close to the astronomical full moon in AD 34 did not deter him any more than the problem of the weekday. After all, no-one could exclude that some undocumented intercalation of a leap day into the Julian calendar after the time of Christ had caused a disturbance of the 28-year cycle. If such a scenario was accepted, it became possible that 25 March had been a Friday in the year of the crucifixion, even though the present-day Julian calendar seemed to signal otherwise.[70]

Paul of Middelburg's approach to the questions surrounding the Passion date was in many ways diametrically opposed to that of his opponent in Louvain, as it relied on the pre- and postdictive powers of mathematical astronomy. Sifting through the times of conjunction

---

[69] Ibid., b2r–4r.

[70] Ibid., c7v: "His ita se habentibus non apparet via commodior ad certe noscendum quis ciclus solaris apud romanos cucurrit temporibus christi." A similar attempt to reconcile the Passion date to the weekday had already been made in the thirteenth century by Alberic of Trois-Fontaines, "Chronica," ed. Paul Scheffer-Boichorst, in *MGH Scriptores*, vol. 23, ed. Georg Heinrich Pertz (Hannover: Hahn, 1874), 679: "Sed quia in nocte dominice resurrectionis ordo temporum immutatus est, et dies qui secundum primam conditionem noctem precedebat modo subsequitur, recte in hoc anno dominicus dies et secundum Dionysium et per tot annos invenitur per C et secundum beatum Augustinum potuit esse B. Item neque conputus paschalis, neque tabula tunc habebatur in ecclesia, et aliter et alio fortisan anno inserebatur tunc bisextum quam modo inseratur."

and opposition for the meridian of Jerusalem according to the Alfonsine tables, Paul, like Alfonso Tostado before him, concluded that 25 March had never come close to either 14 or 15 Nisan during the relevant time-range of AD 30–40. Dropping the traditional date on 25 March, Paul was left with 3 April, AD 33, and 30 March, AD 36, as the only possible candidates, both of which were instances of 14 Nisan in the present-day Jewish calendar. Unlike Tostado, however, Paul refrained from moving the Passion from March to April and instead preferred the unusually late date on 30 March, AD 36. Yet, as he himself knew, this could not be properly reconciled with a lifespan of 33 years and 3 months as long as it was assumed that Jesus had been born in AD 1 or earlier. As a result of these deliberations, and despite the impressive array of source material and astronomical considerations Paul had brought to bear on the matter, he saw himself stuck in an argumentative deadlock. Having "long strayed between these Symplegadian rocks" without any true progress, Paul alleged that he had been finally led on the right track "not by astronomical calculation, but by the compassion and revelation of our saviour Jesus Christ."[71] It was at this very point that Paul's chronological argument took a bizarre and unexpected turn. As he went on to narrate, he had gone to bed after a long night of chronological study, when he was suddenly awoken by the ghostly apparition an elderly man, whose face resembled that of St. Paul, and who delivered the following message:

> Desist, O my Paul…from investigating this problem mathematically and stop spending sleepless nights over it as long as you have not learnt about the errors of the chronographers in numbering the years of Christ our saviour. This is what you shall investigate and try to find, because it is the only reason for the difficulty in determining the day on which Christ was crucified.… And if you really desire to find out this date…turn your gaze upwards and observe and keep in memory what you are about to see, and that which you have long been trying to find in vain by way of calculation, you will now perceive in an instance and see with your own eyes.[72]

[71] Paul of Middelburg, *Paulina*, Pv: "Cunque inter has symplegadas diu versatus essem, & neque errorem chronographorum in annis domini numerandis (propter historiographorum varietatem) demonstrare potuisse, neque diem veneris passioni deputatum adamussim invenire valuissem, ecce divino quodam instinctu per quietem monitus brevi momento explanationem eius didici, non astrologica supputatione, sed Iesu christi salvatoris nostri misericordia, & revelatione."

[72] Ibid., P3v: "Cum dominicae passionis diem multis ac variis lucubrationibus & methodis indagare conatus essem, eiusque supputationem in manus saepe numero

The bishop of Fossombrone went on to describe a highly detailed vision, a woodcut depiction of which was included in the *Paulina*, which he claimed was related to the famous image of the infant Jesus and his mother Mary that, according to medieval legend, had once been revealed to Emperor Augustus by the Tiburtine Sibyl. Apart from the celestial apparition of Jesus and Mary, who were surrounded by the first 35 popes of the Roman Church, Paul also claimed to have witnessed the crucifixion scene at Calvary, accompanied by a solar eclipse, alongside various other biblical episodes. He saw Cyrinus, the governor of Syria, who had allegedly conducted the census at the time of Christ's birth, and King Herod, as he interrogated the three Magi from the East. He also saw the prophet Daniel enumerating the 70 'weeks of years', and St. Paul the Apostle as he preached the Christian message. His strange parade of apparitions even included Julius Caesar and Augustus, who had decreed reforms of the Roman calendar, and the monk Dionysius Exiguus, who had fixed the year of Christ's incarnation. According to Paul's own interpretation, each of these prophetic images belonged to a cycle of 50 divine 'oracles' detailing the date of the crucifixion, the first 25 of which had been designed to point him towards the fact that Jesus's birth had been mistimed.[73]

The most important technical argument that Paul provided in support of this startling revelation was drawn from the chronicle of Eusebius, who mentioned a solar eclipse that had allegedly occurred in the 57th and final year of Augustus's reign. If, as was commonly assumed

adduxissem, & per observationes Ptolemaei aliorumque astrologorum indagare contendissem, cunque nuper hesterno scilicet vesperi in multam & intempestam noctam iam prope auroram calculi supputationes produxissem, tandem defatigato animo lectum petii, ubi postquam paulisper requievissem, ecce visus est adesse mihi senex quidam infulatus, statura procerus, vultu placidus atque decorus, aspectu ipso & habitu sanctitatis insignia divique Pauli apostoli effigiem prae se ferens, qui voce benigna me compellans & manu a somno excitans; desine, inquit, mi Paule calculo haec indagare, desine ob haec noctes insomnes ducere, nisi primum errata chronographorum in annis christi salvatoris numerandis didiceris. Haec investigare satage, haec invenire contende, quoniam in his difficultas indagandi diem quo crucifixus fuit christus constitit, quando quidem & temporum scriptores annos christi non adamussim quadrarunt. Quae si invenire desideras, si chronographorum errata in annis ab incarnatione christi numeratis cognoscere gliscis, si dominicae passionis diem quem tantopere quaeris invenire optas, respice sursum, & visa animadverte ac memoriae commenda, & quod longo iam temporis tractu calculo perscrutari frustra conatus fuisti, nunc brevi momento percipies, & oculata fide conspicies."

[73] Paul of Middelburg, *Paulina*, P3v–7r. The woodcut is on P5r. On the vision of Augustus, see Jacobus de Voragine, *The Golden Legend*, trans. William Granger Ryan, 2 vols. (Princeton, NJ: Princeton University Press, 1993), 1:40.

on the basis of Eusebius's chronology, Jesus had been born in Augustus's 42nd year, this information could be used to date the death of Augustus to AD 15, provided that the birth of Jesus happened in 1 BC, in accordance with the Dionysiac era. The same line of technical argument had been previously pursued by Peter de Rivo, who had enlisted the help of two expert astronomers to identify appropriate candidates for the solar eclipse in question. His first consultant informed him that a solar eclipse had been visible over the Mediterranean on 15 February, AD 17 (Oppolzer no. 2924), but Peter decided that this event was too late still to fall within the year of Augustus's death. Dissatisfied with this solution, Peter consulted a second astronomer, who found that another eclipse had occurred on 21 August, AD 16, which had allegedly been visible over Jerusalem. Not only was this eclipse conveniently close to 19 August, the recorded date of Augustus's death, but its area of visibility was also near to Eusebius's hometown Caesarea, which seemed to explain why the chronographer associated said eclipse with the emperor's demise. As a result, Peter was confident that the death of Rome's first emperor could be dated to AD 16, which in turn implied that the birth of Christ had taken place on 25 December, AD 1.[74]

It was Paul of Middelburg who first noticed that Peter's chronological argument was marred by a serious flaw. Contrary to what Peter's source had claimed, Paul's own calculations indicated that no noteworthy eclipse had been visible over the Mediterranean in AD 16. This is confirmed by modern tables, which show that the path of totality of the eclipse chosen by Peter was restricted to East Asia and the Pacific. As Paul pointed out, it was hardly likely that Eusebius would have cared to record such an unimportant astronomical event, noting that

[74] Peter de Rivo, *Tercus tractatus*, e3v: "De hac eclipsi locutum existimo Eusebium cesariensem, tum quia cesarea civitas sua non longe distat a Iherusalem, ubi visa est dicta eclipsis tante durationis, tum quia eadem eclipsis eo die contigit quo mortuus est Augustus aut eocirca, obiit enim (ut Suetonius scribit) quartodecimo kl septembris qui est dies decimusnonus Augusti. Propter hunc mirabilem concursum tam eclipsis quam obitus Augusti probabile est eandem eclipsim annalibus cesariensium inscriptam esse, ac ob id eclipsis illius et non eclipsis anni sequentis Eusebius in suis cronicis meminisse." See Eusebius, *Chronicle*, GCS 47:171, and also Cassius Dio, *Roman History* 56.29. Since astronomers in Peter's day used a noon-epoch, which precedes our accustomed midnight-epoch by 12 hours, his dates for both eclipses fall one day later than given here on the basis of Oppolzer's canon.

> it is customary among historians and chronographers to record and comment on only those solar eclipses during which the whole sun is obscured and hidden from our vision to such an extent that sometimes the stars become visible and can be counted, which only happens very rarely.[75]

The only eclipse of any significance that Paul, who unlike Peter de Rivo was a skilled astronomer, could find in the relevant range of years was the one on 15 February, AD 17. Peter had rejected this date on the ground that it implied a date of death for Augustus which was one or even two years later than had been generally assumed by previous chronographers. If this year was accepted, it would have meant that the 42nd year of Augustus's reign, in which Jesus had been born according to Eusebius, overlapped with December of AD 2. For Paul, this was a highly welcome result, since it served to corroborate the unusually late crucifixion date on 30 March, AD 36, which he had set out to prove.[76] Paul's deliberations were an important new step in the methodological progress of Christian chronology, seeing how it was the first time a reassessment of a date of Christ's life—and probably of any historical event—was motivated by a calculated eclipse. It is only to be regretted that neither Peter nor Paul were aware of the report in Tacitus (*Annals* 1.16, 28) that a lunar eclipse had occurred shortly after Augustus's death, on 27 September, AD 14 (Oppolzer no. 1884), preventing the Roman legions in Pannonia from mutiny. In this case, they would have had every reason to pre-pone the 42nd year of Augustus to 2 BC, which is the correct year according to our present state of knowledge.[77]

---

[75] Paul of Middelburg, *Paulina*, T2v: "Mos enim historiographis atque chronographis esse solet illas duntaxat eclipses solares in annales referre & annotare, in quibus totus sol a visu nostro occultatur & absconditur, adeo ut interdiu astra caeli stellati videri & numerari possint, quales raro admodum contingunt."

[76] Ibid., T2v–T3r, V2v–V4v. For the paths of visibility of both eclipses in question, see map no. 59 in Oppolzer's *Canon*. Unlike the (annular) eclipse chosen by Peter de Rivo, the (hybrid) elipse on 15 February, AD 17, was indeed visible over North Africa, Sicily, and parts of the Eastern Mediterranean.

[77] The *editio princeps* of the first books of Tacitus's *Annals* only appeared in 1515. See also Salvo De Meis, *Eclipses* (Rome: Istituto Italiano per l'Africa e l'Oriente, 2002), 195–96; Cassius Dio, *Roman History*, 57.4. The Tacitean lunar eclipse was finally taken into account by Heinrich Bünting in 1590. See below, chapter eight.

4. 'Clever Jean' and his Successors: The Continuation of Debate in Sixteenth-Century Italy

Paul of Middelburg's 'prophetic' method of lending support to his chronological arguments was slightly bizarre even in the eyes of some sixteenth-century commentators. His biographer Bernardino Baldi (1553–1617) preferred to leave the mysterious vision, which he summarized in his collection of *Mathematicians' Lives*, and its veracity to the judgment of his readers: "If it were true, it would be accepted as a miracle, if he made it up, it would be attributed to great erudition and ingenuity."[78] Other readers were far less generous. In his twenty-book chronicle, published in 1540, the Flemish scholar Christian Massaeus (ca. 1470–1546) railed against the "embarrassing antics, produced by dreams" (*pudendis somniorum deliramentis*), the "frivolous arguments" (*frivolis rationibus*) and the "singsong" (*naeniis*) by which the bishop of Fossombrone, whom he only referred to as a certain person "writing to Louvain from Italy," tried to back up his correction of the Christian era.[79]

Understandable as Massaeus's indignation may have been, it could not hide the fact that his own chronology of the life of Jesus, too, encountered serious problems. Basing himself on the Dionysiac birth date of Jesus on 25 December, AD 1, he added the canonical lifespan of 33 years and 3 months, arriving at a Passion on 25 March, AD 35. While this date satisfied the condition of being a Friday, the full moon was not even close, 25 March being *luna* 29 according to the traditional 19-year cycle. Moreover, Massaeus's late dating of the Passion carried over to his correlation of the Dionysiac era with Roman chronology in a problematic way. According to his chronicle, Jesus was crucified in the 19th year of Tiberius's reign, in the consulate of Sulpicius Galba and Sulla, two names which normally belong to

[78] Baldi, *Le vite*, 379: "Se questo visione fosse vera, opure se la fingesse, io non saprei; onde non essendo l'historia de gli incerti lascierò, per non esser questo fatto articolo di fede, ciascheduno giudicarne a suo modo. Se fu vera, accettisi per miracolo, se egli se la finse, attribuiscasi a gran dottrina e molto ingegno."

[79] Christian Massaeus, *Chronicorum multiplicis historiae utriusque testamenti libri viginti* (Antwerp, 1540), 91: "Moventur ad haec partim pudendis somniorum deliramentis, partim frivolis rationibus, imo naeniis. Denique, quidam ex ipsis inter multa ridicula, quae Lovaniensibus ex Italia scribit, etiam has sibi revelationes factas coelitus ait."

AD 33.[80] Yet by the time Massaeus composed his work, the basic means to construct a more adequate chronology of the ancient world had in principle already become available. The second half of the fifteenth century saw the appearance in print of a number of crucial sources for the history of Rome, which to that point had been either barley known or whose circulation had been limited to a relatively small number of manuscripts. These included the extant fragments of Livy's *Ab urbe condita* (*editio princeps* in 1469) and the *Roman Antiquities* of Dionysius of Halicarnassus (the Latin translation of which was printed in 1480), but also works of a more encyclopedic nature such as the *Collectanea rerum memorabilium* of C. Iulius Solinus (*editio princeps* in 1473). Among these, no other source was as effective and useful when it came to synchronizing the major eras of ancient history as Censorinus's *De die natali* (21.6–10), a third-century work which was first printed in 1497 by Filippo Beroaldo in Bologna.[81] Drawing on the antiquarian efforts of M. Terentius Varro (116–27 BC), Censorinus offered his readers an extensive correlation of eras for the *annus praesens*, which was designated as the consulate of Ulpius and Pontianus (AD 238):[82]

Olympiad 252.2 (1014)	summer 238–summer 239
Foundation of Rome 991	21 Apr 238–20 Apr 239
Nabonassar 986	25 Jun 238–24 Jun 239
Philippus 562	25 Jun 238–24 Jun 239
Caesar 283	01 Jan 238–31 Dec 238
Augustus (*Egyptian reign*) 267	29 Aug 237–28 Aug 238
Augustus (*Roman reign*) 265	01 Jan 238–31 Dec 238

One important key to unlock these dates could be found in astronomical tables, where the epoch of the era of Nabonassar was clearly given as 26 Feb, 747 BC. From this date, the beginning of the Olympiads could be fairly securely dated to the summer of 776 BC. A look into the chronicle of Eusebius, where the birth of Christ was dated

[80] Ibid., sig. d5r, pp. 92–97.

[81] On the transmission of Censorinus, Livy, and Solinus, see L. D. Reynolds, ed., *Texts and Transmission* (Oxford: Clarendon Press, 1983), 48–50, 203–14, 391–93. For Dionysius of Halicarnassus, see Earnest Cary, trans., *The Roman Antiquities of Dionysius of Halicarnassus*, 7 vols. (Cambridge, MA: Harvard University Press, 1937–50), 1:xli–xlii.

[82] Censorinus, *De die natali* (21.6–10), ed. Sallmann, 52–53.

Ol. 194.3 could thus show that the intended birth year was 2/1 BC. These straightforward conclusions were thwarted by a number of interference signals, which led chronologers in Massaeus's day to assume different correlations. One of the most pervasive sources of error was Eusebius's famous notice according to which Phlegon of Tralles, who was the author of a lost Olympiad chronicle (ca. AD 137), had reported the miraculous solar eclipse that had occurred on the afternoon of Christ's death for the year Ol. 202.4.[83] Phlegon's supposed pagan testimony of the crucifixion-eclipse was of particular value to Christian chroniclers, seeing that "he did know not the Gospels, but was still in agreement with them."[84] Many scholars accordingly believed that they could interpret his Olympiad date on the basis of their already established notions of the Passion date. Thus, while Ol. 202.4, according to the standard correlation, ran from summer AD 32 to summer AD 33, an assumed Passion date in AD 34 or 35 could induce scholars to begin their count of the Olympiads one or even two years too late. A well-known case in point is the chronological chapter in Nicholas Copernicus's *De revolutionibus orbium coelestium*, where the first Olympiad is erroneously marked as having begun 27 Egyptian 'wandering years' and 247 days before the era Nabonassar, thereby implying an epoch of 1 July, 775 BC.[85] Similar problems are encountered in nearly all chronicles and chronological handbooks written during the first half of the sixteenth century, where the displacement of the Olympiads vis-à-vis the Christian era would often carry over to other parts of ancient historiography, in particular the Roman chronology of emperors and

[83] Eusebius, *Chronicle*, GCS 47:174–75: "[Christus] ad passionem venit anno Tiberii XVIII quo tempore etiam in aliis ethnicorum commentariis haec ad verbum scripta repperimus: solis facta defectio, Bithynia terrae motu concussa et in urbe Nicaea aedes plurimae corruerunt. Quae omnia his congruunt, quae in passione Saluatoris acciderant. Scribit vero super his et Flego, qui olympiadarum egregius supputator est, in XIII libro ita dicens: 'quarta autem anno CCII olympiadis magna et excellens inter omnes, quae ante eam acciderant, defectio solis facta. Dies hora sexta ita in tenebrosam noctem versus, ut stellae in caelo uisae sint terraeque motus in Bithynia Nicaenea urbis multas aedes subverterit.' Haec supra dictus vir."

[84] Massaeus, *Chronicorum... libri viginti*, 96: "Sed & Phlegon graecus, olympiadum egregius supputator, Evangelium ignorans & tamen Evangelio consonans, in decimo quarto libro sic ait."

[85] Copernicus, *De revolutionibus*, 76v. See also Noel M. Swerdlow and Otto Neugebauer, *Mathematical Astronomy in Copernicus's De Revolutionibus*, 2 vols. (New York: Springer, 1984), 1:186; Depuydt, "Calendars," 50–51.

consuls. This is particularly obvious in the case of Massaeus, where the consulate of the Gemini (AD 29) is assigned to AD 31.[86]

A more adequate response to Paul of Middelburg's challenge was published in September 1537 in Venice as part of a remarkable collection of chronological tracts, entitled *Opusculum de emendationibus temporum*. According to the dedicatory epistle, dated 15 October, 1535, and addressed to Cardinal Nicholas Schönberg, the pretended author of the *Opus* was a deceased French priest named Johannes Lucidus Samotheus, a man whose work "so elucidated" chronology "that he deservedly ought to be named *Lucidus*" (*ita elucidavit, ut merito Lucidus cognominari debeat*).[87] The epistle was signed by Giovanni Maria Tolosani, a member of the Dominican Order, who is well-known for his harsh criticism of Copernicus's *De revolutionibus*, which shows that the latter only narrowly escaped an early condemnation by the Catholic Church. Tolosani's interest in astronomy is also borne out by his treatise *De correctione calendarii* (1515), which marks his contribution to the discussions that took place in the wake of the Fifth Lateran Council. A version of this treatise also appears in the *Opusculum de emendationibus temporum*, only this time ascribed to the aforementioned Johannes Lucidus Samotheus (whose name may be translated as 'Clever Jean, the Frenchman', 'Samotheus' being a pun on Samothes, the mythical progenitor of the Gauls). In the dedicatory epistle, Tolosani claims that the latter had supplemented his chronology of world history with material from his own *breviloquium temporum*, but in the interest of furnishing the work with greater authority he decided

---

[86] Massaeus, *Chronicorum... libri viginti*, d5r. See also Giacomo Filippo Foresti, *Supplementum chronicarum* (Venice, 1483), Liber octavus, Ar–3r; Gregor Haloander, *Codicis domini Iustiniani sacratissimi principis... libri XII* (Nuremberg, 1530), ggg2v–3r; Paul Constantin Phrygio, *Chronicum* (Basel, 1534), 260–66; Heinrich Glarean, *In omneis, quae quidem extant, T. Livii Patavinii clarissimi historici decadas annotationes* (Basel, 1540), g4r–v; Martin Luther, *Supputatio annorum mundi* (Wittenberg, 1541), Sr–v; Johann Funck, *Chronologia* (Nuremberg, 1545), 65–66. See also the attached *Commentariorum in praecedentem chronologiam liber unus*, 35v–37r. On the genre of early modern 'tabular' chronologies, see Benjamin Steiner, *Die Ordnung der Geschichte* (Cologne: Böhlau, 2008).

[87] Johannes Lucidus Samotheus, *Opusculum*, dedicatory epistle (unpaginated): "Tandem emendationes temporum verissima computatione descripsit dominus Ioannes Lucidus natione Gallus ac professione sacerdos eruditissimus: qui ab orbe condito ad aetatem nostram, tempora iuxta veritatem hebraicm ita elucidavit, ut merito Ludicus cognominari debeat."

to have it all published under Johannes Lucidus's name.[88] As Demetrio Marzi and Edward Rosen have both cogently argued, the simple truth behind Johannes Lucidus Samotheus is that he was a pseudonym for Tolosani to hide behind, as he for some reason did not want to publish his chronological findings under his own name.[89]

This kind of literary camouflage may have been partly motivated by the outspoken criticism of Paul of Middelburg that is included in the treatise *De vero die Passionis Christi ac tempore totius vitae ipsius*, written in 1524.[90] In the *prooemium* of said work, Johannes Lucidus claims to have been approached by two young scholars, a cleric and a layman, who were caught up in an argument about the date of Christ's Passion. While the layman championed the traditional (and chronologically impossible) crucifixion date on 25 March, AD 34, his clerical colleague opted for 3 April, AD 33.[91] The present treatise *De vero die* was alleged to be the outcome of Johannes Lucidus's attempt to adjudicate between the two opinions. As the purported author went on to show, only the latter date (3 April, AD 33) was a wise choice, since it

[88] Ibid.: "Cum autem opus suum, uberius ac perfectius reddere cuperet: ex meo breviloquio temporum in quo & collegeram fideliter emendata tempora, me sibi consensum praebente plura excerpsit, operi suo inseruit. Nam & ego arbitratus sum maioris authoritatis esse opus si sub nomine Ioannis Lucidi quam si sub meo nomine prodiret in lucem. Idcirco sponte assensum praebui, ut mei labores sibi ascriberentur, sicut & fecit Pamphilius martyr, qui suas lucubrationes Eusebio Caesariensi episcopo in suo libro de temporibus condonavit." For the *Epitoma emendationis calendarii Romani*, see ibid., 194r–98r. Tolosani's original treatise is edited in Marzi, *La questione*, 250–54. On his reform proposal, see also Kaltenbrunner, "Die Vorgeschichte," 402–3.

[89] Edward Rosen, "Was Copernicus' *Revolutions* Approved by the Pope?" *Journal of the History of Ideas* 36 (1975): 531–33; Demetrio Marzi, "Giovanni Maria Tolosani e Giovanni Lucido Samoteo," *Miscellanea storica della Valdelsa* 5 (1897): 32–62. See also Marzi, *La questione*, 130–49; Marzi, "Giovanni Maria Tolosani, Alessandro Piccolomini e Luigi Giglio," *Miscellanea storica della Valdelsa* 5 (1897): 202–9; Salvatore I. Camporeale, "Giovanmaria dei Tolosani O.P., 1530–1546: Umanesimo, riforma e teologia controversista," *Memorie Domenicane*, n.s., 17 (1986): 145–252; Miguel A. Granada, "Giovanni Maria Tolosani e la prima reazione Romana di fronte al *De revolutionibus*: La critica di Copernico nell'opusculo *De coelo et elementis*," in *La diffusione del Copernicanesimo in Italia 1543–1610*, ed. Massimo Bucciantini and Maurizio Torrini (Florence: Olschki, 1997), 11–35.

[90] Johannes Lucidus Samotheus, *Opusculum*, 161r–94r. See also the "Liber septimus de vita Christi," ibid., 49v–58r, and the chronological tables, ibid., 113r–v. For the date, see ibid., 171v: "Nam praesenti anno Domini 1524 bisextili currunt litterae CB." Ibid., 166r: "Hoc idem etiam patet in bulla inditionis iubilei pro futuro anno Domini 1525."

[91] Ibid., 161r.

fulfilled all the calendrical and astronomical conditions posed by the four Gospels.[92]

Seizing mainly on Phlegon's eclipse and the data provided by Censorinus, Johannes Lucidus/Tolosani was able to date the crucifixion in AD 33 to the fourth year of the 202nd Olympiad. This correctly set the beginning of the first Olympiad to 776 BC, but some remaining difficulties were incurred in fixing the Roman foundation era (*ab urbe condita* = AUC) on this basis. According to Censorinus, the consulship of AD 238 was the 991st year since the founding of Rome and the 1014th year since the first Olympiad. The difference between both counts was 23 years, indicating to a correlation AUC 1 = Ol. 6.4, but some ambiguity concerning the exact relation between both eras remained. Since the Parilia, the City's birthday feast, took place on 21 April, whereas the Olympic Games were celebrated in summer, such a correlation could be understood to mean either that Rome was founded in the spring of Ol. 6.4, and hence in 752 BC, or in the spring that preceded the start of Ol. 6.4, leading to the Varronian foundation date in 753 BC. While Censorinus evidently intended the latter, Tolosani decided upon the former, namely that the Roman era began only in 752 BC. This misconception seems to have been partly based on the assumption that the Olympiad count began from the autumnal equinox rather than in mid-summer. His source for this idea was Roger Bacon's *Opus maius*, where it was stated that the Greeks began their years in October, although this claim was evidently made with an eye on the Seleucid rather than the Olympiad era.[93] In addition, Tolosani was partly influenced by the testimony of Solinus (*Collectanea* 1.29–30) and Dionysius of Halicarnassus (*Roman Antiquities* 1.75), who both synchronized the beginning of the Roman era with Ol. 7.1, assuming that the City had been founded in the preceding spring.[94]

To have the Roman foundation era begin in 752 BC had the slight advantage of bringing recorded patristic opinion on the birth-year of Christ closer to the Dionysiac era. In his *Historia adversum paganos*, Paulus Orosius had dated the nativity to AUC 752 and the 42nd year of Augustus, which in Tolosani's reading corresponded to 1 BC. This

---

[92] Ibid., 168r–71r, 178v–85r.

[93] Ibid., 175v; Bacon, *Opus majus*, 1:192.

[94] Johannes Lucidus Samotheus, *Opusculum*, 176v–78v. On the Olympiad and AUC-systems of counting the years, see Samuel, *Chronology*, 189–94, 249–53; Bickerman, *Chronology*, 75–78.

implied that Christ had been born in Ol. 194.4, whereas Eusebius's chronicle dated this event to Ol. 194.3. The resulting discrepancy could be easily explained by assuming that Eusebius equated his Olympic years to Roman chronology such that the 42nd year of Augustus, in which Christ was born, began in Ol. 194.3 but also carried over into the first half of Ol. 194.4. However, since Eusebius assigned the crucifixion in the 18th year of Tiberius to Ol. 202.4 = AD 33, this also meant that something had gone wrong in his enumeration of the Roman regnal years. Judging from the latter equation, it followed that Tiberius's reign had begun in the autumn of AD 15, which was the beginning of Ol. 198.3, whereas Eusebius's chronicle assigned the first year of Tiberius to Ol. 198.2. Tolosani hence concluded that Eusebius's chronicle had been corrupted by a scribal error with regard to the duration of Augustus's reign, which ended one year later than indicated in present codices.[95]

The reliability of Eusebius's chronicle was also put into question by the solar eclipse that Paul of Middelburg had seized upon in his attempt to shift the birth of Jesus to the end of AD 2. As Tolosani pointed out, the solar eclipse calculated by Paul for AD 17 actually belonged to the second year of Tiberius and thus could not have preceded the death of Augustus. To disturb the entire chronology of the ancient world on the basis of this "statement of little importance" in the Eusebian chronicle, as the bishop of Fossombrone had done in his *Paulina*, was a methodologically deplorable overreaction. In Tolosani's view, Paul had argued like someone "who, in order to save the fruitless tree in his garden, sets fire to all the fruit-bearing ones." Clearly, the more economical and sensible way of going at things was to assume that Eusebius had erred in a minor detail by claiming that a solar eclipse had preceded Augustus's death, while providing a generally sound chronology for the Roman emperors, rather than vice versa.[96]

As this acute and well-founded criticism indicates, Tolosani was not at all impressed with Paul's alleged prophetic revelations. Having been

---

[95] Johannes Lucidus Samotheus, *Opusculum*, fols. 53, 182.

[96] Ibid., 186v–87r: "Et sic ut defendat & salvet unum dictum parvi momenti Eusebii ante mortem Augusti describentis ipsam eclipsam, destruit omnem computationem annorum Domini magni ponderis, non solum ipsius Eusebii, sed etiam omnium aliorum Chronographorum contra rectam rationem. Inordinante enim ageret, siquis ut salvaret in horto arborem sterilem, incideret omnes alias fructiferas. Secundum autem rectam rationem dicendum est in hoc minimo Eusebium errasse, & in caeteris maximis recte scripsisse, & nullum inde sequitur inconveniens."

unable to bolster his chronological claims by "arguments and calculations" (*rationes et computationes*), the bishop of Fossombrone had resorted to the realm of dreams and visions.[97] That these visions were of less than divine origin had been implicitly acknowledged by Paul of Middelburg himself, who, in the conclusion to his work, admitted to the tenuous character of his arguments, which he wanted his readers to see as mere suggestions and instigation for future research.[98] Accordingly, his vision was little more than a product of literary imagination, comparable to Cicero's *Dream of Scipio*—pleasant to read, but hardly reliable when it came to technical chronology.[99] In the end, the facts compelled Johannes Lucidus "to wash away the error of my friend. Paul, the bishop of Fossombrone, is my friend and master, but truth is my even greater friend."[100] That Tolosani was indeed anxious not to speak too badly of the bishop of Fossombrone, who was still alive when he penned his rebuttal, becomes clear from his further comments:

> And still, we have reason to thank our master Paul, who, with his laborious work, into which he put much sweat, gave us occasion to seek and find the truth, not by way of our ingenuity, but by the special help of God, from whose benevolence everything springs forth.... And since Paul himself is a lover of truth, as I am absolutely sure he is, he will agree with these arguments, once he considers them.[101]

[97] Ibid., 189r: "Cum autem cognosceret dominus Paulus suas rationes et computationes non esse sufficientes ad probandum suam opinionem, convertit se ad somnium vel imaginariam visionem, dum affirmat sibi desuper oraculo demonstratum omnes chronographos in annis Domini in duobus annis errasse."

[98] Paul of Middelburg, *Paulina*, GG3v: "Ut enim divus Hieronymus, non e diffinitio asserimus quaecumque scribimus, sed ea lectoris arbitrio derilinquimus. Sic et nos haec scribimus, ut praestemus diligentiam perquirendi potius quam affirmandi temeritatem." Cited in Johannes Lucidus Samotheus, *Opusculum*, 189r.

[99] Johannes Lucidus Samotheus, *Opusculum*, fol. 189: "Ex quibus verbis patet, quod non ex revelatione divina, sed ex humana adinventione locutus est, quemadmodum loquuntur oratores & poëtae, qui varias adinventiones & similitudines confingunt ut facilius persuadeant quod verum putant humana aestimatione. Unde & iudicio sapientum ipse dominus Paulus in hoc imitatus est Ciceronem, qui in libro de Rep. apte confingit somnium Scipionis, ut suis dictis tanquam divino spiritu prolatis maior fides adhibeatur. Quod quidem somnium eleganter Macrobius exponit referens illud potius ad ingenium Tullij aptissime confingentis quam ad somni veritatem."

[100] Ibid., 173r: "Cogit me veritas nunc amici mei errorem detergere. Amicus meus dominus Paulus episcopus Forosemproniensis, sed magis amica veritas."

[101] Ibid., 189v: "Gratias tamen agimus domino Paulo, qui suo laborioso opere, cui diu insudavit, dedit nobis occasionem indagandi et reperiendi veritatem, non virtute nostri ingenii, sed speciali auxili Dei, a quo bona cuncta procedunt.... Et quoniam ipse praesul dominus Paulus amator est veritatis, ut nobis certissimum est, inspecta

After Paul of Middelburg's claims had been successfully warded off, it was left for 'Johannes Lucidus Samotheus' to respond to the objections by the layman who followed ecclesiastical tradition in dating the crucifixion to 25 March. Tolosani's line of defence, which had been previously tested by Roger Bacon and Alfonso Tostado, was to state boldly that the fathers of the church were as fallible as any other man when it came to chronological dating matters. Since no specific chronological arguments were found in the writings of St. Augustine, who influentially claimed that Jesus had died on 25 March in the consulate of the Gemini, there was no reason to follow him in this question. In fact, a look into Augustine's commentary on the Psalms, where the bishop of Hippo refused to commit himself to the view that the moon reflected the sun's light, showed that he had not been competent enough as an astronomer to make this judgement in the first place. No wonder, then, if Augustine and other ancient church fathers, for all their authority, had fallen prey to an erroneous date for Christ's Passion.[102] In fact, Tolosani believed that he could explain why so many ancient sources dated the crucifixion to March even though astronomical investigation clearly showed that it had taken place in April. The Romans had traditionally equated the Jewish month of Nisan with the March lunation, because March used to be the first month in the Roman calendar and also the Julian month in which Nisan usually began. Since Jesus had died in Nisan, his Passion had therefore quickly become associated

veritate assensum praebebit rationibus ipsam veritatem comprobantibus, quoniam ipse in totius operis suis peroratione dicit."

[102] Johannes Lucidus Samotheus, *Opusculum*, 193r: "Chrysostomus ergo & Augustinus & praedicti reliqui doctores locuti sunt de tempore baptismi & passionis Christi ex opinione simplici, ut vulgo ferebatur, quod patet ex hoc, quod non probaverunt dicta sua per computationes congruas. Nam, ut dicit Ioannes Stofflerinus, antiqui doctores postponebant scientiam Astronomicam, & vacabant aliis doctrinis magis utilibus, quae eo tempore plurimum erant ecclesiae dei neccesariae. Unde Augustinus super psalmum decimum cum proponeret quaestionem de Luna utrum illummetur a Sole, relinquit eam indeterminata tanquam difficilem & parum utilem, dicens ibidem. *In istis obscuritatibus magis negociosis quod fructuosis exercere animum, aut non libet, aut non vacat, aut animus ipse non valet.* Quia ergo antiqui doctores non vacabant Astronomicae disciplinae, quae tradit motum coelestium corporum, & ex consequenti ordinem temporum... quare quandoque errabant. Et quoniam erant viri scientia praestantes ac magnae authoritatis, eos imitati sunt plures recentiores authores absque alia supputationis consyderatione, ideo non est mirum, si etiam ipsi in eundem errorem inciderunt." See Augustine, *Enarrationes in Psalmos* (10.3), CCSL 48: 76.

with March once Christianity had begun to spread from Judaism into the Roman pagan world.[103]

The arguments presented by 'Johannes Lucidus Samotheus' in favour of a Passion on 3 April, AD 33, were well-made and had a considerable impact on the way the Passion date was discussed in sixteenth-century chronological writing. The same date could be subsequently encountered in the work of the Catholic chronologer Alexander Scultetus (1485–1564), who was a friend of Copernicus.[104] Yet just as Alfonso Tostado had found a fierce opponent in his compatriot Juan de Torquemada and Paul of Middelburg had clashed with another scholar from the Low Countries, Tolosani's views were soon to be attacked by a fellow Italian, the legendary astrologer Luca Gaurico (1475/6–1558).[105] The latter's career had made a great leap after he had correctly predicted election to the papacy of Alessandro Farnese, who became Paul III (1534–49). As a reward for his astrological services, Farnese summoned Gaurico to Rome, knighted him, and granted him the privilege of being the pope's table companion. Moreover, in 1539, he was made bishop of Giffoni, a town near his birthplace in Campania. The same year, Gaurico dedicated to his papal benefactor a short work

---

[103] Johannes Lucidus Samotheus, *Opusculum*, 194r. See also ibid., 54v. A similar argument is found in Paul of Middelburg, *Paulina*, F8r–v.

[104] Alexander Scultetus, *Chronographia*, *3v–4r: "Compertum vero est, quod eo anno 18 Tiberii Caesaris, & 33 vitae Christi prima luna, seu novilunium primi mensis verni iuxta rationes astronomicos acciderit in 20 diem Martii ita quod inde 14 Luna acciderit in 2 diem Aprilis, qua Iudaei iuxta legem ad vesperam agnum paschalem comedere debuerant, quemadmodum Christus legi satisfaciens cum discipulis suis fecit." See also ibid., 102, as well as Theodor Bibliander, *De ratione temporum* (Basel, 1551), 253; Bibliander, *Temporum a condito mundo usque ad ultimam ipsius aetatem supputatio, partitioque exactior* (Basel, 1558), 119–21. An implicit commitment to AD 33 may be found in the astronomical still life in Hans Holbein's painting *The Ambassadors* (1533), provided it was the purpose of the depicted instruments to refer to Good Friday in 1533 as the fifteenth centenary of the crucifixion. For an interesting argument in this direction, see John North, *The Ambassadors' Secret* (London: Hambledon and London, 2002).

[105] Baldi, *Le vite*, 426–33; Don Cameron Allen, *The Star-Crossed Renaissance* (New York: Octagon Books, 1973), 51–55; Paola Zambelli, "Da Giulio II a Paolo III: Come l'astrologo provocatore Luca Gaurico divenne vescovo," in *La città dei segreti*, ed. Fabio Troncarelli (Milan: Angeli, 1985), 299–323; Zambelli, "Many Ends for the World: Luca Gaurico Instigator of the Debate in Italy and in Germany," in "*Astrologi hallucinati*", ed. Paola Zambelli (Berlin: de Gruyter, 1986), 239–63; Marco Bertozzi, "Astrologia e potere nel '500: L'avventurosa carriera di Luca Gaurico, vate 'veridicus'," in *Cultura e potere nel Rinascimento*, ed. Luisa Secchi Tarugi (Florence: Cesati, 1999), 543–51.

on the miraculous eclipse that had occurred during the crucifixion.[106] As he made clear in his little treatise, the traditional date of the Passion on 25 March, AD 34, was only two days removed from the opposition of sun and moon (23 March), which fully proved the miraculous nature of the eclipse, since solar eclipses only happened when both luminaries were in conjunction.[107] In addition, it became apparent that the Passion, if it had indeed taken place on 25 March, AD 34, would not have precisely coincided with any full moon, which led Gaurico to reject ordinary astronomical attempts to date the Passion by reference to the lunar phases. On the hypothesis of Paul of Middelburg, who had shifted the Passion to 30 March, AD 36, he noted mockingly:

> I should truly put the greatest trust in his writings, if this date had been revealed to him while he was awake, by some learned and pious man, and not just an oracle by St. Paul the Apostle, or if he had read it in the work of some father of the church.[108]

His treatment of Johannes Lucidus's ideas was similarly scornful:

> I beg this *Lucidus*, who is no doubt a learned man, that he may elucidate me if he ever heard from Apollo's mouth or from some holy man that Christ died on the cross on 3 April and rose from the dead on 5 April, even though this date is nowhere to be found among the chronographers, be they Hebrew, Greek or Latin.[109]

Instead of trying to make the data fit by astronomical experiments, Gaurico professed his confidence in the Latin tradition, which dated the crucifixion on 25 March. He combined this time-honoured

[106] Luca Gaurico, *De eclypsi solis miraculosa in Passione Domini celebrata* (Rome, 1539). The treatise was later reissued by Dominique Jacquinot as *De eclipsi solis miraculosa in Passione Domini observata* (Paris, 1553). In what follows, I shall cite from the 1553 edition. See further Gaurico, *Calendarium ecclesiasticum novum*, 31r–32v, where parts of the same discussion are repeated.

[107] Gaurico, *De eclipsi*, A3r–B2v.

[108] Ibid., B3r: "Ego suis scriptis maximam profecto adhiberem fidem, quando id sibi vigilanti revelatum fuisset ab aliquo viro docto ac religioso, nedum ab oraculo divi Pauli apostoli, vel saltem apud quempiam Ecclesiae doctorem lectitasset Christum ut ipse dicebat, tertio Calendas Aprilis fuisse mortuum in cruce, & Calendis eiusdem a mortuis resurrexisse." See also Gaurico, *Calendarium*, 3r.

[109] Gaurico, *De eclipsi solis*, B3v: "Elucidet mihi obsecro Lucidus ille, vir satis doctus, nunquid etiam ipse ab ore Apollineo, aut a quopiam viro sancto admonitus didicit Christum 3 nonas Arietis, crucis subiisse vexillum, dein nonis eiusdem a mortuis resurrexisse? quum nullibi unquam compertum habeatur apud Chronographos Hebraicos, Graecos, aut Latinos Christum fuisse crucifixum, nisi decimo calendas Aprilis, hoc est, die 23 Martii secundum Graecos, aut 8 calend. Aprilis, id est, 25 Martii secundum orthodoxae fidei doctores."

assumption with the likewise traditional view according to which Jesus had died at the age of 33 years and three months. This life-span, he argued, was backed by the argument of a certain Bernard of Modena, "a most brilliant and upright man, free of all vice and mischief," who had proposed a kabbalistic deduction of the age of Jesus at his death. Since the numerical value of the Hebrew word for 'saviour', *go'el* (3 + 1 + 30 = גאל), was 34, this could only mean that Jesus died in the 34th year of his life.[110]

Yet how could all this be squared with the fact that 25 March had been a Thursday in AD 34? One possible solution was to put the incarnation of Christ in the year AD 1, which was marked with the Sunday letter B. This had the welcome side-effect that the day of the Annunciation (25 March, AD 1), the Adoration of the Magi (6 January, AD 2), and the baptism of Christ (6 January, AD 30) all came to fall on a Friday, whereas the nativity (25 December, AD 1) occurred on a Sunday, which provided all four days with a neat typological connection to the weekdays of Christ's death and resurrection. Since Gaurico assumed that the incarnation and resurrection were separated by 34 years exactly, this also meant that the latter had taken place in AD 35, when 25 March and 27 March had been a Friday and a Sunday respectively.[111] But even if Christ had died in AD 34, a year with the Sunday letter C, Gaurico believed that an explanation of the traditional date was possible. After all, the Jewish day began at sunset and hence six hours earlier than the midnight epoch in use among Christians. As a result, the Sunday of Christ's resurrection had technically begun on 27 March, even though its greater part had fallen on 28 March according to the Julian calendar.[112]

---

[110] Ibid., Cr: "Bernardus Mutinensis vir admodum ingeniosus, atque integer vitae [!] scelerisque purus ex Hebraico idiomate se iampridem didicisse inquit, Christum noluisse crucis subire vexillum ante plenitudinem nominis Goel, quod redemptorem significat. Illudque nomen Goel importare numerum 34 quo anno labente passus fuit Dominus."

[111] Ibid., Cv–C5r. Gaurico's arguments presupposed that the leap year-cycle in the Julian calendar had not been disturbed in the years around the beginning of the Christian era. As Johannes Lucidus Samotheus had already pointed out, however, there had been no intercalation between 5 BC and AD 4, owing to a correction of the calendar implemented by Augustus. As a result, AD 1 historically had the Sunday letter A. See Johannes Lucidus Samotheus, *Opusculum*, 171r–75r.

[112] Gaurico, *De eclipsi*, Cr. See also Gaurico's marginal commentary to the table on sig. C4r: "Labente anno 34 si littera dominicalis fuisset C iam Christus resurrexisset a mortuis post mediam noctem diei 27 in cuius occasu incepisset dies dominicae resurrectionis." See also Gaurico, *Calendarium*, 33r–34v.

While the degree of Gaurico's commitment to either of these two rather implausible scenarios is left unclear by his own words, he seemingly subscribed to the latter in two chronological tables, each of which assigned the conception of Jesus to the year before the beginning of the Christian era, which was a leap year. Instead of designating the latter as 1 BC or the first year *ante Christum*, however, he assigned to it a neutral position '0' and made the retrospective BC-count of the Dionysiac era begin one year earlier than usual. In doing so, Gaurico effectively anticipated the contemporary astronomical method, which makes working with the Christian era more arithmetically convenient by designating 1 BC as year 'zero' and lowering the number of all previous years by 1. As a result, 2 BC becomes year −1, 3 BC becomes −2, and so forth.[113] According to common wisdom, the inventor of this system was Jacques Cassini (1677–1756), but it seems that Gaurico anticipated him by more than 150 years.[114]

His use of the year 'zero' was apparently inspired by the intercalation cycle of the Julian calendar, where every year of the Christian era evenly divisible by four must be a leap year. Due to the lack of a zero-position in the BC/AD timeline, this rule naturally did not apply to years BC, where only those years contained a leap day, whose division by four yielded a remainder of 1. As Gaurico explained towards the end of his short treatise, he was unwilling to accept this doctrine on the leap year rhythm, which had been previously espoused by Paul of Middelburg. Referring to the alleged calendrical decrees of Julius Caesar, he claimed that the year 5116 since the creation of the world and the year 44 'before the incarnation of Christ' were both leap years, "as was the year 5200, in which Christ was conceived through the Holy Spirit."[115] Needless to say, this assertion of a perfectly symmetrical leap year-rule carried the anachronistic presupposition that Julius Caesar already had the Christian era in mind, when he promulgated his

[113] Gaurico, *De eclipsi*, C2v, C6v. See also Gaurico, *Calendarium*, 12r, 40v–41r.

[114] See Jacques Cassini, *Tables Astronomiques* (Paris, 1740), 5. Similar usage of a year 'zero' can already be found in Johannes Kepler, *Tabulae Rudolphinae* (Ulm, 1627); Philippe de la Hire, *Tabulae Astronomicae* (Paris, 1702).

[115] Gaurico, *De eclipsi*, C5v: "Iulius Caesar…statuitque in suis fastis publicoque edicto promulgavit, ut 4. quisque annus efflueret intercalaris, & ex consequenti omnes qui in aequas partes dividi possint. Quamobrem annus 5116 ab orbe condito, & 44 ante Christi conceptionem effluxerat intercalaris, quo sit ut annus 5200 in quo Christus conceptus est de spiritu sancto, fuerit etiam intercalaris. Annus autem 5201 in quo natus est, fuit primus post bisextum, veluti annus 5199 qui erat primus ante conceptum."

calendar. Erroneous as these assumptions were, they had the convenient side-effect of making the year 5200 of the Eusebian world era equivalent to the year 0, which meant that the *anni mundi* obeyed the same leap year-rule as years AD, with bissextile days inserted in AM 5204, 5208, etc.[116]

For all their patent idiosyncrasies, Gaurico's arguments concerning the Passion date were well-received in at least some quarters. The Parisian astronomer Dominique Jacquinot, who reissued Gaurico's *De eclipsi solis* in 1553, told Anselme Petit, the Archdeacon of Langres, in his dedicatory epistle that in the present treatise "everything which I myself had strenuously investigated is presented in such good order and discussed so accessibly" that it seemed to him that "nothing relevant to instruction in these matters could be added or reduced."[117] This favourable view, however, was not shared by Onofrio Panvinio (1529–68), who became the foremost Catholic authority on Roman chronology during the middle years of the sixteenth century. In 1558, he published an extensive commentary on the *Fasti*, the annual lists of magistrates in ancient Rome, which he had already edited the year before. The basis for this edition had been the discovery of the *Fasti Capitolini* under the rubble of the Forum Romanum in 1546. The monument had been originally erected under the reign of Augustus and listed all official magistrates since the beginning of the Roman Republic in 509 BC.[118] All the new data, however, could not save Panvinio from missing the correct correlation between the thread of the Roman consulates and regnal years on the one hand and the Christian era of Dionysius Exiguus on the other. In his excursus *De anno et die passionis D. N. Iesu Christi*, contained in his commentary on the *Fasti*, he dated the consulate of the two Gemini, in which Jesus was baptized, to AD 30.[119] The fixed point of his calculation was once again the notice in Eusebius's chronicle, according to which Phlegon

[116] Ibid., C6v.

[117] Ibid., A2r: "In quo quidem libello, haec omnia quae anxii inquirebamus, ita diserte & aperte discutiuntur, ut nihil eorum quae ad institutum illud pertinent, videatur addi posse aut minui." See also the comparison of Gaurico's and Tolosani's chronologies in Bibliander, *Temporum…supputatio*, 23–24.

[118] On the background, see William McCuaig, "The Fasti Capitolini and the Study of Roman Chronology in the Sixteenth Century," *Athenaeum* (Pavia) 79 (1991): 141–59; Feeney, *Caesar's Calendar*, 167–83.

[119] Panvinio, *Fastorum libri V*, 25–27, 306–12. See also Panvinio, *Fasti et triumphi Rom. a Romulo rege usque ad Carolum V. Caes. Aug.* (Venice, 1557), 39, 44, and Panvinio, *Chronicon Ecclesiasticum* (Louvain, 1573), 5–8.

had recorded the eclipse of Christ's crucifixion for Ol. 202.4. Panvinio followed the 'Roman tradition', once so forcefully evoked by Bede, and equated Ol. 202.4 with AD 33/34, which necessarily meant that his epoch of the Olympiad era belonged to 775 rather than 776 BC.

Turning to the tricky question of how Jesus could have died both on 25 March and in AD 34, Panvinio realized that the traditional date had to be emended to 26 March, which was a Friday in said year. As a result, he chastised Luca Gaurico for believing that 25 March, AD 34, was still a viable option.[120] Panvinio also freely admitted that the full moon in AD 34 had occurred on Tuesday, 23 March, and that there was no coincidence of 25 March with a full moon in the neighbouring years. Yet he also asserted that this only created a problem if the Jews were thought to have always equated 15 Nisan strictly with the day of the opposition of sun and moon. Thanks to his acquaintance with "the books of the ancient Jews," by which he obviously meant the Mishnah or the Talmud, he knew that the Jews of old had once counted their month from the evening of first visibility and sent people on high mountains to report the appearance of the new moon crescent. By Panvinio's logic, 26 March in AD 34 should have been the 15th day from first visibility, bringing the first day of the Jewish month to 12 March. This implied—somewhat implausibly—that the new moon crescent was spotted by the Jews a full three days after the actual day of conjunction (9 March).[121]

Panvinio's idea of shifting the Passion from 25 to 26 March was subsequently also espoused by the Sicilian mathematician Francesco Maurolico (1494–1575), who was abbot of the monastery of Santa Maria del Parto near Castelbuono. Apart from his mathematical achievements,

[120] Panvinio, *Fastorum libri V*, 311: "Quod autem Lucas Gauricus conetur affirmare hoc ipso anno XXXIV aetatis Christi diem Martii XXV fuisse diem Veneris, id repugnat omnino sibi ipsi, qui ait eo anno XXIII die Martii fuisse diem Martis, & XXIV diem Mercurii, item tabulae magnae, cyclo Iulii Caesaris, & aureo numero concilii Nicaeni; Atque his verbis hanc contradictionem tollere sibi persuadet: Incipit Sabbatum iuxta Hebraeorum consuetudinem in occasu Solis diei Veneris, & ita quaelibet feria incipit occasu Solis. Dies igitur Veneris XXV Martii sumpserant initium post occasum Solis diei XXIV. Quod etiam nunc observatur in singulis Italiae urbibus."

[121] Ibid., 310: "Contradictio autem quod Luna XV die Martis XXIII non die Veneris XXVI eo anno fuerit tolli potest, ex antiquis Hebreorum traditionibus, apud quos initium mensis non erat ab ipso statim novilunio, sed ab ipsius Lunae corniculari apparitione. Unde legitur in libris eorum, veteres Iudaeos consuevisse tempore novilunii mittere homines super altissimos montes, qui eis novae Lunae apparitionem nuntiarent, quo mensium initia auspicari possent."

Maurolico is also known for his sharp attacks against those who disagreed with him. In the preface to his *De sphaera liber unus*, printed in 1575 as part of his *Opuscula mathematica*, he famously noted that Copernicus was deserving of "a whip or a scourge rather than a refutation" (*scutica potius, aut flagello, quam reprehensione dignus*) for suggesting that the earth revolved around the sun.[122] The same tome also contained a *Computus ecclesiasticus*, datable to 1568, which deals with the ecclesiastical calendar and its defects. While Maurolico was fully aware of the calendar's errors with respect to the date of the equinox and the full moon, he found it inappropriate to try and change the traditions "which were once handed down to us by our ancestors." Accordingly, he had no sympathy for those chronologers who had in his eyes wasted dozens or hundreds of pages on fanciful proposals to correct the calculation of Easter.[123] He expressed the same contempt for attempts to make the date of Christ's Passion match up with the astronomical full moon. Once again, Maurolico preferred to stick with ecclesiastical tradition, which in his mind prescribed a crucifixion on 26 March, AD 34. Knowing that this had not been the date of a full moon, he nonchalantly noted that

> it is no wonder if the full moon preceded that date by two or three days, for due to the inequality of lunar motion, the 14th day of the lunation does not always correspond exactly to the feast [of Passover], as the curiosity of many would make them believe. And thus, according to the true and simple calculation, Christ suffered in the 34th year of salvation, which has the Sunday letter C and the golden numer 16, on the 26th of March, so that he rose from the dead on the 28th of the same month, which was a Sunday. Mistaken are hence Paul of Middelburg, Johannes

[122] Francesco Maurolico, *Opuscula mathematica* (Venice, 1575), 26. See also Edward Rosen, "Maurolico's Attitude toward Copernicus," *Proceedings of the American Philosophical Society* 101 (1957): 177–94.

[123] Maurolico, *Opuscula*, 42: "Neque obstat, quod tam aequinoctiorum, quam lunationum sedes, ut dictum fuerat, retrocesserint. Cum quibus si paschae locus retraheretur aut, si intercalatis diebus, anni vel mensium exordia transferentur; omnis fastorum, annorum & temporum supputatio confunderetur. Quare, iudicio meo, servandum est semel traditum a maioribus praeceptum: ut retrocessio praedicta per veteres kalendarii notas signata sit antiquitatis tantae venerabile testimonium. Tum etiam, quia certum est, non posse Ecclesiam in hoc animadvertere ac speculari astronomicas minutias, suo tantum aureo numero, suaque dominicali nota contentam. Frustra igitur ac multo curiosius, quam decebat, Paulus de Mildeburgo [!] Forosemproniensis episcopus ingenti volumine, Ioannes Regiomontius, Ioannes Stoflerinus, Ioannes Lucidus, Petrus Pitatus: aliique quotidie super hac re disputant." See ibid., 34, for a reference to 18 December 1567 as the present day.

> Lucidus, and all others who think differently and who busily twist and turn the years according to their views.[124]

Panvinio's and Maurolico's proposals, which, unbeknownst to them, were reminiscent of the works of two forgotten twelfth-century chronologers (Reinher and Constabularius), represented a way of honouring ecclesiastical tradition while retaining a minimum of calendrical plausibility in light of the fact that 25 March fell on a Thursday in AD 34.[125] With a lifespan of 33 years and 3 months, ranging from 25 December, 1 BC, to 26 March, AD 34, Panvinio's assumed dates of Christ's life closely mirrored those found eight centuries earlier in the 47th chapter of Bede's *De temporum ratione*. However, there was one further fly that could be seen as spoiling Panvinio's chronological ointment: since he dated the baptism of Jesus to 6 January in the consulate of the Gemini, which in his reckoning corresponded to AD 30, his Passion date in AD 34 implied that the public ministry had lasted over four years, instead of the usual three years and three months, which were presupposed by Panvinio himself.[126] During this period, there would have been five Passovers (the Passion included), as opposed to the three or four feasts found in the Gospel of John (2:13, 5:1, 6:4, 11:55). This glaring discrepancy compared to traditional views of the chronology of Jesus's life eventually led the geographer Gerhard Mercator to propose a 'five Passover'-theory, which he first outlined in his *Chronologia* (1569) and later repeated in his Gospel harmony, printed in Duisburg in 1592.[127]

---

[124] Ibid., 46: "Notandum, quod secundum Dionysium abbate, eiusque Computum, Christus passus est 26. Martii; cum pridie pascha cum discipulus egisset. Nec mirum est, si plenilunium per biddum, aut per triduum praecesserat. Nam propter inaequalitatem lunaris motus, non adamussim respondebat 14. luna festo. Sicut curiositas multorum credit. Itaque secundum verum & simplicem calculum, Christus passus est anno salutis currente 34. instante litera dominicali C. aureo numero 16. die 26. Martii, ita ut 28. mensus eiusdem, qui fuit dies dominicus, resurrexerit. Falluntur ergo Paulus de Mildeburgo [!], Io. Lucidus & caeteri, qui aliter sentiunt, & annos ad propositum suum ex industria coaptant & torquent."

[125] See also Girolamo Bardi, *Chronologia universale*, 4 vols. (Venice, 1581), 3:10, where the Passion is dated to 26 March, AD 34.

[126] Such criticism was voiced by Nicolas Vignier, *La Bibliothèque Historiale*, 3 vols. (Paris, 1587), 1:695.

[127] Mercator, *Chronologia*, br–d5v; Mercator, *Evangelicae historiae quadripartita monas* (Duisburg, 1592). On the background, see Henk Jan de Jonge, "Sixteenth-Century Gospel Harmonies: Chemnitz and Mercator," in *Théorie et pratique de l'exégèse*, ed. Irena Backus and Francis Higman (Geneva: Droz, 1990), 155–66; Alfred Suhl, "Zu Gerhard Mercators Evangelienharmonie," in *Mercator und Wandlungen der*

The problems entailed by Panvinio's chronological position were also noticed by the Veronese astronomer Pietro Pitati (1490–1567), who is known for publishing a *Compendium super annua solaris* (1560, with subsequent editions in 1564 and 1568). With this work, Pitati seems to have been the first to make the proposal of omitting the Julian leap day in three out of every four centennial years in order to keep the calendar in line with the solar year. According to Pitati, the bissextile day was to be omitted in 1600, even though the latter was a Julian leap year, and in all subsequent centennial years until 1900. This principle was later adopted by the Gregorian reform commission, with the slight difference that the bissextile day is kept in all centennial years evenly divisible by 400 (1600, 2000, 2400, etc.). As a result, 1600 was a leap year, while 1900 was not.[128] Apart from Pitati's views on calendar reform, the *Compendium* also contained a lengthy *Tractatus super incarnationis, nativitatis, atque Passionis D. N. Iesu Christi die, mense et anno*, in which he mainly popularized the arguments and dates that had previously been advanced by Johannes Lucidus Samotheus in the *Opusculum de emendationibus temporum*.[129] In addition, he dedicated an entire chapter to the refutation (*improbatio*) of Luca Gaurico and all others who stubbornly relied on the fathers of the church in dating the Passion to 25 March, AD 34, despite the fact that the latter date defied all chronological probability.[130] The treatise closed with an open letter to Onofrio Panvinio, which took the form of an amicable dialogue between 'Petrus' and 'Onuphrius' and in which

Wissenschaften im 16. und 17. Jahrhundert, ed. Manfred Büttner and René Dirven (Bochum: Brockmeyer, 1993), 43–59; Marijke Hélène de Lang, "De opkomst van de historische en literaire kritiek in de synoptische beschouwing van de Evangliën van Calvijn (1555) tot Griesbach (1774)" (PhD diss., University of Leiden, 1993); de Lang, "The History of the Gospel Synopsis and Gerardus Mercator's Evangelica Historia," in *Gerhard Mercator und die geistigen Strömungen des 16. und 17. Jahrhunderts*, ed. Hans H. Blotevogel and Rienk Vermij (Bochum: Brockmeyer, 1995), 199–208.

[128] Pietro Pitati, *Compendium… super annua solaris, atque lunaris anni quantitate* (Verona, 1560). Further editions appeared in Venice (1564) and Basel (1568). In what follows, I shall cite from the third edition, entitled *Verae Solaris atque lunaris anni quantatis… explicatio*. The crucial proposal ('proposition II') of omitting three days in 400 years can be found on fol. 4v. For biographical details, see Bernardino Baldi, *Le vite*, 489–94. See also Kaltenbrunner, "Die Vorgeschichte," 403–6; Thorndike, *A History*, 5:264–65; Baumgartner, "Popes," 48–49; Noel M. Swerdlow, "The Length of the Year in the Original Proposal for the Gregorian Calendar," *Journal for the History of Astronomy* 17 (1986): 109–18.

[129] Pitati, *Explicatio*, 37r–80r.

[130] Ibid., 65r–74v.

the problem of the chronology of Christ's life was once again tackled. As Pitati pointed out, Panvinio should have dated the Passion to AD 33, since the latter corresponded to the end of the fourth year of the 202nd Olympiad, provided the Olympiads were counted correctly. It followed that Jesus had been baptized three years and three months before his death, on 6 January, AD 30, at which time he had just began the 30th year of his life. Furthermore, he had been born on 25 December, 1 BC, in the fourth year of the 194th Olympiad and not in Ol. 194.3, as Panvinio had supposed.[131]

Although Pitati's chronological treatise did not contain a great number of new ideas, he provided readers with a useful overview of scholarly opinion on the date of Christ's life, indicating with some clarity the existing points of divergence. As his discussion showed, there was a rift in his time between those authors who preferred 3 April, AD 33, for astronomical reasons and those who clung to 25 or 26 March, AD 34, out of respect for patristic tradition. It was the astronomical tradition, which eventually prevailed and dominated chronological opinion in the latter half of the sixteenth century on both halves of the confessional divide.[132] Conversely, the chronologically indefensible notion of 25 March, AD 34 found fewer and fewer adherents as time progressed, although it still cropped up occasionally at the end of the century in works such as Wenceslaus Sturm's German *Chronica*.[133]

---

[131] Ibid., 75r–80r.

[132] See, e.g., Lorenz Codomann, *Supputatio praeteritorum annorum mundi* (Leipzig, 1572), 69–73; Matthaeus Beroaldus, *Chronicum* (Geneva, 1575), D3r–v; Bardi, *Chronologia*, 1:63v–64r; Vignier, *La Bibliothèque*, 1:697–99; Johannes Temporarius, *Chronologicarum demonstrationum libri tres* (Frankfurt, 1596), 169–71; John More, *A Table from the beginning of the world to this day* (Cambridge, 1593), 94; Thomas Pie, *An Houreglasse* (London, 1597), 70–87.

[133] Wenceslaus Sturm, *Chronica* (Leipzig, 1596), 108. A Passion in AD 34, without a calendar date, is also found in Pantaleon Candidus, *Tabulae Chronologicae* (Strasbourg, 1597), 38. The same year was chosen by Cesare Baronio in his *Annales ecclesiastici* (1588). See Baronio, Rinaldi, and Laderchi, *Annales*, 1:109–206.

CHAPTER EIGHT

THE LIFE OF JESUS AND THE EMERGENCE OF SCIENTIFIC CHRONOLOGY

Back in the thirteenth century, Roger Bacon had already committed himself to the important idea that a 'certified' chronology of ancient history had to rely on eclipses of the sun and the moon as independent and precise markers of historical time. It was not until the sixteenth century, however, that this research programme, which in Bacon's work had the character of a promissory note, was fully put into practice. An early trace of eclipse-based chronological argumentation, which I have briefly discussed in the previous chapter, can be found in Peter de Rivo's *Tercius tractatus* and Paul of Middelburg's *Paulina*, where the alleged solar eclipse in the last year of Augustus's reign was used to date the latter's death to AD 16 and 17 respectively. Another significant step in the development of eclipse-dating came in Scaliger's birth year 1540, with the publication of Petrus Apianus's magnificent *Astronomicum Caesareum*, a richly illustrated astronomical handbook in honour of the Emperor Charles V, which included rotating paper volvelles by which readers were invited to compute mechanically the movements of the planets, the sun, and the moon. In the 29th chapter of his work, Apianus, who was a professor of mathematics at the Bavarian university of Ingolstadt, set forth a programmatic discussion of the chronological errors and contradictions found in historical sources and how mathematical astronomy could help to remedy this evil:

> Everybody is aware that those who pass down deeds and histories all but fall out of the heavens (as they say) when it comes to enumerating the years, sometimes counted from the world, sometimes from the foundation of the City. How much darkness, ignorance, doubt, and confusion is introduced by this error every time it comes to evaluating and understanding a work of history, is there to be judged by the scholars. For it therefore happens that, if they want to remind the Christian reader of some memorable event that either preceded or fell not long after the founding of the City, they cannot give a consistent account. Only knowledge of eclipses can amend this great evil and restore it to a better state.

> For eclipses can make it possible to fix all events to precise years, before Christ no less than after him.[1]

Apianus went on to demonstrate the possibility of correcting historical tradition by astronomical means with three very prominent examples from ancient history. The capture of Nicias following the second battle of Syracuse, the battle of Gaugamela, and the Roman victory over the Macedonian king Perseus at Pydna had all been preceded by famous lunar eclipses, which Apianus respectively dated to 8 September, 414 BC (Oppolzer no. 1226), 28 June, 326 BC (Oppolzer no. 1360), and 2 September, 172 BC (Oppolzer no. 1590). In each case a comparison with the years implied by Eusebius's chronicle revealed troubling discrepancies. According to Apianus's own understanding of the chronicle's dates, the years indicated by Eusebius were AM 4782 = 417 BC, AM 4871 = 328 BC, and AM 5034 = 165 BC—a seven-year difference in the case of the battle of Pydna, which marked the end of the Antigonid dynasty and the dismantling of the Macedonian kingdom. The obvious conclusion was that the services of the astronomer in establishing a reliable chronology were urgently needed.[2]

Upon closer inspection, however, Apianus's technical arguments left much room for improvement. His reliance on the Dionysiac era led him to identify Eusebius's year AM 5199 = Ol. 194.3, in which the birth of Christ was thought to have occurred, with the year AD 1/2 instead of 2/1 BC. As a result, his interpretation of the dates found in the chronicle was marred by a systematic error. For example, while Apianus believed that Eusebius had assigned the battle of Gaugamela to AM 4871 = Ol. 112.3 = 328 BC, basing himself on the simple subtraction 5199–4871 = 328, the Olympiad date reveals that Eusebius's

[1] Petrus Apianus, *Astronomicum Caesareum*, I3v: "Nemo ignorat, rerum, historiarumque traditores in enumerandis tum mundi, tum urbis conditae annis, tantum non, ut dicitur, coelo ipso excidere. Qui quidem error quantum tenebrarum, ignorantiae, dubii, confusionisque, in omnia alia historia iudicanda intelligendaque importet, doctorum esto iudicium. Inde enim sit, ut si cuiuspiam rei, memorabilis saltem, lectorem christianum commonefaciant, quae urbem conditam, aut praecesserit, aut non ita multo post, subsequuta sit, ipsis constare, nullo modo possint. Quod quidem tam grande malum sola ecleipsium cognitio emendare et in melius vertere potest. Per ecleipses enim omnia certos in annos reduci possunt, Christum praecedentes, non minus, quam sequentes."

[2] Ibid., I3v–K2r. See also Anthony Grafton, "Petrus Apianus Draws up a Calendar," *Journal for the History of Astronomy* 42 (2011): 55–72; Grafton, *Scaliger*, 2:121–22. For Eusebius's AM dates cited by Apianus, see, e.g., *Chronicon divinum plane opus eruditissimorum autorum* (Basel, 1529), 57r (Nicias), 59r (Gaugamela), 62v–63r (Perseus's reign).

true intended date had been 330 BC. This was close to a lunar eclipse on 20 September, 331 BC (Oppolzer no. 1352), which is now agreed to have been the eclipse of Gaugamela. Instead of remedying the slight deviation from the true date in Eusebius's chronicle, Apianus had actually increased the error by choosing the wrong eclipse. This was partly due to the influence of Plutarch (*Alexander* 31), who set Alexander's victory in the Attic month of Boedromion, for which Apianus chose the wrong Julian equivalent, namely June. The only eclipse in the vicinity that fell in that month was the one in 326 BC (Oppolzer no. 1360). A similar situation occurred in the case of the battle of Pydna, where Apianus was lead astray by the testimony of Livy (*Ab urbe condita* 44.37), who had dated the lunar eclipse in question to the night before 4 September (and the battle itself to the following day). This was a date in the old Roman lunar calendar, which was probably equivalent to 21 June 168 BC (Oppolzer no. 1596) in the proleptic Julian calendar, since this is the correct date of the lunar eclipse yielded by astronomical calculation. Apianus, by contrast, erroneously assumed that the Pydna-eclipse must have fallen in a Julian September, which made him choose an earlier eclipse on 2 September, 172 BC, giving him the impression that Livy had misdated the whole event.[3] Only in the case of Nicias's demise in Sicily did Apianus's approach lead to a slight improvement in accuracy, although his eclipse on 8 September, 414 BC, was still one year too late compared to the lunar eclipse on 27 August, 413 BC (Oppolzer no. 1228), which is now generally associated with this event.

What Apianus's mishaps effectively showed was that the application of astronomy to historical chronology was no guaranteed path to success. Prospective chronologers did not only need to be capable of performing astronomical calculations, but in addition it was necessary to bring a certain amount of philological sensitivity into the equation, if a correct understanding of the historical sources in question was to be achieved. A new wave of fruitful advances in technical chronology, which responded to this challenge, arose in the second half of the sixteenth century and was owed, for the most part, to the efforts of German Protestant scholars.[4] One of these authors was the Lutheran

---

[3] On the background, see Stewart Irvin Oost, "The Roman Calendar in the Year of Pydna (168 B.C.)," *Classical Philology* 48 (1953): 217–30, who likewise denies the reliability of Livy's calendar dates, albeit for different reasons than Apianus.

[4] See Anthony Grafton, "Some Uses of Eclipses in Early Modern Chronology," *Journal of the History of Ideas* 64 (2003): 213–29.

theologian Lorenz Codomann (1529–90), whose *Supputatio praeteritorum annorum mundi* (1572) contains some noteworthy reflections on the task of the chronologer. As Codomann made clear to his readers, the help of astronomy could be fruitfully enlisted in sorting out diverging chronologies and contradictory sources:

> Granted, it is true that astronomy solely by itself cannot show us the beginning or end of any king's reign or historical event and its calculations would rather recede endlessly into the past before it would establish the beginning of the world; but nevertheless, where tried and tested works of history make mention of an eclipse and along with it of the year of some king…the authority of astronomy is such that we can only obtain the right distance of that year from the epoch of Christ, if we find through calculation of the movements of the celestial bodies an eclipse of this sort in the same part of the year. And it is therefore with some justification that big books on chronology are produced, in which the more prominent eclipses are marked at their respective place, to which the author's judgment assigns them, not through some form of divination, but securely determined by astronomical calculation.[5]

Not every chronologer in Codomann's day, however, shared this kind of confidence in the heuristic power of eclipse-dating. Only four years after his confident proclamation, the 'crypto-Calvinist' theologian Leonhard Krentzheim (1532–98), who served as superintendent in the principality of Liegnitz, published a German *Chronologia* (1576), in which he called attention to a systematic one-year discrepancy that he perceived to exist between modern calculations of historical eclipses and the dates for these eclipses given in ancient sources such as the *Almagest*, "in the hope that learned and experienced mathematicians will kindly take notice of my report and be moved to investigate the causes of this discrepancy." Needless to say, such disturbances, which seemed to point to a previously undetected anomaly in the course of

[5] Codomann, *Supputatio*, A3r: "Licet enim Astronomia sola per se nequeat ullius Monarchiae aut rei gestae principium vel finem ostendere, ac in infinitum potius progrederetur, quam certum mundi exordium definiret, tamen ubi in historiis probatis sit mentio Eclipseos, & anni regi alicuius, in quem illa cecidit, vel expresse annotatur, vel ex praecedentibus sequentibus circumstantiis elicitur, simulque si non de die saltem de mense aut anni quadrante constat: ibi tanta est autoritas astronomica, quod annus ille iustam ab Epocha Christi distantiam obtineat, nisi per calculationes motuum coelestium talis Eclipsis in eadem illius anni parte inveniatur. Et iure magni fiunt libri Chronologici, in quibus Eclipses insiginores eo loco annotantur, qui autoris iudicio non tantum divinante, sed per astronomica calculationes certificato, determinatur."

the moon, were bound to cast serious doubts on the reliability of calculated eclipses as indicators of historical dates.[6]

As it turns out, the technical basis of Krentzheim's *Chronologia* was in large parts derived from the findings of the famous geographer Gerhard Mercator, who had published a work of the same name in 1569. Mercator's *Chronologia* was an innovative and much-acclaimed work in its day, not least because of its ample use of astronomical data and elaborate tables, which displayed the timeline of human history with an unprecedented commitment to precision. Onofrio Panvinio, himself an authority in the field, enthusiastically lauded Mercator's skill in a letter that was printed as the *Chronologia*'s foreword, while Joseph Scaliger, who was rarely positive about the efforts of other chronologers, praised the book as a useful resource to his students.[7] Such assessments ignored the fact that the impressive visual façade of Mercator's chronology concealed serious cracks in the technical basement, which threatened to bring the entire house down. Just like a number of previous scholars, including Panvinio and Nicholas Copernicus, Mercator used the traditional 'Bedan' crucifixion year AD 34 in conjunction with the solar eclipse purportedly mentioned by Phlegon of Tralles (Ol. 202.4) as a means of defining the beginning of the Olympic year count. From the equation '(spring of) AD 34 = Ol. 202.4', it naturally—and erroneously—followed that the first Olympiad had begun in the summer of 775 BC. In contrast to Copernicus, however, who had left all other important eras, as they could be found in Censorinus's *De die natali*, in their place, Mercator used

[6] Leonhard Krentzheim, *Chronologia* (Görlitz, 1576), 264v–65r: "Dieses habe ich also . . . vermelden wollen/ weil am allermeisten daran gelegen/ Der tröstlichen zuversicht/ gelehrte und erfahrne *Mathematici*/ werden diese meine erinnerung freundlich vermercken/und darauß ursachen nehmen/ die ursachen dieses mangels gründlicher zuerforschen/ Damit sie dann der gantzen Christenheit/ und dem gemeinen nutz/ grossen Dienst erzeigen/ und dessen beyde für Gott und Menschen ewigen rhum und danck haben." On Krentzheim's book, see also Matthias Pohlig, *Zwischen Gelehrsamkeit und konfessioneller Identitätsstiftung* (Tübingen: Mohr Siebeck, 2007), 216–24.

[7] See *Scaligerana, Thuana, Perroniana, Pithoeana, et Colomesiana*, ed. Pierre des Maizeaux, 2 vols. (Amsterdam, 1740), 2:452: "Sa chronologie bonne ne se trouve plus, bonne & rare. Il y a mis tous les noms de Papes, Roys, Empereurs de Rome, d'Orient, d'Alemagne, des Lombards: je feray le catalogue de ceux des Goths dans mon Eusebe, mais d'autre façon." No adequate study of the contents of Mercator's *Chronologia* exists, but see de Lang, "De opkomst," 68–70; Rienk Vermij, "Gerard Mercator and the Science of Chronology," in Blotevogel and Vermij, *Gerhard Mercator*, 189–97; Vermij, "Mercators Stellung in der Geschichte der Wissenschaft," in *Gerhard Mercator und seine Welt*, ed. Rienk Vermij (Duisburg: Mercator-Verlag, 1997), 90–109.

the Olympiads to derive and re-set the beginning of several further year counts, including the Roman foundation era, which he shifted from 752 BC (Johannes Lucidus Samotheus's estimate) to 751 BC, and the Ptolemaic era of Nabonassar, which now started on 26 February, 746 BC (instead of 747 BC).[8]

This latter move had some dramatic consequences, seeing that the era of Nabonassar was the chronological backbone of Ptolemy's *Almagest*, where it served as the basis for many astronomical datings. As a result of Mercator's displacement of the era of Nabonassar, the eclipses recorded by Ptolemy suddenly occurred in different years than they should have according to modern calculation. Interestingly, this mistake did not always work against the coherence of Mercator's own system. Generations of chronologers had wrestled with the problem that the calculated phases of the moon in the year AD 34 could not accommodate any of the prospective crucifixion dates, as the traditional Easter full moon fell on 21 March (Sunday), while the first opposition in spring was astronomically found to have occurred on 23 March (Tuesday). By extrapolating from the misdated observations found in the *Almagest*, Mercator could instead establish that an opposition had occurred ten days later, on 2 April, AD 34, which was a Friday in that year. Naturally, Mercator was well aware that there was a glaring 12-day discrepancy between his 'Ptolemaic' data and the 'ecclesiastical' phases of the moon, as they could be gleaned from the Dionysiac Easter table—a difference close to the 11-day 'epact shift' by which the lunar phases receded in the Julian calendar from one year to the next. A less audacious chronologer would have taken this as a clear hint that he had placed the crucifixion in the wrong year, realizing that Ptolemy's lunar theory actually supported 3 April, AD 33. Mercator instead went on to put into question the general reliability of astronomical approaches to chronological dating by assuming that some undetected anomaly of the moon's periodic motion caused modern calculations and Ptolemaic observations to disagree.[9]

---

[8] Mercator, *Chronologia*, 52, 54–55, 142, 147.

[9] Ibid., 147: "Posuit autem Dionysius primo anno cycli sui lunam decimam quartam, Nonis Aprilis, ut testatur Beda, unde si quis anno Domini 133 inchoando intervallum ad secundam lunae eclipsim a Ptolomeo, anno 135 observatam colligat, et syzygias luminarium ei convenientes aptet, inveniet lunam decimamquartam dicto 133. anno, diebus circiter 12. prius ex Ptolomei calculo deprehendi, quam ex Dionysiaci cycli continuatione posita intelligatur." See also ibid., d6r–g4r; Grafton, *Scaliger*, 2:131–33; Depuydt, "Calendars," 52.

Mercator's objections did not deter Paul Crusius (1525–72), a Lutheran theologian, former pastor, and professor of mathematics in Jena, from making systematic use of calculated eclipses in his attempt to establish ancient history's chronological benchmarks. In his slim and unimposing *Liber de epochis*, published posthumously in 1578, the German professor used a range of ancient recorded eclipses to provide definite and correct dates for all the major epochs, including the long-debated Olympiads (776 BC), the Roman foundation era (753 BC), and the battle of Gaugamela (331 BC). The result was an astonishing feat of chronological accuracy and argumentative economy, which, as Anthony Grafton has shown, exerted a crucial influence on Joseph Scaliger and on the development of technical chronology in general. In keeping with his straightforward astronomical approach, Crusius expressed his preference for a crucifixion on 3 April, AD 33, while also noting that such a date implied a birth of Jesus earlier than commonly assumed, if the widespread assumption that Jesus had already completed 33 years at the time of his death was retained.[10]

The idea that Jesus had really been born in 2 BC (as opposed to 1 BC or AD 1) had previously already been proposed by Gerhard Mercator, who had been led to this conclusion by an attentive consideration of ancient Rome's regnal and consular chronology.[11] Crusius, probably without having consulted Mercator's book, vindicated this theory via a brilliant application of eclipse-dating. The crucial hint came from Josephus (*Antiquities* 17.167), who reported that a lunar eclipse had occurred shortly before king Herod's death. In principle, there would have been possible dates for this eclipse, on both 13 March, 4 BC (Oppolzer no. 1856), and 9 January, 1 BC (Oppolzer no. 1860). In keeping with an assumed lifespan of Jesus of 33 years and 3 months, Crusius chose the latter eclipse and thus arrived at the conclusion that Jesus had been born a couple of weeks before this astronomical event, in late 2 BC. As a result, he dryly stated that the Dionysiac epoch of Christ's birth, which preceded that of the Passion by 32 years and about three months, "is contrary to the Gospels, for it does not fall

[10] Paul Crusius, *Liber de epochis* (Basel, 1578). On the background, see Grafton, *Scaliger*, 2:109–15, 133–37.

[11] Mercator, *Chronologia*, 142: "Porro hac stante interpretatione & Christi nativitate in annum 42. not autem 43. Augusti admissa, invenimus in annis Christi vulgo numeratis unum deesse, in quo si quis peccatum a nobis putet, expendat argumenta nostra omnia, ponat & reponat, transferat, extendat contrahatve interstitia temporum, & videat si quo modo poterit unum annum excipere, salvo argumentorum consensu."

within Herod's lifetime, but follows upon his death at an interval of roughly nine months."[12] This onslaught against the Dionysiac era was a daring move, not least because it implied that all established dates of world history, which had been dated and memorized according to this era, were in need of correction. Mercator had reacted to this daunting prospect by simply omitting the year of the crucifixion in his count of the years since the birth of Christ. This way, the one-year discrepancy only affected the 34 years of Jesus's life, whereas all years after the crucifixion could be enumerated in harmony with the Dionysiac era—a service which Mercator hoped would minimize confusion among his readers.[13] Crusius by contrast, chose to bypass the use of BC/AD-dating altogether and instead made the year of the Passion of Christ, with an artificial epoch on 1 January, AD 33, into the basis of his astronomical calculations.[14]

No other sixteenth-century chronologer ventured further on Crusius's path of eclipse-based chronology than the Lutheran theologian Heinrich Bünting (1545–1604), whose bulky *Chronologia*, dedicated to the bishop of Halberstadt, appeared in 1590 in the Anhaltian town of Zerbst.[15] Bünting agreed with Crusius in dating the Passion of Christ to 3 April, AD 33, and his birth to the year 2 BC. Where the latter had seized upon the eclipse of 9 January, 1 BC, in his revision of the nativity date, Bünting was able to further corroborate this position by making use of the testimony of Tacitus (*Annals* 1.16, 28), who reported that a lunar eclipse had prevented the Pannonian legions from mutiny

[12] Crusius, *Liber de epochis*, 134: "Hac ergo posita sententia, apparet epochen nativitatis Dominicae, a Dionysio praescriptam, quae passionem Christi annis 32. & mensibus fere tribus antecedit, sacris historiis repugnare: non enim attingit annos Herodis, sed mortem eius sequitur, intervallo fere 9. mensium." See also Temporarius, *Chronologicarum demonstrationum libri tres*, 166–68; Pie, *An Houreglasse*, 70–87.

[13] Mercator, *Chronologia*, 142: "Numeros tamen annorum Christi secundum receptam consuetudinem posui, ne quid historiarum lector turbetur, atque ut commodius id facerem annum passionis Christi innumeratum transsilui."

[14] Crusius, *Liber de epochis*, 136–37: "Cum enim depraehenderim veram epochen incarnationis Christi nonnihil differre, ab epocha nativitatis ipsius Dionysiana, quae in ore omnium celebratissima est, procul dubio futurum esset, ut si alterutram ex iis pro basi totius Chronologiae poneremus, confunderentur animi legentium, & historias uni epochae assignatas, ad alteram transferrent. Quod incommodum vitari potest, si tempus passionis Dominicae ponimus pro epocha principali, quae primo statis aspectu lectore monet, ut cogitet de forma collationis historiarum, ab hac epocha pendentium, ad annos Christi nati, vel iuxta nostram, vel Dionysii computationem."

[15] Heinrich Bünting, *Chronologia* (Zerbst, 1590). See Grafton, *Scaliger*, 2:137–38; Grafton, "Some Uses," 214–17.

shortly after Augustus's death. As his calculations showed, this eclipse had taken place on 26 September, AD 14 (Oppolzer no. 1884), in what had to be the first year of Tiberius's reign. This in turn meant that the 30th year of Jesus's life, which Luke the evangelist linked to the 15th year of Tiberius, overlapped with years 28 and 29 of the Dionysiac era. The resulting correlation clearly showed that Jesus had been born one year earlier than the followers of Dionysius Exiguus believed when they dated the nativity to 25 December, 1 BC. The obvious reason for this error was that Dionysius had not inquired the beginning of the years of Christ "by astronomical calculation, but by use of uncertain and contradictory annals of the Roman consuls."[16] Clearly, an astronomical approach to fixing epochs would have served the ancient chronographers well.

As the example of Apianus shows, however, being able to calculate was not everything. A good chronologer also had to know how to pick the right eclipse. A prominent case in point was Gerhard Mercator, who had tried to fix the death of Augustus by relying on the same solar eclipse mentioned by Eusebius and Cassius Dio (*Roman History* 56.29), which had previously caught the attention of Peter de Rivo and Paul of Middelburg. As Bünting correctly pointed out, no solar eclipse had been properly visible over Roman territory during the relevant period, which he took to stretch from AD 11 to 16.[17] Moreover, he seems to have been the first chronologer to notice that 3 April, AD 33, the putative day of the crucifixion, had also been host to a partial lunar eclipse (Oppolzer no. 1914), which had occurred over Jerusalem in the evening of that day, thereby following the miraculous solar eclipse that

[16] Bünting, *Chronologia*, 207r: "Dionysius Exiguus abbas Romanus, qui primus annos ab incarnatione Christi computare instituit, cum retrorsum per 532 annos, initium annorum Christi, non calculo astronomico, sed ex incertis & discrepantibus annalibus Romanorum consulum inquireret, integro anno a vera Christi nativitate aberravit. Retinenda tamen est, quamvis vitiosa sit, vulgaris illa annorum Domini numerandi ratio, non enim absque magna omnium annalium & historiarum confusione, emendari potest." For Bünting's eclipse calculation, see ibid., 217v–18r.

[17] Ibid., 207r: "Gerardus Mercator autem, uno anno a nostra supputatione discedit, quinque etiam annos ministerii Christi ponit, quod hactenus inauditum fuit. Neglecta enim hac Lunae Eclipsi, quam Cornelius Tacitus annotavit, aliam Solis defectionem ex Dione Cassio & Eusebio citat, quae centum paulo amplius diebus ante mortem Caesaris Augusti facta sit. Sed cum diligenter singulos annos ab undecimo aetatis Christi, usque ad decimum sextum eiusdem, calculo Prutenico examinarem, nullam inveniebam eclipsin Solis, quae in Romano aut nostro hemisphaerio fieri potuisset." See Mercator, *Chronologia*, a3v.

had supposedly accompanied Christ's death on the cross.[18] As a result, Bünting's eclipse-based chronology clearly showed that Mercator had been correct in ante-dating Christ's birth in relation to the Dionysiac era, but obviously wrong in choosing 2 April, AD 34, as a date for the Passion, especially since this date was nowhere close to the astronomical full moon.[19] In exchange for all the worries about the reliability of astronomical calculations of the lunar orbit, which his compatriots Mercator and Krentzheim had raised, Bünting offered a mild taunt:

> The unlearned calculators who do not properly understand the methods of astronomical reckoning can be mistaken, but the periodic motions of the sun and the moon cannot ever go astray. To the inept and unlearned calculators I therefore wish well and instead prefer to rely on the absolutely certain foundations of Ptolemy and all the other highly competent mathematicians.[20]

While these astronomical foundations served to corroborate the Baconian crucifixion date of 3 April, AD 33, they also indicated that Jesus had been born some time before the year 1 BC, when Herod the Great was still among the living. This shift of the nativity raised some complicated new questions concerning the age of Jesus at the time of his baptism, which according to the Greek text of Luke 3:23 was "about 30 years beginning" (ὡσεὶ ἐτῶν τριάκοντα ἀρχόμενος). Earlier in the same chapter, Luke linked the activity of John the Baptist to the 15th year of the reign of Tiberius (3:1), which was the only precise bit of chronological information on the life of Jesus to be found in the entire New Testament. Up until Bünting's time, the dominant stance among Western chronologers had been to assume that the passage meant to state that Jesus had just turned 29, i.e. that he began the course of the 30th year of his life when he was baptized. Assuming a nativity on 25 December, 1 BC, this view would have neatly corresponded with a baptismal date on 6 January, AD 30, which had been influentially chosen by both Johannes Lucidus Samotheus and Onofrio Panvinio.[21] The

---

[18] Bünting, *Chronologia*, 234v, 237r–38r.

[19] Ibid., 238v: "Omnino falsa igitur est, & admodum confusa, illa Geradi Mercatoris supputatio, praesertim cum Christum ad quintum usque annum Ministerii sui pervenisse demonstrare conetur, non tamen satis firmis argumentis."

[20] Ibid.: "Indocti quidem computatores, qui calculum Astronomicum non recte intelligunt, errare possunt, sed Solis ac Lunae periodicae conversiones nunquam errare poterunt. Valeant ergo inepti & indocti computatores, nos Ptolemaei & reliquorum doctissimorum Mathematicorum certissimis nitimur fundamentis."

[21] Johannes Lucidus Samotheus, *Opusculum*, 191v; Panvinio, *Fastorum libri V*, 306.

Genevan preacher and professor of philosophy Matthaeus Beroaldus, who published a chronological handbook in 1575, called for an even more stringent version of this interpretation when he used the duration of the public ministry to shift Jesus's nativity from midwinter to autumn. Since Eusebius's 'prophetic' span of 'three-and-a-half years', if taken literally, implied a baptism at the time of Sukkot (or Tabernacles) in September/October (six months before the crucifixion at Passover), the birth of Christ in 1 BC had to be likewise dated to the time of the autumnal equinox, if it was assumed that Jesus had received baptism at the hands of John very close to his 29th birthday.[22] As Bünting clearly perceived, this interpretation imputed a weight of chronological precision on Luke's testimony, which the text could impossibly sustain. A solution more in tune with the written evidence was to assume that Jesus simply 'began to be 30 years', i.e. was approaching the end of his 30th year. This assumption made it possible to date the baptism of Jesus to the time of Yom Kippur in autumn of AD 29, similar to what Beroaldus had suggested, whilst retaining a birth on 25 December, 2 BC. In Bünting's scheme of things, Jesus's 30th birthday thus still occurred in the 15th year of Tiberius, if the latter's accession to the imperial throne was post-dated to 1 January, AD 15. Moreover, the Passovers of the ensuing three years (AD 30, 31, and 32) could all be found recorded in the Gospel of John (2:13, 5:1, 6:4), with the Passover of AD 33 (11:55) being the one during which the crucifixion had taken place.[23]

As these various examples suggest, a closer look at the chronological literature published during the second half of the sixteenth century clearly invalidates the notion that it was Joseph Justus Scaliger who singlehandedly created the discipline of historical chronology through a series of unprecedented innovations. Quite to the contrary, two of the most crucial developments in the field of chronology—the invention of eclipse-dating and the increasing refinement of chronological considerations with regard to Christ's life, which lead to the abandonment of the Dionysiac era as an indicator of his birth

---

[22] See Beroaldus, *Chronicum*, 186–97, and my "From Sukkot to Saturnalia."

[23] Bünting, *Chronologia*, 207v: "Sed non satisfaciat nobis haec Beroaldi demonstratio, Lucas enim non ita ad amussim tempus aetatis Christi definit, ut certum in anno diem inde colligere possimus.... Ipse Iesus erat quasi incipiens annorum triginta, hoc est, erat ferme annorum triginta, nondum quidem completorum, sed qui mox complendi essent. *Jesus war ohn gefehr von dreissig jaren/ also/ das er bald 30 jar wolt alt werden/* nondum tamen eos compleverat." See also ibid., 207r–13v, 221r–32v.

date—had taken place without any decisive input from the renowned French philologist. In fact, when Scaliger tackled the chronology of the New Testament in the first edition of his *Opus novum de emendatione temporum* (1583), he could in many ways only expand on what his German colleagues had already worked out. Accordingly, he concurred with Crusius that Jesus's birth had to be moved to a date before the lunar eclipse of 9 January, 1 BC, while deploring the error in which Christianity had persevered for so many centuries.[24] Yet instead of fixing the dates of the baptism and crucifixion along the same lines as Crusius, he chose to side with Onofrio Panvinio, who had put the baptism in AD 30 and the crucifixion in AD 34. As a consequence, Jesus's age at his death had to turn out to be higher than 34 years. In order to accommodate this result, Scaliger reinterpreted the crucial passage in Luke 3:23 by claiming that the evangelist's Greek prose contained a previously unnoticed Hebraism, which allowed for the following rendering of Luke's words: "Then Jesus was thirty years old, beginning" (*tunc Iesus erat triginta annorum, incipiens*), meaning that he had already started the 31st year of his life at the time of his baptism.[25] Since Scaliger thought that Jesus had died at the age of 34 in the spring of AD 34, this new reading implied that his baptism had taken place in early AD 30, i.e. shortly after his 30th birthday. This result, however, did not square well with conventional Roman chronology, since Tiberius's 15th year corresponded to AD 28/29. In order to salvage his date for the baptism, Scaliger somewhat arbitrarily declared that Luke had followed a Jewish custom of post-dating regnal years, according to which the years of Roman emperors were only counted from the 1 Nisan that followed their accession. According to this scenario, Tiberius's 15th year had begun in AD 29 and only ended in the spring of AD 30.[26] Similar to Gerhard Mercator, who seems to have had some influence on his arguments, Scaliger also proposed a reconstruction

---

[24] Scaliger, *De emendatione* (1583), 255: "Christus natus est ante obitum Herodis. Ergo natus ante annum primum cycli Lunaris ineuntem. a cuius natali ad hunc annum, quo haec scribebam, annus agitur millesimus quingentesimus octagesimus tertius, quem tamen epilogismus Dionysii Exigui millesimum quingentesimum octagesimum secundum duntxat putat. Tollendus ergo error in posterum: neque vero seccensendum nobis, qui mendum in ratione temporum aperuimus, sed danda opera, ne hostibus Reip. Christianae Iudaeis, & Muhammedistis ludibrium debeamus, qui per tot saecula annum natalis eius ignoravimus, cuius ope & beneficio nos ad gratiam & vitam aeternam renati sumus."

[25] Ibid., 255–56. For further details, see Grafton, *Scaliger*, 2:313–15.

[26] Scaliger, *De emendatione* (1583), 239–40.

of the chronology of Jesus's life which extended his public ministry from three to four complete years, encompassing five Passovers. His main innovation consisted in locating the missing fourth year between chapters 5 and 6 of John's Gospel, assigning to it the events related in chapters 4 to 14 of the Gospel of Matthew.[27]

Joseph Scaliger's take on the dates of Jesus's life, to which he dedicated half of book 6 of his great *Opus de emendatione temporum*, entailed a number of contrived elements that did not escape the criticism of his peers. As the French court historian Nicolas Vignier was to point out only a few years later, Scaliger's approach had been tainted by his decision to date the Passion to 23 April, AD 34, rather than the by now widely accepted 3 April, AD 33. This completely novel calendar date was partly the outcome of Scaliger's research into the history of the Jewish calendar, which brought to the surface a previously unknown 'Palestinian' version of the *molad baharad*, i.e. the epoch of the Jewish world era. Where the European Jews, whose calendar was known from sources such as Sebastian Münster's *Kalendarium Hebraicum*, began to count their years from a conjunction on Monday, 7 October, 3761 BC, the Jews in Palestine had allegedly preserved a more ancient system, which started one year and 20 days later, on 27 October, 3760 BC. Basing himself on some hints in the works of Josephus, Scaliger furthermore assumed that this latter system had already been in use at the time of Jesus Christ, which made it necessary to recalculate the date of the Passion on its basis. As a result, the calendrical coincidence of Friday and 14 Nisan in 3793 JE, which in the mind of Paul of Burgos and many others had supported 3 April, AD 33, led to a crucifixion on 23 April, AD 34, in Scaliger's scheme.[28] Nicolas Vignier was by no means the only observer to remain thoroughly unconvinced by Scaliger's appeal to a 'Palestinian' epoch, which—despite its supposedly ancient origins—had been unknown to all

[27] Ibid., 257–63. See Henk Jan de Jonge, "Joseph Scaliger's Historical Criticism of the New Testament," *Novum Testamentum* 38 (1996): 179–80.

[28] Scaliger, *De emendatione* (1583), 262–63; Vignier, *La Bibliothèque*, 1, i5r: "A toutes lesquelles ie respon en un mot, que puis qu'ils confessent que ce comput là ne revient pas avec celuy que les Iuifs de l'Europe ensuivent auiourd'huy, estant l'un plus tardif en la supputation de leurs annees depuis la creation du monde que l'autre; ie ne suis pas tenu de croire que les Iuifs que precederent la ruine derniere de l'estat de la Iudee sous Vespasian, aient conduit le fait de leur Calendrier selon les mesmes Epoches, & de mesme façon que ceux qui sont venus depuis." See also ibid., 1:696, 698–99.

previous chronologers. Between 1590 and 1594, the German orientalist and astronomer Jacob Christmann (1554–1613), who taught Hebrew at the University of Heidelberg, published no less than four books in which he vociferously impugned both Scaliger's claims concerning the Jewish calendar and the false crucifixion date that had resulted from them. As Christmann was able to show, Scaliger's 'Palestinian' epoch had been a product of his misreading of an obscure Hebrew calendar text, which had been published in 1574/75 by Rabbi Uri ben Simeon and which Christmann himself later translated into Latin.[29]

Scaliger eventually realized his mistakes, which he silently emended in the celebrated second edition of *De emendatione temporum*, printed in Leiden in 1598. Not only did the new edition contain a vastly improved discussion of the history of the Jewish calendar, which was to remain influential for centuries to come, but Scaliger restored the date of Christ's Passion to 3 April, AD 33, whilst also pushing the birth date of Jesus further into the past, to 3 BC. Scaliger's improved *emendatio*, a third edition of which appeared in 1629, along with the publication of his monumental *Thesaurus temporum* (1606^{1}, 1658^{2}), cemented his reputation as the 'founding father' of historical chronology, a discipline which flourished greatly during the early seventeenth century and would later have its sporadic flare-ups during the heyday of German *Geschichtswissenschaft*. At the beginning of the nineteenth century, specialists such as Ludwig Ideler still consulted Scaliger's books at every turn, respectfully referring to his opinions on a variety of subjects, including the history of the Jewish calendar.[30] The only chronologer to achieve a comparable status in the eyes of his contemporaries was the Jesuit Dionysius Petavius (1583–1652), whose *Opus de doctrina temporum* (1627) was written with the express intention of criticizing and improving upon Scaliger's work. On the Protestant side, further noteworthy contributions to technical chronology were made by Seth Calvisius (1556–1615), cantor of the Thomas school in

[29] Jacob Christmann, *Muhamedis Alfragani Arabis chronologica et astronomica elementa* (Frankfurt, 1590); Christmann, *Epistola chronologica ad clarissimum virum Iustum Lipsium* (Heidelberg, 1591); Christmann, *Disputatio de anno, mense et die passionis Dominicae* (Frankfurt, 1593); Christmann, *Calendarium Palaestinorum et Universorum Iudaeorum* (Frankfurt, 1594). For further details, see Grafton, *Scaliger*, 2:188–92, and my "A Sixteenth-Century Debate on the Jewish Calendar: Jacob Christmann and Joseph Justus Scaliger," *Jewish Quarterly Review* (forthcoming).

[30] Ludwig Ideler, *Handbuch der mathematischen und technischen Chronologie*, 2 vols. (Berlin: Rücker, 1825–26), 1:576–83.

Leipzig, whose *Opus chronologicum* (1605), praised by Scaliger, rested on a foundation of no fewer than 271 calculated eclipses, and by the astronomer Johannes Kepler (1571–1630). It was Kepler who made a further decisive step in the growing dissociation of the historical birth year of Jesus from the era of Dionysius Exiguus by realizing that the lunar eclipse mentioned by Flavius Josephus in connection with the last days of Herod had not been that of 9 January, 1 BC, but a previous one on 13 March, 4 BC (Oppolzer no. 1856).[31]

Compared to the results achieved by Kepler, subsequent centuries of research would yield only moderate progress, despite the fact that books on the chronology of Jesus's life kept appearing at an astounding rate. In 1649, the Danish astronomer Villum Lange (1624–82) published a book entitled *De annis Christi*, in which he tackled a great number of technical questions relevant to historical calendars and biblical chronology. Amongst other things, he paid close attention to the fact that the Jews of antiquity had not used any pre-calculated calendar, but based the beginnings of their months on the visibility of the new moon crescent. In order to gain a better understanding of the criteria used by first-century Jewry, Lange relied on a collection of Talmudic and Karaite excerpts on the calendar, published in 1644 by the Hebraist John Selden. One of the insights that seemingly could be drawn from Selden's collection was that the Karaite rules for new moon visibility occasionally allowed for the month to start one day earlier than in the Rabbanite calendar. Lange, who identified the Karaites with the first-century Sadducees, realized that he could use this information to propose a new and possibly more convincing harmonization of the Gospel accounts with regard to the date of the Last Supper. Assuming that Jesus and his disciples had used the Sadducean calendar, he argued that they had begun the month of Nisan (and hence celebrated Passover) one day earlier than the followers of mainstream Pharisean Judaism. His astronomical attempt to determine the date of 15 Nisan in the first-century Sadducean calendar

[31] Pietro Di Rosa, "Denis Petau e la cronologia," *Archivum historicum Societatis Iesu* 29 (1960): 3–54; Detlef Döring, "Sethus Calvisius als Chronolge: Studien zur Entwicklung der Geschichtswissenschaft an der Leipziger Universität am Beginn des 17. Jahrhunderts," *Zeitschrift für Historische Forschung* 21 (1994): 171–202; Kepler, *De Iesu Christi... vero anno natalitio*, 21–23 = KGW 1:375–76; Kepler, *De vero anno*, 77–89 = KGW 5:58–66; Kepler, *Widerholter Ausführlicher Teutscher Bericht* (Strasbourg, 1613), 57–68 = KGW 5:163–69.

subsequently led him to propose the unusual crucifixion date of 2 April, AD 33, which was a Thursday rather than a Friday.[32]

Lange's *De annis Christi* had some influence on the chronological oeuvre of Isaac Newton (1643–1727), whose unpublished manuscript papers contain two sets of drafts on the date of Christ's Passion. Drawing partly from the astronomical visibility criteria described in Maimonides's treatise on the *Laws of the Sanctification of the New Moon*, Newton, at two separate occasions, made calculations for the evening of first visibility at the time of the crucifixion, which caused him to waver between 3 April, AD 33, and 23 April, AD 34, as possible Passion dates.[33] In a posthumously published dissertation *Of the Times of the Birth and Passion of Christ*, Newton finally opted for a crucifixion on 14 Nisan = 23 April, AD 34, the same date which had already been proposed with little success by Joseph Scaliger in 1583. One of his more original arguments in favour of this solution concerned the episode of the plucking of the ears in the Gospel of Luke (6:1). For Newton, the scene implied that the ears had already been ripe at the time of Passover, which hinted at a relatively late date for 14/15 Nisan in that particular year, i.e. in April rather than March. According to his calculations, such a late Passover had occurred in AD 32 on 14 April. Since he also believed that the crucifixion had taken place two years after the scene in Luke 6:1, everything seemed to point towards AD 34 as the historical year of the Passion of Christ.[34]

---

[32] John Selden, *De anno civili & calendario veteris ecclesiae seu Reipublicae Judaicae, dissertatio* (London, 1644); Villum Lange, *De annis Christi libri duo* (Leiden, 1649), 110–28, 392–415. Lange's work was heavily criticized by Hans Wandal in his *De feria Passionis & triduo mortis Domini & Servatoris nostri Jesu Christi* (Leipzig, 1651). On Selden's book, see G. J. Toomer, *John Selden*, 2 vols. (Oxford: Oxford University Press, 2009), 2:626–43.

[33] MS Jerusalem, Jewish National und University Library, Yahuda 24E. See Ari Belenkiy and Eduardo Vila Echagüe, "Stirring Astronomy into Theology: Sir Isaac Newton on the Date of the Passion of Christ," *arXiv*, http://arxiv.org/abs/0810.4358v1.

[34] Isaac Newton, *Observations upon the prophecies of Daniel, and the Apocalypse of St. John* (London, 1733), 167–68: "Thus there remain only the years 33 and 34 to be considered; and the year 33 I exclude by this argument: In the Passover two years before the Passion, when Christ went thro' the corn, and his disciples pluckt the ears, and rubbed them with their hands to eat; this ripeness of the corn shews that the Passover then fell late: and so did the Passover A.C. 32, April 14. But the Passover A.C. 31, March 28th, fell very early. It was not therefore two years after the year 31, but two years after 32 that Christ suffered." On the background, see John P. Pratt, "Newton's Date for the Crucifixion," *Quarterly Journal of the Royal Astronomical Society* 32 (1991): 301–4; Frank E. Manuel, *Isaac Newton Historian* (Cambridge, MA: Harvard University Press, 1963).

While many of the results achieved by early modern chronologers have remained remarkably valid unto this day, Newton's confidence in the transparency and scrutability of biblical chronology, and in particular with regard to the exact dates of the life of Jesus of Nazareth, is not easily shared by those among his successors who work from within the paradigm of modern biblical criticism. Looking back at the deluge of publications which had been produced on the subject by the early twentieth century, Friedrich Karl Ginzel, author of a still-indispensable chronological *Handbuch*, noted both the exhaustiveness with which the subject had been treated and the doubtfulness of ever coming to a satisfactory solution given the sources and tools available.[35] Ginzel's assessment remains valid, even though the stream of speculations concerning the dates of Christ's birth and death has continued unabated, as witnessed by the dedicated work of scholars such as Jack Finegan, Paul Maier, Jerry Vardaman, and the many other participants in two chronological 'Nativity Conferences', which were hosted by Mississippi State University in 1983 and 1992.[36]

That being said, it has not been the purpose of the present study to add any new ideas to the ever-growing canon of theories concerning the chronology of Jesus's life, but rather to see how the early history of these theories has interacted with the development of chronological techniques and ideas. In Chapter 2, I have introduced the term 'computistical chronography' to refer to a specific method in Christian chronological scholarship, which emerged at the beginning of the third century: the retrospective application of Easter cycles to investigate past dates and years. Our most important early witness for this method is the 'canon of Hippolytus', which has come down to us in epigraphic form. As we have seen, the author of this chronological source used a 112-year Easter cycle to generate Julian calendar dates for all the Passovers found in the Bible. These included the birth and death of Jesus Christ, which he respectively assigned to 2 April, 2 BC,

[35] Ginzel, *Handbuch*, 3:185: "Es mag gleich gesagt werden, daß all der Aufwand von Mühe und Scharfsinn, welcher auf jene Probleme verwendet worden ist, zu einer Sicherung der Wahrheit nicht geführt hat. Die Lösung der Fragen wird auch nicht möglich werden, solange die Forschung nur über jene Hilfsmittel verfügt, die ihr gegenwärtig zu Gebote stehen." For a list of pre-twentieth century works, beginning with Johannes Kepler, see ibid., 3:382.

[36] Jerry Vardaman and Edwin M. Yamauchi, eds., *Chronos, Kairos, Christos* (Winona Lake, IN: Eisenbrauns, 1989); Jerry Vardaman, ed., *Chronos, Kairos, Christos II* (Macon, GA: Mercer University Press, 1998).

and 25 March, AD 29, dates which neatly agreed with the assumptions current among early Christian exegetes. A comparison of the Hippolytan canon with other late antique chronological texts has made it plausible that the widespread Western tradition of dating the Passion on 25 March, AD 29, did not simply emerge "from symbological, allegorical, or typical interpretations,"[37] but was crucially influenced by technical considerations and calculations, based on primitive lunisolar cycles. The importance of computistical chronography, as I have construed it in the present study, lies chiefly in the fact that it exerted a strong influence on the development of some of the chronological traditions that we encounter in late antique Christian liturgy and chronography—a fact still underappreciated even though the development of these traditions cannot be adequately understood if the role of the computus is ignored.

An appreciation of the computistical mechanisms involved should also make us susceptible for the profound changes that awaited Latin chronography, once the 19-year cycle of the Alexandrians made its way to Rome and the West, where it soon collided with existing chronological traditions. In Chapter 3, I have argued that the widespread adoption of the 19-year cycle eventually led to a rift in the fabric of Christian chronology in the West, which is closely linked to the spread of the *Annus Domini* as a means of identifying years in Easter tables. Thanks to the chronological doctrines of Dionysius Exiguus and the Venerable Bede, two pivotal points on the timeline of salvation—the creation of the world and the crucifixion of Christ—effectively ceased to be properly datable events. What had once been shining cornerstones of Christian chronology had dissolved into a state of chronological confusion, whose symptoms can still be detected in ninth-century sources such as the Carolingian 'Seven-Book-Computus' and Claudius of Turin's world chronicle. If I have dedicated considerable space in the present book to pointing out the existence of this early medieval crisis of computistical chronography, then this is because I think that its discovery amounts to more than just a marginal contribution to Carolingian intellectual history. The real importance of this crisis becomes apparent once we put it into wider historical perspective: ever since the work of Hippolytus in the early third century, the application of the Easter computus to chronographic problems had been an essential

[37] Jones, *Bedae Opera*, 6.

element of Christian chronological thought, which influenced the choice of dates and years for crucial events: the creation of the world as well as the incarnation, birth, and crucifixion of Jesus. At first glance, the westward expansion of the Alexandrian 19-year cycle should have led to a stabilization or even unification of chronological traditions in East and West. In the Byzantine world, for instance, the search for a viable chronological structure of world history, based on the 19-year cycle, had by the seventh century led to a remarkably coherent system, which found its most palpable expression in the incipient use of the Byzantine world era (starting from 5509/8 BC) and with AD 31 as the year of Christ's death and resurrection. In Egypt and Ethiopia, on the other hand, chronographers and Churches preferred to work with an even older system, which set the world's creation in 5493/92 BC and the crucifixion in AD 42.

It is quite conceivable that a similar state of stability could have been reached in Rome and the West, if the Victorian Easter table had continued to enjoy papal sanction and had managed to emerge as the standard of reckoning in the kingdom of the Franks (we might then be designating the present year 2011 as AP or *Annus Passionis* 1984). Owing to the aforementioned crisis, however, the development of Christian chronology in the Latin-speaking world instead took a distinctly different turn. Beginning with Abbo of Fleury around the year 1000, a new generation of computists felt the need to react to the ongoing state of disarray and began to draw some of the consequences that their predecessors had shunned. Embracing the natural rhythms of the heavenly luminaries and the words of the Gospels as their only sure guides, they boldly put aside the eras of Bede and Dionysius and undertook to restore coherence to chronology by re-anchoring the events of the world's creation and Christ's birth and death. In the hands of these 'critical computists' of the eleventh and twelfth centuries, the Easter computus turned into a razor-sharp tool with which the weightiest chronographic traditions could be mercilessly cut down. Chroniclers such as Marianus Scottus perfected this approach by showing how chronological shifts could be made historically plausible, for instance by strategically inserting additional years into the succession of the Roman emperors. Yet for all the ingenuity they invested into their systems, the critical computists failed to make a lasting impact on the chronological habits of their contemporaries, who were simply not prepared to give up their received count of the years in favour of a novel system that bluntly contradicted the timeline of Roman history as it had been inherited from Eusebius, Orosius, and others. More importantly, by the eleventh century the Dionysiac era had established itself as the preferred instrument of dating in annals, charters, inscriptions, etc., and no alternative incarnation era could hope to replace it merely by the force of computistical reasoning.

At the same time, however, and despite their outward lack of success, the endeavours of the critical computists served to heighten appreciation in their readers for the importance of astronomical cycles as a check upon chronographic data. This new sensibility could become the basis for serious methodological progress: if the search for Christ's vital dates continued to fail despite all the care invested into fixing them by computistical means, one could be tempted to question the reliability of the computus when it came to squaring the relevant data for the first century AD. Under these circumstances, medieval computists were all the more prepared to soak up the wave of new astronomical data that hit Western Europe during the twelfth century, owing to the import of translated Arabic texts from Iberia. By the end of the century, computistical experts such as Reinher of Paderborn clearly perceived that the ecclesiastical calendar was hopelessly out of tune with the natural phenomena and that only its reform could save Christianity from embarrassing ridicule by Muslim and Jewish onlookers, whose own methods of time reckoning seemed far more sophisticated. This new scientific awareness naturally also affected the long-standing preoccupation with the dates of Jesus Christ's life. After centuries of unsuccessful friction between computus and chronography, scholars came to the realization that it was not only possible, but essential to tackle these long-standing problems with the new and improved tools that were provided by Arab astronomy. Not only did the computational basis change dramatically, but the underlying historical problem, that of the Jewish calendar during the first century, was treated with newly found rigour and insight. Where the critical computists had more or less equated the lunisolar calendar of the Hebrews with that of the Alexandrian church, scholars gradually began to treat the Jewish calendar as an entity of its own, whose history and specifics had to be studied carefully if one hoped to finally solve the obstinate problem of dating Christ's Passion. A decisive step was taken in about 1267, when the English Franciscan Roger Bacon sent his famous *Opus maius* to pope Clement IV. Not only did Bacon's knowledge of the Jewish calendar and his use of astronomical tables enable him to suggest a completely new date for the crucifixion—3 April, AD 33, a date which is in fact still widely accepted by scholars today—but he explicitly and programmatically acknowledged the importance of astronomy as a means of reconstructing the timeline of the distant past. Basing himself on the work of Claudius Ptolemy, Bacon singled out eclipses of the sun and the moon as the most valuable astronomical benchmarks, which could be used to purge chronographic tradition of its many errors and ambiguities or, as Bacon himself put it, "to reach certainty about the times" (*certificare de temporibus*).

Similar to the case of the calendar, which was eventually reformed by Pope Gregory XIII in 1582, it took Western scholarship some three centuries to fully implement Bacon's ambitious program of an eclipse-based chronology of ancient history. While fifteenth-century discussions of the Passion date, from Alfonso Tostado to Paul of Middelburg and his opponent Peter de Rivo, show an ever-increasing level of attention to detail and technical sophistication, it was only in the sixteenth century that the potential of astronomical dating for the solution of chronological problems was fully realized. After 1578, chronologers all over Europe could turn to Paul Crusius's *Liber de epochis*, which provided dated eclipses to establish the major chronological benchmarks of Greek and Roman history. The timeline that emerged from these calculations was already very recognizably that of our contemporary history books and Crusius did in fact serve as a basis for Joseph Justus Scaliger's *De emendatione temporum* (1583), which has often been described as the founding work of scientific chronology. Like many scholars in his day, Scaliger was not inclined to credit medieval authors with much importance for the discipline that he himself allegedly invented. But even a superficial look at his treatment of the Passion date reveals startling continuities with several centuries' worth of medieval discussion on the same topic. Upon closer examination, the common notion of scientific chronology as a typical invention of the Renaissance is thus in need of serious qualification. As it turns out, one of the discipline's central ideas, namely that of applying astronomical calculations to the dating of historical events, had been a part of Christian scholarship ever since the third century. Not only can the roots of an important component of technical chronology be thus found in pre-modern Christianity, but it also becomes clear that the methodological developments described in this present study hinged upon a number of contingent factors, some of which—such as the early medieval crisis of computistical chronography—were specific to the Latin-speaking Christian world. These factors may thus go some way towards explaining why the turn towards scientific chronology happened first and foremost within the context of Western civilization.

In the present study, I could do little more than sketch the bare outlines of the medieval contribution to historical chronology, a contribution which was richer and more variegated than the present narrow focus on the life of Jesus would suggest. Much work remains to be done and many alleys of medieval scholarship are yet to be explored until anything like a comprehensive history of this field can be written. It is my contention that such a history could provide valuable insights into the minds of those medieval writers, who, with their relentless search for chronological accuracy—their "obsession de la date," as Bernard Guenée put it—have contributed to shaping the timeline that we still work with today.[38] Dealing with this 'obsession' in greater depth may give weight to an ongoing project in the historiography of science and intellectual culture: that of replacing a traditional narrative, according to which we became 'modern' only because we managed to overcome the period of the Middle Ages, with a different and possibly sounder one, which suggests that we did so, because we inherited it and successfully built upon its legacy.

[38] Guenée, *Histoire*, 147.

APPENDIX

PROMINENT ATTEMPTS TO DATE CHRIST'S BIRTH AND DEATH (200–1600)

Note: 25 December is here given as the default date even in cases where no specific calendar date for the nativity is mentioned in the source text.

Author	Birth	Lifespan	Passion
Hippolytus	02 Apr 2 BC	30y	25 Mar 29
De pascha computus	28 Mar 4 BC	31y	09 Apr 28
Victorius of Aquitaine	—	—	26 Mar 28
Bede the Venerable	25 Dec 1 BC	33y 3m	25 Mar 34
Claudius of Turn	25 Dec 1 BC	33y 3m	21 Mar 34
Abbo of Fleury	25 Dec 21 BC	32y 3m	25 Mar 12
Heriger of Lobbes	25 Dec AD 9	32y 3m	23 Mar 42
Gerland	25 Dec AD 8	33y 3m	23 Mar 42
Marianus Scottus	25 Dec 22 BC	33y 3m	25 Mar 12
Heimo of Bamberg	25 Dec 33 BC	33y 3m	25 Mar 01
Reinher of Paderborn	25 Dec AD 1	32y 3m	26 Mar 34
Albert the Great	—	—	25 Mar 34
Roger Bacon	25 Dec AD 1	31y 3m	03 Apr 33
Robert of Leicester	—	—	23 Mar 42
Jean des Murs	25 Dec AD 1	31y 3m	03 Apr 33
Alfonso Tostado	25 Dec 1 BC	32y 3m	03 Apr 33
Paul of Middelburg	25 Dec AD 2	33y 3m	30 Mar 36
Joh. Lucidus Samotheus	25 Dec 1 BC	32y 3m	03 Apr 33
Christian Massaeus	25 Dec AD 1	33y 3m	25 Mar 35
Onofrio Panvinio	25 Dec 1 BC	33y 3m	26 Mar 34
Gerhard Mercator	25 Dec 2 BC	34y 3m	02 Apr 34
Matthaeus Beroaldus	Sep/Oct 1 BC	32y 6m	03 Apr 33
Paul Crusius	25 Dec 2 BC	33y 3m	03 Apr 33
J. J. Scaliger (1583)	25 Dec 2 BC	34y 3m	23 Apr 34
J. J. Scaliger (1598)	Sep/Oct 3 BC	34y 6m	03 Apr 33

BIBLIOGRAPHY

Manuscripts

[Astronomical tables for Novara]. "Composui hanc tabulam ad inveniendum diem." Munich, Bayerische Staatsbibliothek, Clm 17703, 27v–31r.

Bacon, Roger. "Compotus." London, British Library, Royal 7.F.VIII, 99r–163v.

——. "Opus majus." London, British Library, Cotton, Tiberius C.5, 49r–119r.

Claudius of Turin. "Chronicle." Madrid, Biblioteca Nacional, 9605, 103r–11r, 112r–16v; Monza, Biblioteca Capitolare, c-9/69, 66ra–83vb; Paris, Bibliothèque Nationale, lat. 5001, 1r–8v.

"Compotus Constabularii." London, British Library, Cotton, Vitellius A.XII, 87ra–97va.

"De comparatione epactarum Dionysii et Victorii." Cologne, Erzbischöfliche Diözesan- und Dombibliothek, 83^{II}, 176v–78r.

Gerland [Garlandus]. "Compotus." Oxford, Bodleian Library, Auct. F.1.9, 12vb–26v.

Jean des Murs. "Autores kalendari…" Vienna, Österreichische Nationalbibliothek, CVP 5273, 91r–102r; CVP 5292, 199r–209v.

——. "Sermo de regulis computistarum." Brussels, Bibliothèque Royale, 1022–1047, 40ra–vb, 203r–4v; London, British Library, Royal 12.C.XVII, 213ra–vb.

"Liber ysagogarum Alchorismi." Munich, Bayerische Staatsbibliothek, Clm 13021, 27r–31v; Paris, Bibliothèque Nationale, lat. 16208, 67ra–71ra.

Reinher of Paderborn. "Compotus emendatus." Admont, Stiftsbibliothek, 442, 38r–51v; Hannover, Niedersächsische Landesbibliothek, IV 373, 3v–10v; Leiden, Universiteitsbibliotheek, B. P. L. 191 E, 129r–40r; Oxford, Bodleian Library, Can. Misc. 561, 82r–95r; Prague, Národní Knihovna, XIII.F.8, 105r–27r.

Robert of Leicester. "De compoto Hebraeorum." Erfurt, Universitätsbibliothek, Amploniana, Quart. 361, 80rb–85rb.

Roger of Hereford. "Compotus." Cambridge, University Library, Kk.1.1., 222v–39r.

Walcher of Malvern. "Qui facultatem computandi eligant dicunt lunam." Oxford, Bodleian Library, Auct. F.1.9, 86ra–96ra.

——. "Sententia Petri Ebrei, cognomento Anphus, de dracone." Oxford, Boldeian Library, Auct. F.1.9, 96ra–99ra.

Zoestius, Hermann. "Calendarium hebraicum novum." Munich, Bayerische Staatsbibliothek, Clm 18470, 19v–27v; Clm 24868, 23v–32r.

——. "De fermento et azymo." Munich, Bayerische Staatsbibliothek, Clm 3564, 145ra–55ra; Clm 18536, 201r–18v.

——. "Phaselexis." Munich, Bayerische Staatsbibliothek, Clm 3564, 138ra–43brb; Clm 18470, 4r–19r; Clm 24868, 1r–23v.

Early Modern Printed Books

Apianus, Petrus. *Astronomicum Caesareum*. Ingolstadt, 1540.

Bardi, Girolamo. *Chronologia universale*. 4 vols. Venice, 1581.

Beroaldus, Matthaeus. *Chronicum, Scripturae Sacrae autoritate constitutum*. Geneva, 1575.

Bianchini, Francesco. *De Kalendario et cyclo Caesarisac De paschali canone S. Hippolyti martyris dissertationes duae*. Rome, 1703.

Bibliander, Theodor. *De ratione temporum, Christianis rebus & cognoscendis & explicandis accomodata, liber unus*. Basel, 1551.

——. *Temporum a condito mundo usque ad ultimam ipsius aetatem supputatio, partitioque exactior*. Basel, 1558.

Bünting, Heinrich. *Chronologia hoc est, omnium temporum et annorum series*. Zerbst, 1590.

Candidus, Pantaleon. *Tabulae Chronologicae, continentes seriem annorum mundi*. Strasbourg, 1597.

Cassini, Jacques. *Tables Astronomiques du soleil, de la lune, des planets, des étoiles fixes, et des satellites de Jupiter et de Saturne*. Paris, 1740.

Christmann, Jacob. *Muhamedis Alfragani Arabis chronologica et astronomica elementa*. Frankfurt, 1590.

——. *Epistola chronologica ad clarissimum virum Iustum Lipsium*. Heidelberg, 1591.

——. *Disputatio de anno, mense et die passionis Dominicae*. Frankfurt, 1593.

——. *Calendarium Palaestinorum et Universorum Iudaeorum, ad annos quadraginta supputatum*. Frankfurt, 1594.

Codomann, Lorenz. *Supputatio praeteritorum annorum mundi, et septuaginta Hebdomadarum Danielis*. Leipzig, 1572.

Copernicus, Nicholas. *De revolutionibus orbium coelestium, libri VI*. Nuremberg, 1543.

Crusius, Paul. *Liber de epochis seu aeris temporum et imperiorum omnium facultatum studiosis utilissimus*. Basel, 1578.

d'Ailly, Pierre. *Tractatus et sermones*. Strasbourg, 1490. Reprint, Frankfurt: Minerva, 1971.

de la Hire, Philippe. *Tabulae Astronomicae*. Paris, 1702.

Eusebius of Caesarea et al. *Chronicon divinum plane opus eruditissimorum autorum, repetitum ab ipso mundi initio, ad annum usque salutis MDXII*. Basel, 1529.

Foresti, Giacomo Filippo. *Supplementum chronicarum*. Venice, 1483.

Funck, Johann. *Chronologia, hoc est. Omnium temporum et annorum ab initio mundi usque ad resurrectionem domini nostri Iesu Christi, computatio*. Nuremberg, 1545.

Gaurico, Luca. *Calendarium ecclesiasticum novum*. Venice, 1552.

——. *De eclipsi solis miraculosa in Passione Domini observata*. Paris, 1553.

Glarean, Heinrich. *In omneis, quae quidem extant, T. Livii Patavinii clarissimi historici decadas annotationes*. Basel, 1540.

Haloander, Gregor. *Codicis domini Iustiniani sacratissimi principis... libri XII*. Nuremberg, 1530.

Hugh of St. Cher. *Opera omnia in universum Vetus & Novum Testamentum*. 8 vols. Lyons, 1669.

Johannes de Sacrobosco. *Libellus de Sphaera. Eiusdem autoris libellus, cuius titulus est Computus, eruditißimam anni & mensium descriptionem continens*. Edited by Philipp Melanchthon. Wittenberg, 1538.

John of Ashenden. *Summa astrologiae iudicialis de accidentibus mundi*. Venice, 1489.

Kepler, Johannes. *De Iesu Christi Servatoris Nostri vero anno natalitio*. Frankfurt, 1606.

——. *De stella nova in pede Serpentarii, et qui sub ejus exortum de novo iniit, trigono igneo*. Prague, 1606.

——. *Widerholter Ausführlicher Teutscher Bericht/ Das unser Herz und Hailand Jesus Christus nit nuhr ein Jahr vor dem anfang unserer heutiges tags gebreuchigen Jahrzahl geboren sey...sondern fünff gantzer Jahr*. Strasbourg, 1613.

——. *De vero anno quo aeternus Dei filius humanam naturam in utero benedictae Virginis Mariae assumpsit*. Frankfurt, 1614.

——. *Eclogae Chronicae*. Frankfurt, 1615.

——. *Tabulae Rudolphinae*. Ulm, 1627.

Krentzheim, Leonhard. *Chronologia. Das ist Gründtliche und fleissige JahrRechnung*. Görlitz, 1576.

Labbé, Philippe. *Novae Bibliothecae Manuscriptorum Librorum*. 2 vols. Paris, 1657.
Lange, Villum. *De annis Christi libri duo*. Leiden, 1649.
Lucidus Samotheus, Johannes. *Opusculum de emendationibus temporum ab orbe condito ad usque hanc aetatem nostram iuxta veram ac rectam chronographiam ex antiquis ac probatissimis authoribus*. Venice, 1537.
Luther, Martin. *Supputatio annorum mundi*. Wittenberg, 1541.
Mandel, Christoph. *Beweisung aus der Juden Gesatz*. Neuburg an der Donau, 1557.
Massaeus, Christian. *Chronicorum multiplicis historiae utriusque testamenti libri viginti*. Antwerp, 1540.
Maurolico, Francesco. *Opuscula mathematica*. Venice, 1575.
Mercator, Gerhard. *Chronologia. Hoc est temporum demonstratio exactissima, ab initio mundi, usque ad annum Domini MDLXVIIII*. Cologne, 1569.
——. *Evangelicae historiae quadripartia monas, sive Harmonia quatuor evangelistarum*. Duisburg, 1592.
More, John. *A Table from the beginning of the world to this day*. Cambridge, 1593.
Münster, Sebastian. *Kalendarium Hebraicum*. Basel, 1527.
——. תורת המשיח *Evangelium secundum Matthaeum in Lingua Hebraica, cum versione latina atque succinctis annotationibus*. Basel, 1537.
Newton, Isaac. *Observations upon the prophecies of Daniel, and the Apocalypse of St. John*. London, 1733.
Panvinio, Onofrio. *Fasti et triumphi Rom. a Romulo rege usque ad Carolum V. Caes. Aug*. Venice, 1557.
——. *Fastorum Libri V a Romulo Rege usque ad Imp. Caesarem Carolum V. Austrium Augustum. Eiusdem in Fastorum Libros Commentarii*. Venice, 1558.
——. *Chronicon Ecclesiasticum*. Louvain, 1573.
Paul of Middelburg. *Paulina de recta Paschae celebratione: et de die passionis domini nostri Iesu Christi*. Fossombrone, 1513.
Petavius, Dionysius. *Uranologion sive systema variorum authorum, qui de sphaera, ac sideribus, eorumque motibus Graece commentati sunt*. Paris, 1630.
——. *De doctrina temporum*. 3 vols. Antwerp, 1703.
Peter de Rivo. *Opus responsivum ad Epistolam apologeticam M. Pauli de Middelburgo de anno, die et feria dominicae passionis*. Louvain, 1488.
——. *Tercius tractatus de anno, die et feria dominice passionis atque resurrectionis*. Louvain, 1492.
Phrygio, Paul Constantin. *Chronicum regum regnorumque omnium catalogum, et perpetuum ab exordio mundi temporum, seculorumque seriem complectens*. Basel, 1534.
Pie, Thomas. *An Houreglasse contayning I: A Computation from the Beginning of Time to Christ by X Articles. II: A Confirmation of the same for the times controversed before Christ. As also that there wanteth a yeare after Christ, in the usuall Computation*. London, 1597.
Pitati, Pietro. *Compendium…super annua solaris, atque lunaris anni quantitate, Paschalis item solennitatis iuxta veters Ecclesiae canones recognitione, Romanique Calendarii instauratione, deque vero passionis dominicae die*. Verona, 1560.
——. *Verae Solaris atque lunaris anni quantatis aliarumque rerum ad calendarii Romani emendationem pertinentium…explicatio*. Basel, 1568.
Ptolemy, Claudius. *Almagestum*. Venice, 1515.
Regiomontanus, Johannes. *Calendarium*. Venice, 1476.
Scaliger, Joseph Justus. *Opus novum de emendatione temporum in octo libros tributum*. Paris, 1583.
——. *Hippolyti Episcopi Canon Paschalis*. Leiden, 1595.
——. *Opus de emendatione temporum: Castigatus et multis partibus auctius, ut novum videri possit*. Leiden, 1598.
Scaligerana, Thuana, Perroniana, Pithoeana, et Colomesiana. Edited by Pierre des Maizeaux. 2 vols. Amsterdam, 1740.

Scultetus, Alexander. *Chronographia sive annales omnium fere regum, principum, & potentatuum, ab orbe condito, usque ad hunc annum Domini MDXLV*. Rome, 1546.

Selden, John. *De anno civili & calendario veteris ecclesiae seu Reipublicae Judaicae, dissertatio*. London, 1644.

Smetius, Martin. *Inscriptionum antiquarum quae passim per Europam, liber*. Leiden, 1588.

Sturm, Wenceslaus. *Chronica: Von Anfangk der Welt biß auffs Jahr Christi unsers Erlösers MDXCVI*. Leipzig, 1596.

Temporarius, Johannes. *Chronologicarum demonstrationum libri tres* Frankfurt, 1596.

Textus biblie: Quintus pars huius operis in se continens glosam ordinariam cum expositione lyre literali et morali necnon additionibus ac Replicis. Basel, 1507.

Tostado, Alfonso. *Opera omnia*. 27 vols. Venice, 1728.

Vignier, Nicolas. *La Bibliothèque Historiale*. 3 vols. Paris, 1587.

Vincent of Beauvais. *Speculum quadruplex sive Speculum maius*. 4 vols. Douai, 1624. Reprint, Graz: Akademische Druck- u. Verlagsanstalt, 1964–65.

Vossius, Gerhard Johannes. *Dissertatio Gemina: Una De Iesu Christi Genealogia; Altera De Annis, Quibus Natus, Baptizatus, Mortuus*. Amsterdam, 1643.

Wandal, Hans. *De feria Passionis & triduo mortis Domini & Servatoris nostri Jesu Christi... Diatribe historico-theologica, novis ac paradoxis opinionibus Wilhelmi Langii libro de Annis Christi contentis, opposita*. Leipzig, 1651.

Source Editions and Secondary Literature (after 1800)

Abraham bar Ḥiyya. *Sefer ha-Ibbur*. Edited by Herschell Filipowski. London: Longman, Brown, Green, and Longmans, 1851.

Abulafia, Anna Sapir. *Christians and Jews in the Twelfth-Century Renaissance*. London: Routledge, 1995.

——. "The Ideology of Reform and Changing Ideas Concerning Jews in the Works of Rupert of Deutz and Hermannus quondam Iudeus." *Jewish History* 7 (1993): 3–23.

Abū Maʿshar. *On Historical Astrology: The Book of Religions and Dynasties (On the Great Conjunctions)*. Edited by Keiji Yamamoto and Charles Burnett. 2 vols. Islamic Philosophy, Theology, and Science 33–34. Leiden: Brill, 2000.

Adler, William. *Time Immemorial: Archaic History and its Sources in Christian Chronography from Julius Africanus to George Syncellus*. Dumbarton Oaks Studies 26. Washington, DC: Dumbarton Oaks, 1989.

Adler, William, and Paul Tuffin. *The Chronography of George Synkellos: A Byzantine Chronicle of Universal History from the Creation*. Oxford: Oxford University Press, 2002.

Alberic of Trois-Fontaines. "Chronica." Edited by Paul Scheffer-Boichorst. In *MGH Scriptores*, vol. 23, edited by Georg Heinrich Pertz, 631–950. Hannover: Hahn, 1874.

Albertus Magnus. *Opera Omnia*. Edited by Auguste Borgnet. 38 vols. Paris: Vivès, 1890–99.

——. *Super Dionysii Epistulas*. Edited by Paulus Simon. Alberti Magni Opera Omnia 37.2. Münster: Aschendorff, 1978.

Al-Bīrūnī. *The Chronology of Ancient Nations*. Translated by C. Edward Sachau. London: Allen, 1879.

Alexander of Hales. *Quaestiones disputatae "antequam esset frater"*. 3 vols. Bibliotheca Franciscana Scholastica Medii Aevi 19–21. Quaracchi: Collegium S. Bonaventurae, 1960.

Al-Farghānī. *Il "Libro dell'aggregazione delle stelle"*. Edited by Romeo Campiani. Città di Castello: Lapi, 1910.

Allard, André. "The Arabic Origins and Development of Latin Algorisms in the Twelfth Century." *Arabic Sciences and Philosophy* 1 (1991): 233–83.

——. *Muhammad Ibn Mūsā Al-Khwārizmī: Le calcul indien (Algorismus)*. Paris: Blanchard, 1992.

Allen, Don Cameron. *The Star-Crossed Renaissance: The Quarrel about Astrology and its Influence in England*. New York: Octagon Books, 1973.

Allen, Michael Idomir. "The Chronicle of Claudius of Turin." In *After Rome's Fall: Narrators and Sources of Early Medieval History*, edited by Alexander Callander Murray, 288–319. Toronto: University Press of Toronto, 1998.

Andrei, Osvalda. "Dalle Chronographiai di Giulio Africano alla Synagoge di 'Ippolito': Un dibattito sulla scrittura cristiana del tempo." In Wallraff, *Julius Africanus*, 113–45.

Anheim, Étienne, Benoît Grévin, and Martin Morard. "Exégèse judéo-chrétienne, magie et linguistique: Un recueil de *Notes* inédites attribuées à Roger Bacon." *Archives d'Histoire Doctrinale et Littéraire du Moyen Age* 68 (2001): 95–154.

Anselm of Canterbury. *Opera Omnia*. Edited by Franciscus S. Schmitt. 6 vols. Stuttgart-Bad Cannstatt: Frommann, 1968.

Arduini, Maria Lodovica. *Ruperto di Deutz e la controversia tra Cristiani ed Ebrei nel secolo XII*, Studi storici 119–121. Rome: Istituto storico italiano per il medio evo, 1979.

Aufgebauer, Peter. "'Ein vielfaches Ärgernis in der Kirche': Probleme der christlichen Zeitrechnung im Mittelalter." In *Herrschaftspraxis und soziale Ordnungen im Mittelalter und in der frühen Neuzeit: Ernst Schubert zum Gedenken*, edited by Peter Aufgebauer and Christine van den Heuvel, 73–85. Hannover: Hahn, 2006.

Augustine of Hippo. *On Christian Teaching*. Translated by R. P. H. Green. Oxford: Oxford University Press, 1997.

Avvakumov, Georgij. *Die Entstehung des Unionsgedankens: Die lateinische Theologie des Hochmittelalters in der Auseinandersetzung mit dem Ritus der Ostkirche*. Veröffentlichungen des Grabmann-Institutes 47. Berlin: Akademie Verlag, 2002.

Azariah de' Rossi. *The Light of the Eyes*. Translated by Joanna Weinberg. New Haven, CT: Yale University Press, 2001.

Bacon, Roger. *Compendium studii philosophiae*. In *Opera quaedam hactenus inedita*, 393–519.

——. *Compotus*. Edited by Robert Steele. Opera hactenus inedita Rogeri Baconi 6. Oxford: Clarendon Press, 1926.

——. *Metaphysica*. Edited by Robert Steele. Opera hactenus inedita Rogeri Baconi 1. London: Moring, 1909.

——. *Opus majus*. Edited by John Henry Bridges. 3 vols. Oxford: Clarendon Press, 1897–1900.

——. *Opus tertium*. In *Opera quaedam hactenus inedita*, 3–310.

——. *Opera quaedam hactenus inedita*. Vol. 1. Edited by John Sherren Brewer. Rerum Britannicarum Medii Aevi Scriptores (Rolls Series) 15. London: Longman, Green, Longman, and Roberts, 1859.

Baer, Yitzhak. *A History of the Jews in Christian Spain*. Translated by Louis Schoffman. 2 vols. Philadelphia: Jewish Publication Society of America, 1961–66.

Bainton, Roland. "Basilidian Chronology and New Testament Interpretation." *Journal of Biblical Literature* 42 (1923): 81–134.

Baldi, Bernardino. *Le vite de' matematici*. Edited by Elio Nenci. Milan: Angeli, 1998.

Baldovin, John F. "Hippolytus and the *Apostolic Tradition*: Recent Research and Commentary." *Theological Studies* 64 (2003): 520–42.

Baron, Salo W. *A Social and Religious History of the Jews*. 2nd ed. 18 vols. New York: Columbia University Press, 1952–83.

Baronio, Cesare, Odorico Rinaldi, and Jacopo Laderchi. *Annales ecclesiastici: Denuo excusi et ad nostra usque tempora perducti*. Edited by Augustin Theiner. 37 vols. Bar-le-Duc: Guérin, 1864–83.

Barr, James. "Luther and Biblical Chronology." *Bulletin of the John Rylands University Library of Manchester* 72 (1990): 51–69.

——. "Why the World was Created in 4004 BC: Archbishop Ussher and Biblical Chronology." *Bulletin of the John Rylands University Library of Manchester* 67 (Spring 1985): 575–608.

Bartlett, Robert. *The Natural and the Supernatural in the Middle Ages*. The Wiles Lectures 2006. Cambridge: Cambridge University Press, 2008.

Bauer, Adolf. *Die Chronik des Hippolytos im Matritensis Graecus 121*. Texte und Untersuchungen zur Geschichte der altchristlichen Literatur 14.1. Leipzig: Hinrichs, 1905.

Baumgartner, Frederic J. "Popes, Astrologers, and Early Modern Calendar Reform." In *History Has Many Voices*. Edited by Lee Palmer Wandel, 41–56. Kirksville, MO: Truman State University Press, 2003.

Bayer, Axel. *Spaltung der Christenheit: Das sogenannte Morgenländische Schisma von 1054*. Beihefte zum Archiv für Kulturgeschichte 53. Cologne: Böhlau, 2002.

Beaucamp, Joëlle, R. Bondoux, J. Lefort, Marie-France Rouan, and Irène Sorlin. "Temps et histoire I: Le prologue de la Chronique Pascale." *Travaux et mémoires du Centre de recherche d'histoire et civilisation de Byzance* 7 (1979): 223–301.

Beckwith, Roger T. *Calendar and Chronology, Jewish and Christian: Biblical, Intertestamental and Patristic Studies*. Arbeiten zur Geschichte des antiken Judentums und des Urchristentums 33. Leiden: Brill, 1996.

——. *Calendar, Chronology, and Worship: Studies in Ancient Judaism and Early Christianity*. Arbeiten zur Geschichte des antiken Judentums und des Urchristentums 61. Leiden: Brill, 2005.

Belenkiy, Ari. "A Unique Feature of the Jewish Calendar—Deḥiyot." *Culture and Cosmos* 6 (2002): 3–22.

Belenkiy, Ari, and Eduardo Vila Echagüe. "Stirring Astronomy into Theology: Sir Isaac Newton on the Date of the Passion of Christ." *arXiv*. http://arxiv.org/abs/0810.4358v1.

Bellagente, Elisabetta. "La *Chronica de sex aetatibus* di Claudio vescovo di Torino." *Aevum* 73 (1999): 237–46.

Belloso Martín, Nuria. "Sobre la guerra y la paz en Alfonso de Madrigal." *La Corónica* 33, no. 1 (2004): 17–38.

Beltrán de Heredia, Vicente. *Cartulario de la Universidad de Salamanca (1218–1600)*. 6 vols. Salamanca: Universidad, 1970–73.

Benjamin, Francis S., Jr., and G. J. Toomer. *Campanus of Novara and Medieval Planetary Theory*. Publications in Medieval Science 16. Madison: The University of Wisconsin Press, 1971.

Bergdolt, Klaus. *Das Auge und die Theologie: Naturwissenschaften und "Perspectiva" an der päpstlichen Kurie in Viterbo (ca. 1260–1285)*. Nordrhein-Westfälische Akademie der Wissenschaften: Vorträge G 413. Paderborn: Schöningh, 2007.

Berger, Samuel. *Quam notitiam linguae Hebraicae habuerint Christiani medii aevi in Gallia*. Paris: Hachette, 1893.

Berggren, J. Lennart. "Medieval Arithmetic: Arabic Texts and European Motivations." In Contreni and Casciani, *Word, Image, Number*, 351–65.

Bergmann, Werner. "Chronographie und Komputistik bei Hermann von Reichenau." In *Historiographia medaevalis: Studien zur Geschichtsschreibung und Quellenkunde des Mittelalters*, edited by Dieter Berg and Hans-Werner Goetz, 103–17. Darmstadt: Wissenschaftliche Buchgesellschaft, 1988.

Bernays, Jacob. *Joseph Justus Scaliger*. Berlin: Hertz, 1855.

Bertozzi, Marco. "Astrologia e potere nel '500: L'avventurosa carriera di Luca Gaurico, vate 'veridicus'." In *Cultura e potere nel Rinascimento*, edited by Luisa Secchi Tarugi, 543–51. Florence: Cesati, 1999.

Bezdechi, Stefan. "Un projet de réforme du calendrier par Nicéphore Grégoras." In *Mélanges d'Histoire Générale*, edited by Constantin Marinescu, 2 vols., 1:68–74. Cluj: Cartea românească, 1927–38.

Bickerman, Elias J. *Chronology of the Ancient World*. Rev. ed. London: Thames & Hudson, 1980.

Bischoff, Bernhard. *Mittelalterliche Studien: Ausgewählte Aufsätze zur Schriftkunde und Literaturgeschichte*. 3 vols. Stuttgart: Hiersemann, 1966–81.

Blackburn, Bonnie J., and Leofranc Holford-Strevens. *The Oxford Companion to the Year: An Exploration of Calendar Customs and Time-Reckoning*. Oxford: Oxford University Press, 1999.

Blázquez Hernández, Joaquín. "El Tostado alumno graduado y profesor de la Universidad de Salamanca." In *XV Semana española de Teologia (19–24 Sept. 1955)*, 411–47. Madrid: Consejo Superior de Investigaciones Científicas, 1956.

Blotevogel, Hans H., and Rienk Vermij, eds. *Gerhard Mercator und die geistigen Strömungen des 16. und 17. Jahrhunderts*. Duisburger Mercator-Studien 3. Bochum: Brockmeyer, 1995.

Bonaventure. *Opera omnia*. 10 vols. Quaracchi: Collegium S. Bonaventurae, 1882–1901.

Borst, Arno. "Alkuin und die Enzyklopädie von 809." In *Science in Western and Eastern Civilization in Carolingian Times*, edited by Paul Leo Butzer and Dietrich Lohrmann, 53–75. Basel: Birkhäuser, 1993.

——. *Astrolab und Klosterreform an der Jahrtausendwende*. Sitzungsberichte der Heidelberger Akademie der Wissenschaften, phil.-hist. Kl., 1989.1. Heidelberg: Winter, 1989.

——. *Das Buch der Naturgeschichte: Plinius und seine Leser im Zeitalter des Pergaments*. 2nd ed. Heidelberg: Winter, 1995.

——. "Computus: Zeit und Zahl im Mittelalter." *Deutsches Archiv* 44 (1988): 1–82.

——. "Ein Forschungsbericht Hermanns des Lahmen." *Deutsches Archiv* 40 (1984): 379–477.

——. *Die karolingische Kalenderreform*. MGH Schriften 46. Hannover: Hahn, 1998.

——. *Der karolingische Reichskalender und seine Überlieferung bis ins 12. Jahrhundert*. 3 vols. MGH Antiquitates 3/Libri memoriales 2. Hannover: Hahn, 2001.

——. *The Ordering of Time: From the Ancient Computus to the Modern Computer*. Translated by Andrew Winnard. Chicago: University of Chicago Press, 1993.

——. *Schriften zur Komputistik im Frankenreich von 721 bis 818*. 3 vols. MGH Quellen zur Geistesgeschichte des Mittelalters 21. Hannover: Hahn, 2006.

Boulhol, Pascal. *Claude de Turin: Un évêque iconoclaste dans l'Occident Carolingien*. Paris: Institut d'Études Augustiniennes, 2002.

Bradshaw, Paul F., and Lawrence A. Hoffman, eds. *Passover and Easter*. 2 vols. Two Liturgical Traditions 5–6. Notre Dame, IN: University of Notre Dame Press, 1999.

Brent, Allen. *Hippolytus and the Roman Church in the Third Century: Communities in Tension before the Emergence of a Monarch-Bishop*. Supplements to Vigiliae Christianae 31. Leiden: Brill, 1995.

Brown, Raymond E. *The Birth of the Messiah: A Commentary on the Infancy Narratives in Gospels of Matthew and Luke*. 2nd ed. New York: Doubleday, 1993.

——. *The Death of the Messiah: From Gethsemane to the Grave; A Commentary on the Passion Narratives in the Four Gospels*. 2 vols. New York: Doubleday, 1994.

Bryer, Anthony. "Chronology and Dating." In *The Oxford Handbook of Byzantine Studies*, edited by Elizabeth Jeffreys, John Haldon, and Robin Cormack, 31–37. Oxford: Oxford University Press, 2008.

Buchner, Max. "Die Areopagitika des Abtes Hilduin von St. Denis und ihr kirchenpolitischer Hintergrund." Pt. 2.2. *Historisches Jahrbuch* 58 (1938): 55–96.

Buhler, Stephen M. "Marsilio Ficino's *De stella magorum* and Renaissance Views of the Magi." *Renaissance Quarterly* 43 (1990): 348–71.

Burke, Robert Belle, trans. *The Opus Majus of Roger Bacon*. 2 vols. 1928. Reprint, New York: Russell & Russell, 1962.

Burke-Gaffney, W. "Kepler and the Star of Bethlehem." *Journal of the Royal Astronomical Society of Canada* 31 (1937): 417–25.

Burnett, Charles, ed. *Adelard of Bath: An English Scientist and Arabist of the Early Twelfth Century*. Warburg Institute Surveys and Texts 14. London: The Warburg Institute, 1987.

——. "The Astrological Categorization of Religions in Abū Maʿshar, the *De vetula*, and Roger Bacon." In *Language of Religion—Language of the People: Medieval Judaism, Christianity and Islam*, edited by Ernst Bremer, Jörg Jarnut, Michael Richter, and David J. Wasserstein, 127–38. Munich: Fink, 2006.

——. "Catalogue: The Writings of Adelard of Bath and Closely Associated Works, together with the Manuscripts in Which They Occur." In Burnett, *Adelard of Bath*, 163–96.

——. *The Introduction of Arabic Learning into England*. The Panizzi Lectures 1996. London: The British Library, 1997.

——. "Mathematics and Astronomy in Hereford and its Region in the Twelfth Century." In *Medieval Art, Architecture and Archaeology at Hereford*, edited by David Whitehead, 50–59. Leeds: British Archaeological Association, 1995.

——. "The Semantics of Indian Numerals in Arabic, Greek and Latin." *Journal of Indian Philosophy* 34 (2006): 15–30.

——. "The Works of Petrus Alfonsi: Questions of Authenticity." *Medium Aevum* 66 (1997): 42–79.

Camporeale, Salvatore I. "Giovanmaria dei Tolosani O.P., 1530–1546: Umanesimo, riforma e teologia controversista." *Memorie Domenicane*, n.s., 17 (1986): 145–252.

Candal, Manuel. "El 'De fermento et azymo' del Ottoboniano Latino 718." In *Mélanges Eugène Tisserant*, 7 vols., 6:289–311 (Vatican City: Biblioteca Apostolica Vaticana, 1964).

Cantera Burgos, Francisco. "La conversión del célebre talmudista Salomón Levi (Pablo de Burgos)." *Boletin de la biblioteca de Menéndez Pelayo* l5 (1933): 419–48.

Capitularia regum Francorum. Vol. 1. Edited by Alfred Boretius. MGH Leges 2.1. Hannover: Hahn, 1883.

Carlebach, Elisheva. *Palaces of Time: Jewish Calendar and Culture in Early Modern Europe*. Cambridge, MA: Harvard University Press, 2011.

Carmody, Francis J. *The Astronomical Works of Thabit b. Qurra*. Berkeley: University of California Press, 1960.

Cary, Earnest, trans. *The Roman Antiquities of Dionysius of Halicarnassus*. 7 vols. Cambridge, MA: Harvard University Press, 1937–50.

Caspar, Max. *Johannes Kepler*. 4th ed. Stuttgart: Verlag für Geschichte der Naturwissenschaften und der Technik, 1995.

Censorinus. *De die natali liber ad Q. Caerellium*. Edited by Nicolaus Sallmann. Leipzig: Teubner, 1983.

Cerrato, John A. *Hippolytus between East and West: The Commentaries and the Provenance of the Corpus*. Oxford: Oxford University Press, 2002.

Chadwick, Henry. *East and West: The Making of a Rift in the Church; From Apostolic Times until the Council of Florence*. Oxford: Oxford University Press, 2003.

Chaîne, Marius. *La Chronologie des temps chrétiens de l'Égypte et de l'Éthiopie*. Paris: Geuthner, 1925.

Charles-Edwards, Thomas M. *Early Christian Ireland*. Cambridge: Cambridge University Press, 2000.

Chronica minora saec. IV. V. VI. VII. Edited by Theodor Mommsen. 3 vols. MGH Auctores Antiquissimi 9, 11, 13. Berlin: Weidmann, 1892–98.

Chupungco, Anscar J. *The Cosmic Elements of Christian Passover*. Studia Anselmiana 72. Rome: Editrice Anselmiana, 1977.

Clough, Cecil H. "Federigo da Montefeltro's Patronage of the Arts, 1468–1482." *Journal of the Warburg and Courtauld Institutes* 36 (1973): 129–44.

Cochrane, Louise. *Adelard of Bath: The First English Scientist*. London: British Museum Press, 1994.

Contreni, John J. "Counting, Calendars, and Cosmology: Numeracy in the Early Middle Ages." In Contreni and Casciani, *Word, Image, Number*, 43–83.

Contreni, John J., and Santa Casciani, eds. *Word, Image, Number: Communication in the Middle Ages*. Micrologus' Library 8. Florence: SISMEL, 2002.

Cordoliani, Alfred. "Abbon de Fleury, Hériger de Lobbes et Gerland de Besançon sur l'ère de l'incarnation de Denys le Petit." *Revue d'histoire ecclésiastique* 44 (1949): 463–87.

——. "L'activité computistique de Robert, évêque de Hereford." In *Mélanges offerts à Réné Crozet*, edited by Pierre Gallais and Yves-Jean Riou, 2 vols., 1:333–40. Poitiers: Société d'études médiévales, 1966.

——. "Le comput de Gerland de Besançon." *Revue du moyen âge latin* 2 (1946): 309–13.

——. "Le computiste Hermann de Reichenau." *Miscellanea storica ligure* 3 (1963): 165–90.

Corning, Caitlin. *The Celtic and Roman Traditions: Conflict and Consensus in the Early Medieval Church*. New York: Palgrave Macmillan, 2006.

Crowley, Theodore. *Roger Bacon: The Problem of the Soul in his Philosophical Commentaries*. Louvain: Éditions de l'Institut supérieur de philosophie, 1950.

Cuesta, Luisa. "La edición de las obras del Tostado, empresa de la Corona Española." *Revista de archivos, bibliotecas y museos* 56 (1950): 321–34.

Cullen, Olive M. "A Question of Time or a Question of Theology: A Study of the Easter Controversy in the Insular Church." PhD diss., St. Patrick's College, Maynooth, 2007.

Cumont, Franz. "Regula Philippi Arrhidaei." *Isis* 26 (1936): 8–12.

Curtze, Maximilian. "Der Briefwechsel Regiomontans mit Giovanni Bianchini, Jacob von Speier, und Christian Roder." *Abhandlungen zur Geschichte der Mathematischen Wissenschaften* 12 (1902): 189–336.

Cyril of Scythopolis. *Vita Sabae*. Edited by Eduard Schwartz. Texte und Untersuchungen zur Geschichte der altchristlichen Literatur 49.2. Leipzig: Hinrichs, 1939.

Dachowski, Elizabeth. *First among Abbots: The Career of Abbo of Fleury*. Washington: The Catholic University of America Press, 2008.

Dales, Richard C. "The Computistical Works Ascribed to Robert Grosseteste." *Isis* 80 (1989): 74–79.

de Ceuleneer, Adolf. "Paulus van Middelburg en de Kalenderhervorming." In *Handelingen van het eerste Vlaamsch Taal- en Geschiedkundig Congres, gehouden te Antwerpen, den 17-18-19 September 1910*, 276–89. Antwerp: De Vos en Van der Groen, 1910.

Declercq, Georges. *Anno Domini: The Origins of the Christian Era*. Turnhout: Brepols, 2000.

——. "Dionysius Exiguus and the Introduction of the Christian Era." *Sacris Erudiri* 41 (2002): 165–246.

de Jongh, Henri. *L'Ancienne faculté de théologie de Louvain au premier siècle de son existence (1432–1540)*. Louvain: Revue d'histoire ecclésiastique, 1911.

Dekkers, Eligius. *Clavis Patrum Latinorum*. 3rd ed. Steenbrugge: Brepols, 1995.

de Lang, Marijke Hélène. "The History of the Gospel Synopsis and Gerardus Mercator's Evangelica Historia." In Blotevogel and Vermij, *Gerhard Mercator*, 199–208.

——. "De opkomst van de historische en literaire kritiek in de synoptische beschouwing van de Evangliën van Calvijn (1555) tot Griesbach (1774)." PhD diss., University of Leiden, 1993.

De Meis, Salvo. *Eclipses: An Astronomical Introduction for Humanists*. Serie Orientale Roma 96. Rome: Istituto Italiano per l'Africa e l'Oriente, 2002.

Déprez, Eugène. "Une tentative de reforme du calendrier sous Clement VI: Jean des Murs et la Chronique de Jean de Venette." *Mélanges d'archéologie et d'histoire* 19 (1899): 131–43.

Depuydt, Leo. "Calendars and Years in Ancient Egypt: The Soundness of Egyptian and West Asian Chronology in 1500–500 BC and the Consistency of the Egyptian 365-Day Wandering Year." In *Calendars and Years: Astronomy and Time in the Ancient Near East*, edited by John M. Steele, 35–81. Oxford: Oxbow Books, 2007.

——. "The Date of Death of Jesus of Nazareth." *Journal of the American Oriental Society* 122 (2002): 466–80.

——. "'More Valuable than All Gold': Ptolemy's Royal Canon and Babylonian Chronology." *Journal of Cuneiform Studies* 47 (1995): 97–117.

de Rijk, Lambert Marie, ed. *Garlandus Compotista: Dialectica*. Wijsgerige teksten en studies 3. Assen: Van Gorcum, 1959.

Derolez, René. *Runica Manuscripta: The English Tradition*. Rijksuniversiteit te Gent: Werken uitgegeen door de faculteit van de wijsbegeerte en letteren 118. Bruges: De Tempel, 1954.

de Smet, Antoine. "Savants humanistes et astrologie." In *Acta Conventus Neo-Latini Lovaniensis*, edited by J. I. Jsewijn and E. Keßler, 191–97. Louvain: Leuven University Press, 1973.

de Solan, Olivier. "Les propositions de réforme du calendrier au XVe siècle." *École Nationale des Chartes: Positions des thèses* (1998): 275–81.

——. "La réforme du calendrier dans une question quodlibétique d'Henri de Runen (1444)." *Bibliothèque de l'École des Chartes* 157 (1999): 171–220.

Diacu, Florin. *The Lost Millennium: History's Timetables Under Siege*. Toronto: Vintage Canada, 2006.

Dickey, Bruce George. "Adelard of Bath: An Examination Based on heretofore Unexamined Manuscripts." PhD diss., University of Toronto, 1982.

Diekamp, Franz. "Der Mönch und Presbyter Georgios." *Byzantinische Zeitschrift* 9 (1900): 14–51.

Di Rosa, Pietro. "Denis Petau e la cronologia." *Archivum historicum Societatis Iesu* 29 (1960): 3–54.

Dondaine, Antoine. "'Contra Graecos': Premiers écrits polémiques des Dominicains d'Orient." *Archivum Fratrum Praedicatorum* 21 (1951): 320–446.

——. "Nicolaus de Cotrone et les sources du Contra errores Graecorum de Saint Thomas." *Divus Thomas*, 3rd ser., 28 (1950): 313–40.

Dondaine, Hyacinthe. "Le Contra Errores Graecorum et le IVe livre du Contra Gentiles." *Les sciences philosophiques et théologiques* 1 (1941–42): 156–62.

Döring, Detlef. "Sethus Calvisius als Chronolge: Studien zur Entwicklung der Geschichtswissenschaft an der Leipziger Universität am Beginn des 17. Jahrhunderts." *Zeitschrift für Historische Forschung* 21 (1994): 171–202.

Dowd, Matthew F. "Astronomy and Computus at Oxford University in the Early Thirteenth Century: The Works of Robert Grosseteste." PhD diss., University of Notre Dame, 2003.

Dugdale, William. *Monasticon Anglicanum*. Edited by John Caley, Henry Ellis, and Bulkeley Bandinel. 6 vols. London: March, 1817–30.

Duhem, Pierre. *Le système du monde: Histoire des doctrines cosomologiques de Platon à Copernic*. 10 vols. Paris: Hermann, 1913–59.

Easton, Stewart C. *Roger Bacon and His Search for a Universal Science*. Oxford: Blackwell, 1952.

Eastwood, Bruce S. "The Astronomy of Macrobius in Carolingian Europe: Dungal's Letter of 811 to Charles the Great." *Early Medieval Europe* 3 (1994): 117–34.

Edson, Evelyn. "World Maps and Easter Tables: Medieval Maps in Context." *Imago Mundi* 48 (1996): 25–42.

Einhard. *Vita Karoli Magni*. Edited by Georg Waitz. 6th ed. MGH Scriptores Rerum Germanicarum 25. Hannover: Hahn, 1911.

Elliott, James Keith. *The Apocryphal New Testament: A Collection of Apocryphal Christian Literature in an English Translation Based on M. R. James*. Oxford: Clarendon Press, 1993.

Emanuel, Simcha. "A Jewish-Christian Debate—France, 1110." [In Hebrew.] *Zion* 63 (1998): 143–55.

Emden, Alfred B. *A Biographical Register of the University of Oxford to A.D. 1500*. 3 vols. Oxford: Clarendon Press, 1957–59.

Englisch, Brigitte. "Artes und Weltsicht bei Roger Bacon." In *Artes im Mittelalter*, edited by Ursula Schaefer, 53–67. Berlin: Akademie Verlag, 1999.

——. "Karolingische Reformkalender und die Fixierung der christlichen Zeitrechnung." In Warntjes and Ó Cróinín, *Computus*, 238–58.

Epistolae Pontificae de rebus in Concilio Florentino annis 1438–1439 gestis. Edited by Georg Hofmann. Concilium Florentinum: Documenta et Scriptores, ser. A, vol. 2. Rome: Pontificium Institutum Orientalium Studiorum, 1944.

Erickson, John H. "Leavened and Unleavened: Some Theological Implications of the Schism of 1054." *St. Vladimir's Theologial Quarterly* 14 (1970): 155–76.

Ernst, Germana. *Religione, ragione e natura: Ricerche su Tommaso Campanella e il tardo Rinascimento*. Milan: Angeli, 1991.

Evans, James. *The History and Practice of Ancient Astronomy*. New York: Oxford University Press, 1998.

Evans, Jonathan. "Peter de Rivo and the Problem of Future Contingents." *Carmina Philosophiae* 10 (2001): 39–55.

Faracovi, Ornella Pompeo. *Gli oroscopi di Cristo*. Venice: Marsilio Editori, 1999.

Feeney, Denis. *Caesar's Calendar: Ancient Time and the Beginnings of History*. Berkeley: University of California Press, 2007.

Feldman, William Moses. *Rabbinical Mathematics and Astronomy*. 3rd ed. New York: Hermon Press, 1978.

Ferrari, Mirella. "Note su Claudio di Torino 'Episcopus ab ecclesia damnatus'." *Italia medioevale e umanistica* 16 (1973): 291–308.

Ferrari-d'Occhieppo, Konradin. "Die Osterberechnung als Kalenderproblem von der Antike bis Regiomontanus." In *Regiomontanus-Studien*, edited by Günther Hamann, 91–108 (Vienna: Verlag der Österreichschen Akademie der Wissenschaften, 1980).

Finegan, Jack. *Handbook of Biblical Chronology: Principles of Time Reckoning in the Ancient World and Problems of Chronology in the Bible*. Rev. ed. Peabody, MA: Hendrickson, 1998.

First, Mitchell. *Jewish History in Conflict: A Study of the Major Discrepancy between Rabbinic and Conventional Chronology*. Northvale, NJ: Aronson, 1997.

Floëri, Fernand, and Pierre Nautin, eds. *Homélies pascales*. Vol. 3, *Une homélie anatolienne sur la date de Paques en l'an 387*. Sources Chrétiennes 48. Paris: Éditions du Cerf, 1957.

Folkerts, Menso. "Adelard's Version of Euclid's *Elements*." In Burnett, *Adelard of Bath*, 55–68.

Fomenko, Anatoly T. *Empirico-Statistical Analysis of Narrative Material and its Applications to Historical Dating*. 2 vols. Dordrecht: Kluwer, 1994.

——. *History: Fiction or Science?* 2nd ed. 7 vols. Paris: Delamere, 2003–6.

Fotheringham, John K. "The Evidence of Astronomy and Technical Chronology for the Date of the Crucifixion." *Journal of Theological Studies* 35 (1934): 146–62.

Frank, Edgar. *Talmudic and Rabbinical Chronology: The Systems of Counting Years in Jewish Literature*. New York: Feldheim, 1956.

Friedländer, Michael. "Ibn Ezra in England." *Transactions of the Jewish Historical Society of England* 2 (1894–95): 47–75.

Gack-Scheiding, Christine. *Johannes de Muris: Epistola super reformatione antiqui kalendarii. Ein Beitrag zur Kalenderreform im 14. Jahrhundert*. MGH Studien und Texte 11. Hannover: Hahn, 1995.

García, Santiago. "La competencia hebraica de Alfonso de Madrigal." *La Corónica* 33, no. 1 (2004): 85–98.

Garin, Eugenio. *Lo zodiaco della vita: La polemica sull'astrologia dal Trecento al Cinquecento*. Rome: Laterza, 1976.

Garrastachu, Jaquinto M. "Los manuscritos del Cardenal Torquemada en la Biblioteca Vaticana." *La Ciencia Tomista* 22 (1930): 188–217, 291–322.

Gelzer, Heinrich. *Sextus Julius Africanus und die byzantinische Chronographie*. 2 vols. Leipzig: Teubner, 1880–98.

Georgius Syncellus. *Ecloga chronographica*. Edited by Alden A. Mosshammer. Leipzig: Teubner, 1984.

Gerlach, Karl. *The Antenicene Pascha: A Rhetorical History*. Liturgia condenda 7. Louvain: Peeters, 1998.

Germann, Nadja. "À la recherche de la structure du temps: Abbon de Fleury et le comput." In *Abbon, un abbé de l'an mil*, edited by Annie Dufour and Gillette Labory, 153–76. Turnhout: Brepols, 2008.

——. *De temporum ratione: Quadrivium und Gotteserkenntnis am Beispiel Abbos von Fleury und Hermanns von Reichenau*. Studien und Texte zur Geistesgeschichte des Mittelalters 89. Leiden: Brill, 2006.

——. "Zwischen *veritas naturae* und *fides historiae*: Zeit und Dauer bei Abbo von Fleury." In *Das Sein der Dauer*, edited by Andreas Speer and David Wirmer, 171–95. Berlin: de Gruyter, 2008.

Gibbon, Edward. *The History of the Decline and Fall of the Roman Empire*. Edited by J. B. Bury. 7 vols. London: Methuen, 1909–14.

Gill, Joseph. *The Council of Florence*. Cambridge: Cambridge University Press, 1959.

Ginzel, Friedrich Karl. *Handbuch der mathematischen und technischen Chronologie: Das Zeitrechnungswesen der Völker*. 3 vols. Leipzig: Hinrichs, 1906–14.

Glorieux, Palémon. "Autour du Contra Errores Graecorum: Suggestions chronologiques." In *Autour d'Aristote: Recueil d'études de philosophie ancienne et médiévale offert à Monseigneur A. Mansion*, 497–512. Louvain: Publications Universitaires de Louvain, 1955.

Goldberg, Sylvie Anne. *La Clepsydre: Essai sur la pluralité des temps dans le judaïsme*. Paris: Michel, 2000.

Goldstein, Bernard R. "Ancient and Medieval Values for the Mean Synodic Month." *Journal for the History of Astronomy* 34 (2003): 65–74.

Goldstine, Hermann H. *New And Full Moons, 1001 B.C. to A.D. 1651*. Memoirs of the American Philosophical Society 94. Philadelphia: American Philosophical Society, 1973.

Golubovich, Hieronymus. "Disputatio Latinorum et Graecorum seu Relatio Apocrisariorum Gregorii IX de gestis Nicaeae in Bithynia et Nymphaeae in Lydia 1234." *Archivum Franciscanum Historicum* 12 (1919): 418–70.

Gorman, Michael. "The Commentary on Genesis of Claudius of Turin and Biblical Studies under Louis the Pious." *Speculum* 72 (1997): 279–329.

Gosselin, Edward A. "A Listing of the Printed Editions of Nicolaus de Lyra." *Traditio* 26 (1970): 399–426.

Grabmann, Martin. *Mittelalterliches Geistesleben: Abhandlungen zur Geschichte der Scholastik und Mystik*. 3 vols. Munich: Hueber, 1926–56.

Grafton, Anthony. "The Antiquary as Pariah: The Strange Case of Jean Hardouin." *Journal of the Warburg and Courtauld Institutes* 62 (1999): 241–67.

——. *Cardano's Cosmos: The Worlds and Works of a Renaissance Astrologer*. Cambridge, MA: Harvard University Press, 1999.

——. "Chronology and Its Discontents in Renaissance Europe: The Vicissitudes of a Tradition." In *Time: Histories and Ethnologies*, edited by Diane Owen Hughes and Thomas R. Trautmann, 139–66. Ann Arbor: University of Michigan Press, 1995.

——. "Chronology, Controversy, and Community in the Republic of Letters: The Case of Kepler." Chap. 6 in *Worlds Made by Words: Schlarship and Community in the Modern West*. Cambridge, MA: Harvard University Press, 2009.

——. "The Chronology of the Flood." In *Sintflut und Gedächtnis: Erinnern und Vergessen des Ursprungs*, edited by Martin Mulsow and Jan Assmann, 65–82. Munich: Fink, 2006.

——. "From *De die natali* to *De emendatione temporum*: The Origins and Settings of Scaliger's Chronology." *Journal of the Warburg and Courtauld Institutes* 48 (1985): 100–43.

——. "Joseph Scaliger and Historical Chronology: The Rise and Fall of a Discipline." *History and Theory* 14 (1975): 156–85.

——. *Joseph Scaliger: A Study in the History of Classical Scholarship*. 2 vols. Oxford: Clarendon Press, 1983–93.

——. "Kircher's Chronology." In *Athanasius Kircher: The Last Man Who Knew Everything*, edited by Paula Findlen, 171–87. New York: Routledge, 2004.

——. "Petrus Apianus Draws up a Calendar." *Journal for the History of Astronomy* 42 (2011): 55–72.

——. "Some Uses of Eclipses in Early Modern Chronology." *Journal of the History of Ideas* 64 (2003): 213–29.

——. "Tradition and Technique in Historical Chronology." In *Ancient History and the Antiquarian: Essays in Memory of Arnaldo Momigliano*, edited by M. H. Crawford and C. R. Ligota, 15–31. London: The Warburg Institute, 1995.

Grafton, Anthony, and Henk Jan de Jonge. "Joseph Scaliger: A Bibliography 1850–1993." In *The Scaliger Collection*, edited by Rijk Smitskamp, i–xxx. Leiden: Smitskamp Oriental Antiquarium, 1993.

Grafton, Anthony, and Noel Swerdlow. "Calendar Dates and Ominous Days in Ancient Historiography." *Journal of the Warburg and Courtauld Institutes* 51 (1988): 14–42.

——. "Greek Chronography in Roman Epic: The Calendrical Date of the Fall of Troy in the Aeneid." *Classical Quarterly*, n.s., 36 (1986): 212–18.

——. "The Horoscope of the Foundation of Rome." *Classical Philology* 81 (1986): 148–53.

——. "Technical Chronology and Astrological History in Varro, Censorinus and Others." *Classical Quarterly*, n.s., 35 (1985): 454–65.

Grafton, Anthony, and Joanna Weinberg. *"I Have Always Loved the Holy Tongue": Isaac Casaubon, the Jews, and a Forgotten Chapter in Renaissance Scholarship*. Cambridge, MA: Harvard University Press, 2011.

Granada, Miguel A. "Giovanni Maria Tolosani e la prima reazione Romana di fronte al *De revolutionibus*: La critica di Copernico nell'opusculo *De coelo et elementis*." In *La diffusione del Copernicanesimo in Italia 1543–1610*, edited by Massimo Bucciantini and Maurizio Torrini, 11–35. Florence: Olschki, 1997.

Granada, Miguel A., and Dario Tessicini. "Copernicus and Fracastoro: The Dedicatory Letters to Pope Paul III, the History of Astronomy, and the Quest for Patronage." *Studies in History and Philosophy of Science* 36 (2005): 431–76.

Grévin, Benoît. "L'hébreu des franciscains: Nouveaux éléments sur la conaissance de l'hébreu en milieu chrétien au XIIIe siècle." *Médiévales* 41 (2001): 65–82.

Grumel, Venance. *La chronologie*. Traité d'Études Byzantines 1. Paris: Presses universitaires de France, 1958.

——. "Les premières ères mondiales." *Revue des Études Byzantines* 10 (1952): 93–108.

——. "Le problème de la date pascale aux IIIe et IVe siècles." *Revue des Études Byzantines* 18 (1960): 163–78.

Guarducci, Margherita. *Epigrafia greca*. 4 vols. Rome: Istituto Poligrafico dello Stato, 1967–78.

——. "La statua di 'sant'Ippolyto' in Vaticano." *Rendiconti della Pontificia Accademia Romana di Archeologia* 47 (1974–75): 163–90.

Guenée, Bernard. *Entre l'Église et l'État: Quatre vies de prélats français à la fin du Moyen Âge (XIIIe–XVe siècle)*. Paris: Gallimard, 1987.

——. *Histoire et culture historique dans l'occident medieval*. Paris: Éditions Aubier-Montaigne, 1980.

Guilland, Rodolphe. *Essai sur Nicéphore Grégoras*. Paris: Geuthner, 1926.

Guttmann, Jacob. *Die Scholastik des dreizehnten Jahrhunderts in ihren Beziehungen zum Judenthum und zur jüdischen Literatur*. Breslau: Marcus, 1902.

Hackett, Jeremiah, ed. *Roger Bacon and the Sciences: Commemorative Essays*. Studien und Texte zur Geistesgeschichte des Mittelalters 57. Leiden: Brill, 1997.

——. "The Published Works of Roger Bacon." *Vivarium* 35 (1997): 315–20.

Haeusler, Martin. *Das Ende der Geschichte in der mittelalterlichen Weltchronistik*. Beihefte zum Archiv für Kulturgeschichte 13. Cologne: Böhlau, 1980.

Hanssens, Jean Michel. *La liturgie d'Hippolyte: Ses documents—son titulaire, ses origines et son caractère*. Orientalia Christiana Analecta 155. Rome: Pontificium Institutum Orientalium Studiorum, 1959.

Harrison, Kenneth. "Easter Cycles and the Equinox in the British Isles." *Anglo-Saxon England* 7 (1978): 1–8.

Haskins, Charles Homer. *Studies in the History of Mediaeval Science*. Harvard Historical Studies 27. Cambridge, MA: Harvard University Press, 1924.

Heil, Johannes. "Claudius von Turin—eine Fallstudie zur Geschichte der Karolingerzeit." *Zeitschrift für Geschichtswissenschaft* 45 (1997): 385–412.

Heine, Ronald E. "Hippolytus, Ps.-Hippolytus and the Early Canons." In *The Cambridge History of Early Christian Literature*, edited by Frances Young, Lewis Ayres, and Andrew Louth, 142–51. Cambridge: Cambridge University Press, 2004.

Herold, Werner. "Der *computus emendatus* des Reinher von Paderborn." In *Kulturarbeit und Kirche: Festschrift Msgr. Dr. Paul Mai zum 70. Geburtstag*, edited by Werner Chobrak and Karl Hausberger, 39–47. Regensburg: Verlag des Vereins für Regensburger Bistumsgeschichte, 2005.

Heydeck, Kurt. *Die mittelalterlichen Handschriften der Universitätsbibliothek Rostock*. Wiesbaden: Harrassowitz, 2001.

Hirsch, Samuel A. "Early English Hebraists: Roger Bacon and His Predecessors." *Jewish Quarterly Review* 12 (1899–1900): 34–88.

——. "Roger Bacon and Philology." In *Roger Bacon: Essays Contributed by Various Writers on the Occasion of the Commemoration of the Seventh Centenary of his Birth*, edited by Andrew G. Little, 101–51. Oxford: Clarendon Press, 1914.

Hirzel, Oskar. *Abt Heriger von Lobbes 990–1007*. Beiträge zur Kulturgeschichte des Mittelalters und der Renaissance 8. Leipzig: Teubner, 1910.

Hofmann, Heinz. "Literary Culture at the Court of Urbino during the Reign of Federico da Montefeltro." *Humanistica Lovaniensia* 57 (2008): 5–59.

Holford-Strevens, Leofranc. "Church Politics and the Computus: From Milan to the Ends of the Earth." In Warntjes and Ó Cróinín, *Easter Controversy* (forthcoming).

——. *The History of Time: A Very Short Introduction*. Oxford: Oxford University Press, 2005.

——. "Paschal Lunar Calendars up to Bede." *Peritia* 20 (2008): 165–208.

Honecker, Martin. "Die Entstehung der Kalenderreformschrift des Nikolaus von Cues." *Historisches Jahrbuch* 60 (1940): 581–92.

Honselmann, Klemens. "Magister Reinher: Schrittmacher für die Kalenderreform und die moderne Rechenkunst." In *Von der Domschule zum Gymnasium Theodorianum in Paderborn*, edited by Klemens Honselmann, 107–26. Paderborn: Verein für Geschichte und Altertumskunde Westfalens, 1962.

Hotson, Howard. *Paradise Postponed: Johann Heinrich Alsted and the Birth of Calvinist Millenarianism*. Archives Internationales d'Histoire des Idées 172. Dordrecht: Kluwer, 2000.
Hughes, Jeremy. *Secrets of the Times: Myth and History in Biblical Chronology*. Journal for the Study of the Old Testament, Supplement Series 66. Sheffield: JSOT Press, 1990.
Humphreys, Colin J. *The Mystery of the Last Supper: Reconstructing the Final Days of Jesus*. Cambridge: Cambridge University Press, 2011.
Humphreys, Colin J., and W. Graeme Waddington. "Astronomy and the Date of the Crucifixion." In Vardaman and Yamauchi, *Chronos, Kairos, Christos*, 165–81.
——. "Crucifixion Date." *Nature* 348 (1990): 684.
——. "Dating the Crucifixion." *Nature* 306 (1983): 743–46.
Humphries, Mark. "Chronicle and Chronology: Prosper of Aquitaine, his Methods and the Development of Early Medieval Chronography." *Early Medieval Europe* 5 (1996): 155–75.
Ideler, Ludwig. *Handbuch der mathematischen und technischen Chronologie*. 2 vols. Berlin: Rücker, 1825–26.
Illig, Heribert. *Das erfundene Mittelalter: Die größte Zeitfälschung der Geschichte*. Düsseldorf: Econ, 1996.
Jacobus de Voragine. *The Golden Legend: Readings on the Saints*. Translated by William Granger Ryan. 2 vols. Princeton, NJ: Princeton University Press, 1993.
James, Montague Rhodes. *The Ancient Libraries of Canterbury and Dover*. Cambridge: Cambridge University Press, 1903.
Jaritz, Gerhard, and Gerson Moreno-Riaño, eds. *Time and Eternity: The Medieval Discourse*. International Medieval Reserach 9. Turnhout: Brepols, 2003.
Jenkins, R. M. "The Star of Bethlehem and the Comet of AD 66." *Journal of the British Astronomical Association* 114 (2004): 336–43.
Jerome of Stridon. *De viris illustribus*. Edited by Aldo Ceresa-Gastaldo. Florence: Nardini, 1988.
Jones, Charles W. *Bedae Opera de Temporibus*. The Medieval Academy of America Publication No. 41. Cambridge, MA: The Medieval Academy of America, 1943.
——. "Bede's Place in Medieval Schools." In *Famulus Christi: Essays in Commemoration of the Thirteenth Centenary of the Birth of the Venerable Bede*, edited by Gerald Bonner, 261–85. London: SPCK, 1976.
——. "An Early Medieval Licensing Examination." *History of Education Quarterly* 3 (1963): 19–29.
——. "A Legend of St. Pachomius." *Speculum* 18 (1943): 198–210.
——. "The Victorian and Dionysiac Paschal Tables in the West." *Speculum* 9 (1934): 408–21.
Jonge, Henk Jan de. "Joseph Scaliger's Historical Criticism of the New Testament." *Novum Testamentum* 38 (1996): 176–93.
——. "Sixteenth-Century Gospel Harmonies: Chemnitz and Mercator." In *Théorie et pratique de l'exégèse*, edited by Irena Backus and Francis Higman, 155–66. Geneva: Droz, 1990.
Jordan, Mark D. "Theological Exegesis and Aquinas's Treatise 'against the Greeks'." *Church History* 56 (1987): 445–56.
Juan de Torquemada. *Apparatus super Decretum Florentinum Unionis Graecorum*. Edited by Emmanuel Candal. Concilium Florentinum: Documenta et Scriptores, ser. B, vol. 2.1. Rome: Pontificium Institutum Orientalium Studiorum, 1942.
Kaltenbrunner, Ferdinand. "Die Vorgeschichte der gregorianischen Kalenderreform." *Sitzungsberichte der kaiserlichen Akademie der Wissenschaften (Wien)*, phil.-hist. Kl., 82 (1876): 289–414.
Kanter, Shamai. *Rabban Gamaliel II: The Legal Traditions*. Brown Judaic Studies 8. Chico, CA: Scholars Press, 1980.

Kennedy, Edward S. "The World-Year Concept in Islamic Astrology." In *Actes du Dixième Congrès International d'Histoire des Sciences*, 2 vols., 1:23–43. Paris: Hermann, 1964.

Kennedy, Edward S., and David Pingree. *The Astrological History of Māshā'allāh.* Cambridge, MA: Harvard University Press, 1971.

Kennedy, Edward S., and B. L. van der Waerden. "The World-Year of the Persians." *Journal of the American Oriental Society* 83 (1963): 315–27.

Klempt, Adalbert. *Die Säkularisierung der universalgeschichtlichen Auffassung: Zum Wandel des Geschichtsdenkens im 16. und 17. Jahrhundert.* Göttinger Bausteine zur Geschichtswissenschaft 31. Göttingen: Musterschmidt, 1960.

Klopsch, Paul. *Pseudo-Ovidius De Vetula: Untersuchungen und Text.* Mittellateinische Studien und Texte 2. Leiden: Brill, 1967.

Knowles, Dom David, C. N. L. Brooke, and Vera C. M. London. *The Heads of Religious Houses, England and Wales.* 2nd ed. Vol. 1, *940–1216.* Cambridge: Cambridge University Press, 2001.

Koch, Hugo. "Tertullianisches." *Theologische Studien und Kritiken* 101 (1929): 458–74.

Kohl, Wilhelm. *Die Zisterzienserabtei Marienfeld.* Germania Sacra. Dritte Folge 2: Die Bistümer der Kirchenprovinz Köln; Das Bistum Münster 11. Berlin: de Gruyter, 2010.

Kokkinos, Nikos. "Crucifixion in A.D. 36: The Keystone for Dating the Birth of Jesus." In Vardaman and Yamauchi, *Chronos, Kairos, Christos*, 133–63.

Krojer, Franz. *Die Präzision der Präzession: Illigs mittelalterliche Phantomzeit aus astronomischer Sicht.* Munich: Differenz-Verlag, 2003.

Krusch, Bruno. "Das älteste fränkische Lehrbuch der dionysianischen Zeitrechnung." In *Mélanges offerts à M. Émile Chatelain par ses élèves et ses amis*, 232–42. Paris: Champion, 1910.

——. "Die Chronicae des sogenannten Fredegar." Pt. 2. *Neues Archiv der Gesellschaft für ältere deutsche Geschichtskunde* 7 (1882): 423–516.

——. *Neue Bruchstücke der Zeitzer Ostertafel vom Jahre 447.* Berlin: de Gruyter, 1933.

Lacarra María Jesús, ed. *Estudios Sobre Pedro Alfonso de Huesca.* Collección de Estudios Altoaragonenses 41. Huesca: Instituto de Estudios Altoaragonenses, 1996.

Lazzarato, Damiano. *Chronologia Christi seu discordantium fontium concordantia ad juris normam.* Naples: M. d'Auria, 1952.

Leclercq, Henri. "Hippolyte (statue et cimetière de saint)." In *Dictionnaire d'archéologie chrétienne et de liturgie*, 15 vols., 6:2419–83. Paris: Letouzey et Ané, 1903–51.

Lederer, Stephan. *Der spanische Cardinal Johann von Torquemada: Sein Leben und seine Schriften.* Freiburg i. Br.: Herder, 1879.

Lejbowicz, Max. "Des tables pascales aux tables astronomiques et retour: Formation et réception du comput patristique." *Methodos: Savoirs et textes* 6 (2006). http://methodos.revues.org/ documents538.html.

——. "Les Pâques baptismales d'Augustin d'Hippone, une étape contournée dans l'unification des partiques computistes latines." In Warntjes and Ó Cróinín, *Computus*, 1–39.

——. "Présentation de Jean de Murs, 'observateur et calculateur sagace et laborieux'." In *Méthodes et statut des sciences à la fin du Moyen Âge*, edited by Christophe Gerrard, 159–80. Villeneuve d'Ascq: Presses Universitaires du Septentrion, 2004.

Lemay, Richard. "The Hispanic Origin of our Present Numeral Forms." *Viator* 8 (1977): 435–62.

Lindberg, David C. "Lines of Influence in Thirteenth Century Optics: Bacon, Witelo, and Pecham." *Speculum* 46 (1971): 66–83.

——. *Roger Bacon's Philosophy of Nature: A Critical Edition, with English Translation, Introduction, and Notes, of* De multiplication specierum *and* De speculis comburentibus. Oxford: Clarendon Press, 1983.

——. *Roger Bacon and the Origins of* Perspectiva *in the Middle Ages: A Critical Edition and English Translation of Bacon's* Perspectiva *with Introduction and Notes*. Oxford: Clarendon Press, 1996.
Lindhagen, Arvid. "Die Neumondtafel des Robertus Lincolniensis." *Arkiv för Matematik, Astronomi och Fysik* 11, no. 2 (1916): 1–41.
Little, Andrew G. "Annual Lecture on a Master Mind: Roger Bacon." *Proceedings of the British Academy* 14 (1928): 265–96.
——. *The Grey Friars in Oxford*. Oxford: Oxford Historical Society, 1892.
——. *Part of the Opus tertium of Roger Bacon, including a Fragment Now Printed for the First Time*. Aberdeen: The University Press, 1912.
Loenertz, Raymond J. "Autour du traité de Fr. Barthélemy de Constantinople Contre les Grecs." *Archivum Fratrum Praedicatorum* 6 (1936): 361–71.
Loi, Vincenzo. "Il 25 Marzo data pasquale e la cronologia giovannea della passione in età patristica." *Ephemerides Liturgicae* 85 (1971): 48–69.
López Martínez, Nicolás. "Pablo de Santa María y el sentido literal bíblico en las controversias con los judíos," In *Biblia, exégesis y cultura: estudios en honor del prof. José María Casciaro*, edited by Gonzalo Aranda Pérez, Claudio Basevi, and J. Chapa, 475–83. Pamplona: Ediciones Universidad de Navarra, 1994.
Love, Rosalind. "Bede and John Chrysostom." *Journal of Medieval Latin* 17 (2007): 72–86.
Lutz, Edward. *Roger Bacon's Contribution to Knowledge*. Franciscan Studies 17. New York: Wagner, 1936.
Mahler, Eduard. *Handbuch der jüdischen Chronologie*. Leipzig: Fock, 1916.
Malcolm, Noel. *Aspects of Hobbes*. Oxford: Clarendon Press, 2002.
——. "*Leviathan*, the Pentateuch, and the Origins of Modern Biblical Criticism." In *Leviathan after 350 Years*, edited by Tom Sorell and Luc Foisneau, 241–64. Oxford: Clarendon Press, 2004.
Mansi, Giovanni Domenico. *Sacrorum conciliorum nova et amplissima collectio*. 53 vols. Paris: Welter, 1901–27.
Manuel, Frank E. *Isaac Newton Historian*. Cambridge, MA: Harvard University Press, 1963.
Marcos Rodríguez, Florencio. "Los manuscritos de Alfonso de Madrigal conservados en la Biblioteca Universitaria de Salamanca." *Salmanticensis* 4 (1957): 3–50.
Marcovich, Miroslav. "Note on Hippolytus' *Refutatio*." *Journal of Theological Studies*, n.s., 15 (1964): 69–74.
Marzi, Demetrio. "Giovanni Maria Tolosani, Alessandro Piccolomini e Luigi Giglio." *Miscellanea storica della Valdelsa* 5 (1897): 202–9.
——. "Giovanni Maria Tolosani e Giovanni Lucido Samoteo." *Miscellanea storica della Valdelsa* 5 (1897): 32–62.
——. "Nuovi studi e ricerche intorno alla questione del calendario durante i secoli XV e XVI." In *Atti del congresso internazionale di scienze storiche (Roma, 1–9 Aprile 1903)*, 12 vols., 3:637–50. Rome: Accademia dei Lincei, 1904–7.
——. *La questione della riforma del calendario nel Quinto Concilio Lateranense (1512–1517)*. Florence: Carnesecchi, 1896.
Massa, Eugenio. *Ruggero Bacone: Etica e poetica nella storia dell' "Opus maius"*. Uomini e dottrine 3. Rome: Edizioni di Storia e Letteratura, 1955.
McCarthy, Daniel. "Bede's Primary Source for the Vulgate Chronology in his Chronicles in *De Temporibus* and *De Temporum Ratione*." In Warntjes and Ó Cróinín, *Computus*, 159–89.
——. "The Emergence of *Anno Domini*." In Jaritz and Moreno-Riaño, *Time and Eternity*, 31–53.
——. *The Irish Annals: Their Genesis, Evolution and History*. Dublin: Four Courts Press, 2008.

McCarthy, Daniel and Aidan Breen. *The Ante-Nicene Christian Pasch*—De ratione paschali: *The Paschal Tract of Anatolius, Bishop of Laodicea*. Dublin: Four Courts Press, 2003.

——. "An Evaluation of Astronomical Observations in Irish Annals." *Vistas in Astronomy* 41 (1997): 117–38.

McCluskey, Stephen C. *Astronomies and Cultures in Early Medieval Europe*. Cambridge: Cambridge University Press, 1998.

——. "Changing Contexts and Criteria for the Justification of Computistical Knowledge and Practice." *Journal for the History of Astronomy* 34 (2003): 201–17.

McCuaig, William. "The Fasti Capitolini and the Study of Roman Chronology in the Sixteenth Century." *Athenaeum* (Pavia) 79 (1991): 141–59.

Mentz, Arthur. *Beiträge zur Osterfestberechnung bei den Byzantinern*. Königsberg: Leupold, 1906.

Merchavia, Ch. "The Talmud in the Additiones of Paul of Burgos." *Journal of Jewish Studies* 16 (1965): 115–34.

Mercier, Raymond. "Astronomical Tables in the Twelfth Century." In Burnett, *Adelard of Bath*, 87–118.

——. "The Dates in Syriac Martyr Acts." *Analecta Bollandiana* 117 (1999): 47–66.

——. "East and West Contrasted in Scientific Astronomy." In *Occident et Proche-Orient: Contacts scientifiques au temps de Croisades*, edited by Isabelle Draelants, Anne Tihon, and Baudouin van den Abeele, 325–42. Turnhout: Brepols, 2000.

Metzger, Marcel, ed. *Les constitutions apostoliques*. 3 vols. Sources Chrétiennes 320, 329, 336. Paris: Éditions du Cerf, 1985–87.

Meyer, Otto. "Weltchronistik und Computus im hochmittelalterlichen Bamberg." *Jahrbuch für fränkische Landesforschung* 19 (1959): 241–60.

Milikowsky, Chaim. "*Seder 'Olam* and Jewish Chronography in the Hellenistic and Roman Periods." *Proceedings of the American Academy for Jewish Research* 52 (1985): 115–39.

Millar, Fergus. "Reflections on the Trials of Jesus." In *A Tribute to Geza Vermes: Essays on Jewish and Christian Literature and History*, edited by Philip R. Davies and Richard T. White, 335–81. Sheffield: JSOT Press, 1990.

Millás Vallicrosa, José María. "La aportación astronómica de Pedro Alfonso." *Sefarad* 3 (1943): 65–105. Translated by David Assouline as "Pedro Alfonso's Contribution to Astronomy." With postscript by Charles Burnett. *Aleph* 10 (2010): 139–68.

——. *El libro de los fundamentos de las Tablas astronómicas de R. Abraham Ibn Ezra*. Madrid: Casa Provincial de Caridad de Barcelona, 1947.

Minnis, Alastair J. "Fifteenth-Century Versions of Thomistic Literalism: Girolamo Savonarola and Alfonso de Madrigal." In *Neue Richtungen in der hoch- und spätmittelalterlichen Bibelexegese*, edited by Robert E. Lerner, 163–80. Munich: Oldenbourg, 1996.

Moesgaard, Kristian Peder. "Thābit ibn Qurra between Ptolemy and Copernicus: Analysis of Thābit's Solar Theory." *Archive for the History of Exact Sciences* 12 (1974): 199–216.

Möller, Astrid. "Epoch-Making Eratosthenes." *Greek, Roman, and Byzantine Studies* 45 (2005): 245–60.

Mommsen, Theodor. *Über den Chronographen aus dem Jahre 354*. Leipzig: Weidmann, 1850.

——. *Zeitzer Ostertafel vom Jahre 447*. Berlin: Dümmler, 1863.

Moreton, Jennifer. "Before Grosseteste: Roger of Hereford and Calendar Reform in Eleventh- and Twelfth-Century England." *Isis* 86 (1995): 562–86.

——. "The *Compotus* of 'Constabularius' (1175): A Preliminary Study." In *Langage, Sciences, Philosophie au XII*[e] *siècle*, edited by Joël Biard, 61–82. Paris: Vrin, 1999.

——. "Doubts about the Calendar: Bede and the Eclipse of 664." *Isis* 89 (1998): 50–65.

——. "John of Sacrobosco and the Calendar." *Viator* 25 (1994): 229–44.

——. "On Not Editing Grosseteste." In *Editing Robert Grosseteste*, edited by Evelyn A. Mackie and Joseph Goering, 167–84. Toronto: University of Toronto Press, 2003.

——. "Robert Grosseteste and the Calendar." In *Robert Grosseteste: New Perspectives on His Thought and Scholarship*, edited by James McEvoy, 77–88. Turnhout: Brepols, 1995.

Mosshammer, Alden A. *The Chronicle of Eusebius and Greek Chronographic Tradition*. Lewisburg: Bucknell University Press, 1979.

——. "The Computus of 455 and the Laterculus of Augustalis, with an Appendix on the Fractional Method of Agriustia." In Warntjes and Ó Cróinín, *Easter Controversy* (forthcoming).

——. *The Easter Computus and the Origins of the Christian Era*. Oxford: Oxford University Press, 2008.

Müller, Tom. *"Ut reiecto paschali errore veritati insistamus": Nikolaus von Kues und seine Konzilsschrift* De reparatione kalendarii. Buchreihe der Cusanus-Gesellschaft 17. Münster: Aschendorff, 2010.

Nautin, Pierre. *Hippolyte et Josipe: Contribution à l'histoire de la littérature chrétienne du troisième siècle*. Études et textes pour l'historie du dogme de la trinité 1. Paris: Éditions du Cerf, 1947.

Neugebauer, Otto. "Abū-Shāker and the Ethiopic Hasāb." *Journal of Near Eastern Studies* 42 (1983): 55–58.

——. *Abu Shaker's "Chronography": A Treatise of the 13th Century on Chronological, Calendrical, and Astronomical Matters, Written by a Christian Arab, Preserved in Ethiopic*. Sitzungsberichte der Österreichischen Akademie der Wissenschaften, phil.-hist. Kl., 498. Vienna: Verlag der Österreichischen Akademie der Wissenschaften, 1988.

——. *The Astronomical Tables of Al-Khwārizmī*. Kgl. Danske Vidensk. Selskab, Hist.-filos. Skr., 4.2. Kopenhagen: Munksgaard, 1962.

——. *Chronography in Ethiopic Sources*. Sitzungsberichte der Österreichischen Akademie der Wissenschaften, phil.-hist. Kl., 512. Vienna: Verlag der Österreichischen Akademie der Wissenschaften, 1989.

——. *Ethiopic Astronomy and Computus*. Sitzungsberichte der Österreichischen Akademie der Wissenschaften, phil.-hist. Kl., 347. Vienna, Verlag der Österreichischen Akademie der Wissenschaften, 1979.

——. *A History of Ancient Mathematical Astronomy*. 3 vols. Studies in the History of Mathematics and Physical Sciences 1. Berlin: Springer, 1975.

——. "Regula Philippi Arrhidaei." *Isis* 50 (1950): 477–78.

——. "Thâbit ben Qurra 'On the Solar Year' and 'On the Motion of the Eighth Sphere'." *Proceedings of the American Philosophical Society* 106 (1962): 264–99.

Neugebauer, Otto, and Richard A. Parker. *Egyptian Astronomical Texts*. Vol. 3, *Decans, Planets, Constellations, and Zodiacs*. Providence, RI: Brown University Press, 1969.

Neusner, Jacob. *The Tosefta: Translated from the Hebrew with a New Introduction*. 2 vols. Peabody, MA: Hendrickson, 2002.

Nicephorus Gregoras. *Byzantina Historia*. Edited by Ludwig Schopen. 3 vols. Bonn: Weber, 1829–55.

Nicholas of Cusa. *Die Kalenderverbesserung (De correctione kalendarii)*. Edited by Viktor Stegemann. Heidelberg: Kerle, 1955.

Nolan, Edmond, and Samuel A. Hirsch. *The Greek Grammar of Roger Bacon and a Fragment of his Hebrew Grammar*. Cambridge: Cambridge University Press, 1902.

North, John. *The Ambassadors' Secret: Holbein and the World of the Renaissance*. London: Hambledon and London, 2002.

——. "Astrology and the Fortunes of Churches." *Centaurus* 24 (1980): 181–211.

——. "Astronomy and Mathematics." In *The History of the University of Oxford*, edited by Trevor H. Aston, 8 vols., 2:103–74. Oxford: Clarendon Press, 1984–2000.

——. "Chaucer, Holbein, Libra, and the Crucifixion." In *Sic Itur ad Astra: Studien zur Geschichte der Mathematik und Naturwissenschaften; Festschrift für den Arabisten Paul Kunitzsch zum 70. Geburtstag*, edited by Menso Folkerts and Richard Lorch, 461–73. Wiesbaden: Harrassowitz, 2000.

——. *Cosmos: An Illustrated History of Astronomy and Cosmology*. Chicago: University of Chicago Press, 2008.

——. "The Crucifixion and the Heavens." In *Seventh Centenary of the Teaching of Astronomy in Bologna, 1297–1997*, edited by Pierluigi Battistini, Fabrizio Bònoli, Alessandro Braccesi, and Dino Buzzetti, 17–35. Bologna: CLUEB, 2001.

——. *Horoscopes and History*. Warburg Institute Surveys and Texts 13. London: The Warburg Institute, 1986.

——. "Roger Bacon and the Saracens." In *Filosofia e scienza classica, arabo-latina medievale e l'età moderna*, edited by Graziella Federici Vescovini, 129–60. Louvain-la-Neuve: Fédération Internationale des Instituts d'Études Médiévales, 1999.

——. "The Western Calendar—'Intolerabilis, Horribilis, et Derisibilis': Four Centuries of Discontent." In *Gregorian Reform of the Calendar*, edited by George V. Coyne, Michael A. Hoskin, and Olaf Pedersen, 75–113. Vatican City: Specola Vaticana, 1983.

Nothaft, C. Philipp E. "Between Crucifixion and Calendar Reform: Medieval Christian Perception of the Jewish Lunisolar Calendar." In *Living the Lunar Calendar*, edited by Jonathan Ben-Dov, Wayne Horowitz, and John Steele. Oxford: Oxbow Books, (forthcoming).

——. "From Sukkot to Saturnalia: The Attack on Christmas in Sixteenth-Century Chronological Scholarship." *Journal of the History of Ideas* 72 (2011): 503–22.

——. "A Sixteenth-Century Debate on the Jewish Calendar: Jacob Christmann and Joseph Justus Scaliger." *Jewish Quarterly Review* (forthcoming).

O'Connell, D. J. K. "Copernicus and Calendar Reform." *Studia Copernicana* 13 (1975): 189–202.

Ó Cróinín, Dáibhí. *Early Irish History and Chronology*. Dublin: Four Courts Press, 2003.

Offenberg, Adri K. "The First Use of Hebrew in a Book Printed in the Netherlands." *Quaerendo* 4 (1974): 44–54.

Ogg, George. *The Chronology of the Public Ministry of Jesus*. Cambridge: Cambridge University Press, 1940.

——. "Hippolytus and the Introduction of the Christian Era." *Vigiliae Christianae* 16 (1962): 2–18.

——. "A Note on *Stromateis* 1.144.1–146.4." *Journal of Theological Studies* 46 (1945): 59–63.

——. *The Pseudo-Cyprianic De Pascha Computus*. London: S. P. C. K., 1955.

——. "The Year of the Exodus in the Pseudo-Cyprianic *De Pascha Computus*." *The Expository Times* 60 (1948–49): 226.

Ohashi, Masako. "The Annus Domini and the Sexta Aetas: Problems in the Transmission of Bede's De Temporibus." In Warntjes and Ó Cróinín, *Computus*, 190–203.

——. "'Sexta aetas continent annos praeteritos DCCVIIII' (Bede, *De temporibus*, 22): A Scribal Error?" In Jaritz and Moreno-Riaño, *Time and Eternity*, 55–61.

Oost, Stewart Irvin. "The Roman Calendar in the Year of Pydna (168 B.C.)." *Classical Philology* 48 (1953): 217–30.

Oppert, Gustav. "Über die Entstehung der Aera Dionysiana und den Ursprung der Null." *Verhandlungen der Berliner Gesellschaft für Anthropologie, Ethnologie und Urgeschichte* 32 (1900): 102–36.

Orlando, Thomas A. "Roger Bacon and the *Testimonia gentilium de secta christiana*." *Recherches de théologie ancienne et médiévale* 43 (1976): 202–18.

Overgaauw, Eef. "Auseinandersetzung um den Ostertermin: Die Berliner und Zeitzer Fragmente der Zeitzer Ostertafel (447)." In *Handschriften und frühe Drucke aus der Zeitzer Stiftsbibliothek*, edited by Frank-Joachim Stewing, 14–17. Petersberg: Imhof, 2009.

Overgaauw, Eef, and Frank-Joachim Stewing. *Die Zeitzer Ostertafel aus dem Jahre 447*. Petersberg: Imhof, 2005.

Palmer, James T. "Computus after the Paschal Controversy of AD 740." In Warntjes and Ó Cróinín, *Easter Controversy* (forthcoming).

Panti, Cecilia. *Moti, virtù e motori celesti nella cosmologia di Roberto Grossatesta: Studio ed edizione dei trattati De sphera, De cometis, De motu supercelestium*. Corpus Philosophorum Medii Aevi: Testi e Studi 16. Florence: SISMEL, 2001.

Parilla, Carmen. "*Qui scit, docere debet*: Acerca de Alfonso Tostado, el Tostado." *Archivum: Revista de la Facultad de Filología* 54–55 (2004–5): 367–90.

Parker, Richard A. *The Calendars of Ancient Egypt*. The Oriental Institute of Chicago: Studies in Ancient Oriental Civilization 26, Chicago: University of Chicago Press, 1950.

Parry, G. J. R. "Trinity College Dublin MS 165: The Study of Time in the Sixteenth Century." *Historical Research* 62 (1989): 15–33.

Patrides, C. A. "Renaissance Estimates of the Year of Creation." *Huntington Library Quarterly* 26 (1963): 315–22.

Pattison, Mark. "Joseph Scaliger." Chap. 6 in *Essays by the Late Mark Pattison, Sometime Rector of Lincoln College*. Edited by Henry Nettleship. Vol. 1. Oxford: Clarendon Press, 1889.

Pedersen, Fritz S. *The Toledan Tables: A Review of the Manuscripts and the Textual Versions with an Edition*. 4 vols. Kgl. Danske Vidensk. Selskab, Hist.-filos. Skr., 24.1–4. Kopenhagen: Reitzel, 2002.

Pedersen, Olaf. "The Ecclesiastical Calendar and the Life of the Church." In *Gregorian Reform of the Calendar*, edited by George V. Coyne, Michael A. Hoskin, and Olaf Pedersen, 17–74. Vatican City: Specola Vaticana, 1983.

——. "In Quest of Sacrobosco." *Journal for the History of Astronomy* 16 (1985): 175–221.

——. *A Survey of the Almagest*. Odense: Odense University Press, 1974.

Petrus Alfonsi. *Dialogue against the Jews*. Translated by Irven M. Resnick. The Fathers of the Church: Mediaeval Continuation 8. Washington: The Catholic University of America Press, 2006.

Photius of Constantinople. *Bibliotheca*. Edited by René Henry. 9 vols. Paris: Société d'Édition Les Belles Lettres, 1959–91.

Pillonel-Wyrsch, Roland-Pierre. *Le calcul de la date de Pâques au Moyen Âge: Analyse et commentaires sur "De Temporum Ratione" de Bède*. Fribourg: Academic Press, 2004.

Pingree, David. "Astronomy and Astrology in India and Iran." *Isis* 54 (1963): 229–46.

——. "The Indian and Pseudo-Indian Passages in Greek and Latin Astronomical and Astrological Texts." *Viator* 7 (1976): 141–95.

——. *Preceptum Canonis Ptolomei*. Corpus des Astronomes Byzantins 8. Louvain-la-Neuve: Academia-Bruylant, 1997.

——. *The Thousands of Abū Maʿshar*. Studies of the Warburg Institute 30. London: The Warburg Institute, 1968.

Piper, Ferdinand. "Der erste Tag der Welt." In *Königlich Preussischer Staats-Kalender für das Jahr 1856*, 6–35. Berlin: Verlag der Königlichen Geheimen Ober-Hofbuchdruckerei, 1856.

——. *Die Kalendarien und Martyrologien der Angelsachsen*. Berlin: Verlag der Königlichen Geheimen Ober-Hofbuchdruckerei, 1862.

Plassard, Joël. "Projets de réforme du calendrier à Paris au début du XIVe siècle: Textes édités et commentés." *École Nationale des Chartes: Positions de thèses* (1975): 175–81.

Pohlig, Matthias. *Zwischen Gelehrsamkeit und konfessioneller Identitätsstiftung: Lutherische Kirchen- und Universalgeschichtsschreibung 1546–1617*. Tübingen: Mohr Siebeck, 2007.

Porrer, Sheila M. *Jacques Lefèvre d'Étaples and the Three Maries Debates: Introduction, Latin Text, English Translation, and Annotation*. Geneva: Droz, 2009.

Power, Amanda. "'In the last days at the end of the world': Roger Bacon and the Reform of Christendom." *Canterbury Studies in Franciscan History* 1 (2008): 135–51.

——. "A Mirror for Every Age: The Reputation of Roger Bacon." *English Historical Review* 121 (2006): 657–92.

Power, Matthew A. "Nisan Fourteenth and Fifteenth in Gospel and Talmud: A Study in Jewish Camouflage." *American Journal of Theology* 24 (1920): 252–76.

Pratt, John P. "Newton's Date for the Crucifixion." *Quarterly Journal of the Royal Astronomical Society* 32 (1991): 301–4.

Preuschen, Erwin. "Todesjahr und Todestag Jesu." *Zeitschrift für die neutestamentliche Wissenschaft* 5 (1904): 1–17.

Price, B. B. "The Use of Astronomical Tables by Albertus Magnus." *Journal for the History of Astronomy* 22 (1991): 221–40.

Prijs, Joseph. *Die Basler hebräischen Drucke (1492–1866)*. Edited by Bernhard Prijs. Olten: Graf, 1964.

Ptolemy, Claudius. *Almagest*. Translated by G. J. Toomer. New York: Springer, 1984.

Quicherat, Jules. *Histoire des Règnes de Charles VII et de Louis XI par Thomas Basin*. 4 vols. Ouvrages publiés par la Société de l'histoire de France 29. Paris: Lahure, 1855–59.

——. "Thomas Basin, sa vie et ses écrits." *Bibliothèque de l'École des Chartes* 3 (1841–42): 313–76.

Rauner, Erwin. *Katalog der lateinischen Handschriften der Bayerischen Staatsbibliothek München: Die Handschriften aus Augsburger Bibliotheken*. Vol. 1, *Stadtbibliothek: Clm 3501–3661*. Wiesbaden: Harrassowitz, 2007.

Recio, Roxana, and Antonio Cortijo Ocaña. "Alfonso de Madrigal 'El Tostado': Un portavoz único de la intelectualidad castellana del siglo XV." *La Corónica* 33, no. 1 (2004): 7–15.

Reeves, Eileen. *Painting the Heavens: Art and Science in the Age of Galileo*. Princeton, NJ: Princeton University Press, 1997.

Reich, Karin. "Problems of Calendar Reform from Regiomontanus to the Present." In Zinner, *Regiomontanus*, 345–62.

Reinhardt, Klaus, and Horacio Santiago-Otero. *Biblioteca bíblica ibérica medieval*. Medievala et Humanistica 1. Nomenclator de autores medievales hispanos 1. Madrid: Consejo Superior de Investigaciones Científicas, 1986.

Renaker, David. "The Horoscope of Christ." *Milton Studies* 12 (1978): 213–33.

Reynolds, L. D., ed. *Texts and Transmission: A Survey of the Latin Classics*. Oxford: Clarendon Press, 1983.

Richard, Marcel. "Comput et chronographie chez saint Hippolyte." Pts. 1 and 2. *Mélanges de science religieuse* 7 (1950): 237–68; 8 (1951): 19–50.

——. "Le comput pascal par octaétéris." *Le Muséon* 87 (1974): 307–39.

——. "Notes sur le comput des cent-douze ans." *Revue des Études Byzantines* 24 (1966): 257–77.

Rosenberg, Daniel, and Anthony Grafton. *Cartographies of Time: A History of the Timeline*. New York: Princeton Architectural Press, 2010.

Rosen, Edward. "Galileo's Misstatements about Copernicus." *Isis* 49 (1958): 319–30.

——. "Maurolico's Attitude toward Copernicus." *Proceedings of the American Philosophical Society* 101 (1957): 177–94.

——. "Was Copernicus' *Revolutions* Approved by the Pope?" *Journal of the History of Ideas* 36 (1975): 531–42.

Roth, Norman. *Conversos, Inquisition, and the Expulsion of the Jews from Spain*. Madison: The University of Wisconsin Press, 1995.

Ruggles, Clive. "The Moon and the Crucifixion." *Nature* 345 (1990): 669–70.

Rühl, Franz. "Der Ursprung der jüdischen Weltära." *Deutsche Zeitschrift für Geschichtswissenschaft*, n.s., 2 (1897–98): 185–202.

Russell, Josiah C. *Dictionary of Writers of Thirteenth Century England*. London: Longmans, Green, 1936.

——. "Hereford and Arabic Science in England about 1175–1200." *Isis* 18 (1932): 14–25.

Sachs, A. J., and C. B. F. Walker. "Kepler's View of the Star of Bethlehem and the Babylonian Almanac for 7/6 B.C." *Iraq* 46 (1984): 43–55.

Salmon, George. "The Commentary of Hippolytus on Daniel." *Hermathena* 8 (1893): 161–90.

——. "Some Notes on the Chronology of Hippolytus." *Hermathena* 1 (1873): 82–128.

Sambin, Paolo. "Il dottorato padovano in medicina di Paolo da Middelburgo (1480)." *Quaderni per la storia dell'Università di Padova* 9–10 (1976–77): 252–56.

Samuel, Alan E. *Greek and Roman Chronology: Calendars and Years in Classical Antiquity*. Handbuch der Altertumswissenschaft, Abt. 1, vol. 7. Munich: C. H. Beck, 1972.

——. *Ptolemaic Chronology*. Münchener Beiträge zur Papyrusforschung und antiken Rechtsgeschichte 43. Munich: C. H. Beck, 1962.

Santiago-Otero, Horacio. *Manuscritos de autores medievales hispanos*. Vol. 1. Medievalia et Humanistica 3. Madrid: Consejo Superior de Investigaciones Científicas, 1987.

Sato, Masayuki. "Comparative Ideas of Chronology." *History and Theory* 30 (1991): 275–301.

Schabel, Chris. "*Ad correctionem calendarii*... The Background to Clement VI's Initiative?" *Cahiers de l'Institut du Moyen-Âge Grec et Latin* 68 (1998): 13–34.

——. "Jean des Murs and Firmin of Beauval's Letter and Treatise on Calendar Reform for Clement VI." *Cahiers de l'Institut du Moyen-Âge Grec et Latin* 66 (1996): 187–215.

——. "Peter de Rivo and the Quarrel over Future Contingents at Louvain: New Evidence and New Perspectives." *Documenti e studi sulla tradizione filosofica medievale* 6 (1995): 363–473; 7 (1996): 369–435.

——. *Theology at Paris, 1316–1345: Peter Auriol and the Problem of Divine Foreknowledge and Future Contingents*. Aldershot: Ashgate, 2000.

Schaefer, Bradley E. "Lunar Visibility and the Crucifixion." *Quarterly Journal of the Royal Astronomical Society* 31 (1990): 53–67.

Schwartz, Eduard. *Christliche und jüdische Ostertafeln*. Abhandlungen der Königlichen Gesellschaft der Wissenschaften zu Göttingen, phil.-hist. Kl., n.s., vol. 8.6. Berlin: Weidmann, 1905.

——. "Osterbetrachtungen." *Zeitschrift für die neutestamentliche Wissenschaft* 7 (1906): 1–33.

Screech, M. A. "The Magi and the Star (Matthew 2)." In *Histoire de l'exégèse au XVIe siècle*, edited by Olivier Fatio and Pierre Fraenkel, 385–409. Genva: Droz, 1978.

Sela, Shlomo. *Abraham Ibn Ezra and the Rise of Medieval Hebrew Science*. Brill's Series in Jewish Studies 32. Leiden: Brill, 2003.

——. "Contactos científicos entre judíos y cristianos en el siglo XII: El caso del *Libro de las Tablas Astronómicas* de Abraham Ibn Ezra en su versión latina y hebrea." *Miscelánea de estudios árabes y hebraicos*, Sección de Hebreo, 45 (1996): 185–222.

Serruys, Daniel. "Des quelques ères usitées chez les chroniqueurs byzantins." *Revue de Philologie*, n.s., 31 (1907): 151–89.

——. "Les Canons d'Eusèbe, d'Annianos et d'Andronicus d'après Élie de Nisibe." *Byzantinische Zeitschrift* 22 (1913): 1–36.

Shuger, Deborah Kuller. *The Renaissance Bible: Scholarship, Sacrifice, and Subjectivity*. Berlekey: University of California Press, 1994.

Shumaker, Wayne. *Renaissance Curiosa*. Medieval and Renaissance Texts and Studies 8. Binghamton: Center for Medieval and Early Renaissance Studies, 1982.

Simon, Marcel. *Verus Israel: A Study of the Relations between Christians and Jews in the Roman Empire (AD 135–425)*. London: The Littman Library of Jewish Civilization, 1996.

Siniscalco, Paolo. "Le età del mondo in Beda." *Romanobarbarica* 3 (1978): 297–331.

Smith, Mahlon H., III. *And Taking Bread…: Cerularius and the Azyme Controversy of 1054*. Paris: Éditions Beauchesne, 1978.

Smoller, Laura Ackerman. *History, Prophecy, and the Stars: The Christian Astrology of Pierre d'Ailly, 1350–1420*. Princeton, NJ: Princeton University Press, 1994.

Southern, Richard W. *Medieval Humanism and Other Studies*. Oxford: Basil Blackwell, 1970.

Spier, Arthur. *The Comprehensive Hebrew Calendar*. New York: Behrman House, 1952.

Springsfeld, Kerstin. *Alkuins Einfluß auf die Komputistik zur Zeit Karls des Großen*. Sudhoffs Archiv: Beihefte 48. Stuttgart: Steiner, 2002.

——. "Eine Beschreibung der Handschrift St. Gallen Stiftsbibliothek, 225." In Warntjes and Ó Cróinín, *Computus*, 204–37.

Stacey, Robert C. "Jews and Christians in Twelfth-Century England: Some Dynamics of a Changing Relationship." In *Jews and Christians in Twelfth-Century Europe*, edited by Michael A. Signer and John Van Engen, 340–54. Notre Dame: University of Notre Dame Press, 2001.

Starke, Ronald. *Niemand hat an der Uhr gedreht! Die Phantomzeittheorie auf dem Prüfstand*. Munich: Differenz-Verlag, 2009.

Stegmüller, Friedrich. "Pedro de Osma: Ein Beitrag zur spanischen Universitäts-, Konzils- und Ketzergeschichte." *Römische Quartalschrift* 43 (1935): 205–66.

Steiner, Benjamin. *Die Ordnung der Geschichte: Historische Tabellenwerke in der Frühen Neuzeit*. Norm und Struktur 34. Cologne: Böhlau, 2008.

Stern, Sacha. *Calendar and Community: A History of the Jewish Calendar, Second Century BCE–Tenth Century CE*. Oxford: Oxford University Press, 2001.

——. "Fictitious Calendars: Early Rabbinic Notions of Time, Astronomy and Reality." *Jewish Quarterly Review*, n.s., 87 (1996): 103–29.

——. "Near Eastern Lunar Calendars in the Syriac Martyr Acts." *Le Muséon* 117 (2004): 447–72.

Stern, Sacha, and Piergabriele Mancuso. "An Astronomical Table by Shabbetai Donnolo and the Jewish Calendar in Tenth-Century Italy." *Aleph* 7 (2007): 13–41. With corrigenda in *Aleph* 8 (2008): 343–44.

Stevenson, Jane. *The "Laterculus Malalianus" and the School of Archbishop Theodore*. Cambridge: Cambridge University Press, 1995.

——. "Theodore and the *Laterculus Malalianus*." In *Archbishop Theodore: Commemorative Studies on his Life and Influence*, edited by Michael Lapidge, 204–21. Cambridge: Cambridge University Press, 1995.

Stevenson, W. H. "A Contemporary Description of the Domesday Survey." *English Historical Review* 22 (1907): 72–84.

Stevens, Wesley M. *Cycles of Time and Scientific Learning in Medieval Europe*. Variorum Collected Studies Series 482. Aldershot: Variorum, 1995.

Strobel, August. *Texte zur Geschichte des frühchristlichen Osterkalenders*. Liturgiewissenschaftliche Quellen und Forschungen 64. Münster: Aschendorff, 1984.

——. *Ursprung und Geschichte des frühchristlichen Osterkalenders*. Texte und Untersuchungen zur Geschichte der altchristlichen Literatur 121. Berlin: Akademie Verlag, 1977.

——. "Weltenjahr, große Konjunktion und Messiasstern: Ein themageschichtlicher Überblick." In *Aufstieg und Niedergang der Römischen Welt. Teil II: Principat*, vol. 20.2, edited by Wolfgang Haase, 998–1187. Berlin: de Gruyter, 1987.

Struik, Dirk Jan. "Paolo di Middelburg e il suo posto nella storia delle scienze esatte." *Periodico di Matematiche*, 4th ser., 5 (1925): 337–47.

——. "Paulus van Middelburg (1445–1533)." *Mededeelingen van het Nederlandsch Historisch Instituut te Rome* 5 (1925): 79–118.

——. "Sull'opera matematica di Paolo di Middelburg." *Rendiconti della Reale Accademia Nazionale dei Lincei: Classe di scienze fisiche, matematiche e naturali* 1 (1925): 305–8.

Suhl, Alfred. "Zu Gerhard Mercators Evangelienharmonie." In *Mercator und Wandlungen der Wissenschaften im 16. und 17. Jahrhundert*, edited by Manfred Büttner and René Dirven, 43–59. Bochum: Brockmeyer, 1993.

Suter, Heinrich, ed. *Die astronomischen Tafeln des Muhammed ibn Mūsā al-Khwārizmī in der Bearbeitung des Maslama ibn Ahmed al-Madjrītī und der lateinischen Übersetzung des Athelhard von Bath*, Kgl. Danske Vidensk. Selskab, Skrifter, 7th ser., Hist. og filos. Afd., 3.1. Copenhagen: A. F. Høst & Søn, 1914.

Swerdlow, Noel M. "The Length of the Year in the Original Proposal for the Gregorian Calendar." *Journal for the History of Astronomy* 17 (1986): 109–18.

Swerdlow, Noel M., and Otto Neugebauer. *Mathematical Astronomy in Copernicus's* De Revolutionibus. 2 vols. Studies in the History of Mathematics and Physical Sciences 10. New York: Springer, 1984.

Szpiech, Ryan. "Scrutinizing History: Polemic and Exegesis in Pablo de Santa María's *Siete edades del mundo*." *Medieval Encounters* 16 (2010): 96–142.

Talley, Thomas J. "Further Light on the Quartodeciman Pascha and the Date of the Annunciation." *Studia Liturgica* 33 (2003): 151–58.

——. *The Origins of the Liturgical Year*. 2nd ed. Collegeville, MN: The Liturgical Press, 1991.

Tatian. *Oratio ad Graecos and Fragments*. Edited by Molly Whittaker. Oxford: Clarendon Pres, 1982.

Theophanes Confessor. *Chronographia*. Edited by Carl de Boor. 2 vols. Leipzig: Teubner, 1883–85.

Thomas Aquinas. *Opera Omnia Iussu Impensaque Leonis XIII*. 50 vols. Rome: Typographia Polyglotta S. C. de Propaganda Fide, 1882–.

——. *Scriptum super libros Sententiarum magistri Petri Lombardi episcopi Parisiensis*. Edited by P. Mandonnet and M. F. Moos. 4 vols. Paris: Lethielleux, 1929–47.

——. *Super Evangelium S. Matthaei lectura*. Edited by R. Cai. 5th ed. Turin: Marietti, 1951.

——. *Super Evangelium S. Ioannis lectura*. Edited by R. Cai. 5th ed. Turin: Marietti, 1952.

Thompson, James Westfall. "The Introduction of Arabic Science into Lorraine in the Tenth Century." *Isis* 12 (1929): 184–93.

Thorndike, Lynn. "Computus." *Speculum* 29 (1954): 223–38.

——. *A History of Magic and Experimental Science*. 6 vols. New York: Columbia University Press, 1923–58.

Thorndike, Lynn, and Pearl Kibre. *A Catalogue of Incipits of Mediaeval Scientific Writings in Latin*. Rev. ed. The Medieval Academy of America Publication no. 29. London: The Medieval Academy of America, 1963.

Thornton, Timothy C. G. "Problematical Passovers: Difficulties for Diaspora Jews and Early Christians in Determining Passover Dates During the First Three Centuries A.D." *Studia Patristica* 20 (1989): 402–8.

Timmer, David E. "Biblical Exegesis and the Jewish-Christian Controversy in the Early Twelfth Century." *Church History* 58 (1989): 309–21.

Tolan, John. *Petrus Alfonsi and his Medieval Readers*. Gainesville: University Press of Florida, 1993.

Tönsmeyer, Josef. "Hermann Zoestius von Marienfeld: Ein Vertreter der konziliaren Theorie am Konzil von Basel." *Westfälische Zeitschrift* 87 (1930): 114–91.

Toomer, G. J. *John Selden: A Life in Scholarship*. 2 vols. Oxford: Oxford University Press, 2009.

——. "A Survey of the Toledan Tables." *Osiris* 15 (1968): 5–174.

Topper, Uwe. *Kalender-Sprung: Falsche Geschichtsschreibung bestimmt die Zukunft; Europas Religionswechsel um 1500*. Tübingen: Grabert, 2006.

Tränkle, Hermann. *Q. S. F. Tertulliani Adversus Iudaeos: Mit Einleitung und kritischem Kommentar*. Wiesbaden: Steiner, 1964.

Tuckerman, Bryant. *Planetary, Lunar, and Solar Positions 601 B.C. to A.D. 1 at Five-Day and Ten-Day Intervals*. Memoirs of the American Philosophical Society 56. Philadelphia: American Philosophical Society, 1962.

van Bladel, Kevin. *The Arabic Hermes: From Pagan Sage to Prophet of Science*. Oxford: Oxford University Press, 2009.

Vanden Broecke, Steven. *The Limits of Influence: Pico, Louvain, and the Crisis of Renaissance Astrology*. Medieval and Early Modern Science 4. Leiden: Brill, 2003.

——. "Paulus von Middelburg, *Paulina*." In *De geleerde wereld van Keizer Karel*, edited by Tineke Padmos and Geerd Vanpaemel, 291–93. Louvain: Leuven University Press, 2000.

Van Engen, John H. *Rupert of Deutz*. Publications of the UCLA Center for Medieval and Renaissance Studies 18. Berkeley: University of California Press, 1983.

van Wijk, Walter Émile. "Un Comput de la fin du douzième siècle." *Archives Internationales d'Histoire des Sciences* 4 (1951): 867–73.

——. *Le comput emendé de Reinherus de Paderborn (1171)*. Verhandelingen der Koninklijke Nederlandse Akademie van Wetenschappen, Afd. Letterkunde, n.s., vol. 57.3. Amsterdam: North-Holland, 1951.

Vardaman, Jerry, ed. *Chronos, Kairos, Christos II: Chronological, Nativity, and Religious Studies in Memory of Ray Summers*. Macon, GA: Mercer University Press, 1998.

Vardaman, Jerry, and Edwin M. Yamauchi, eds. *Chronos, Kairos, Christos: Nativity and Chronological Studies Presented to Jack Finegan*. Winona Lake, IN: Eisenbrauns, 1989.

Verbist, Peter. "Abbon de Fleury et l'ère chrétienne vers l'an mil: Un esprit critique vis-à-vis d'une tradition erronée." In *Abbon de Fleury: Philosophie, sciences et comput autour de l'an mil*, edited by Barbara Obrist, 61–93. Paris: Publications universitaires Denis Diderot, 2004.

——. "Abbo of Fleury and the Computational Accuracy of the Christian Era." In Jaritz and Gerson Moreno-Riaño, *Time and Eternity*, 63–80.

——. *Duelling with the Past: Medieval Authors and the Problem of the Christian Era, c. 990–1135*. Studies in the Early Middle Ages 21. Turnhout: Brepols, 2010.

——. "De Epistola ad Hugonem van Heriger van Lobbes († 1007): Een belangrijk chronologisch traktaat." *Millennium: Tijdschrift voor middeleeuwse studies* 12 (1998): 30–42.

——. "Reconstructing the Past: The Chronicle of Marianus Scottus (d. 1082)." *Peritia* 16 (2002): 284–334.

Vermij, Rienk. "Gerard Mercator and the Science of Chronology." In Blotevogel and Vermijk *Gerhard Mercator*, 189–97.

——. "Mercators Stellung in der Geschichte der Wissenschaft." In *Gerhard Mercator und seine Welt*, edited by Rienk Vermij, 90–109. Duisburg: Mercator-Verlag, 1997.

Vernarecci, Augusto. *Fossombrone dai tempi antichissimi ai nostri*. 2 vols. Fossombrone: Monacelli, 1907–14.

Vogtherr, Thomas. "100 Jahre Forschungen zur Chronologie des Mittelalters und der Neuzeit." *Archiv für Diplomatik* 54 (2008): 223–47.

Voigt, Ulrich. "Geminos von Rhodos und die Oktaëteris." *Beiträge zur Astronomiegeschichte* 10 (2010): 7–34.

——. "Über das Jahr 1 des Alexander Severus auf der Passatafel des Hippolytus." *Scholion* 5 (2008): 99–121.

von Bezold, Friedrich. "Astrologische Geschichtsconstruction im Mittelalter." *Deutsche Zeitschrift für Geschichtswissenschaft* 8 (1892): 29–72.

von den Brincken, Anna-Dorothee. "Beobachtungen zum Aufkommen der retrospektiven Inkarnationsära." *Archiv für Dipolmatik* 25 (1979): 1–20.

——. "Marianus Scottus als Universalhistoriker iuxta veritatem Evangelii." In *Die Iren und Europa im früheren Mittelalter*, edited by Heinz Löwe, 2 vols., 2:970–1009. Stuttgart: Klett-Cotta, 1982.

——. "Marianus Scottus: Unter besonderer Berücksichtigung der nicht veröffentlichten Teile seiner Chronik." *Deutsches Archiv* 17 (1961): 191–238.

——. *Studien zur lateinischen Weltchronistik bis in das Zeitalter Ottos von Freising.* Düsseldorf: Triltsch, 1957.

——. "Welt- und Inkarnationsära bei Heimo von St. Jakob: Kritik an der christlichen Zeitrechnung durch Bamberger Komputisten in der ersten Hälfte des 12. Jahrhunderts." *Deutsches Archiv* 16 (1960): 155–94.

von Döllinger, Johann Joseph Ignaz. *Hippolytus und Kallistus; oder die römische Kirche in der ersten Hälfte des dritten Jahrhunderts.* Regensburg: Manz, 1853.

von Stuckrad, Kocku. *Das Ringen um die Astrologie: Jüdische und christliche Beiträge zum antiken Zeitverständnis.* Religionsgeschichtliche Versuche und Vorarbeiten 49. Berlin: de Gruyter, 2000.

——. *Locations of Knowledge in Medieval and Early Modern Europe: Esoteric Discourse and Western Identities.* Brill's Studies in Intellectual History 186. Leiden: Brill, 2010.

Wacholder, Ben Zion, and David B. Weisberg. "Visibility of the New Moon in Cuneiform and Rabbinic Sources." *Hebrew Union College Annual* 42 (1971): 227–42.

Wagenaar, Jan A. *Origin and Transformation of the Ancient Israelite Festival Calendar.* Beihefte zur Zeitschrift für Altorientalische und Biblische Rechtsgeschichte 6. Wiesbaden: Harrassowitz, 2005.

Wallis, Faith. *Bede: The Reckoning of Time.* Translated Texts for Historians 29. Liverpool: Liverpool University Press, 1999.

——. "The Church, the World, and the Time: Prolegomena to a History of the Medieval Computus." In *Normes et pouvoir à la fin du moyen âge*, edited by Marie-Claude Déprez-Masson, 15–29. Montreal: Editions CERES, 1990.

Wallraff, Martin, ed. *Julius Africanus und die christliche Weltchronistik.* Texte und Untersuchungen zur Geschichte der altchristlichen Literatur 157. Berlin: de Gruyter, 2006.

Walmsley, Conrad. "Two Long Lost Works of William Woodford and Robert of Leicester." *Archivum Franciscanum Historicum* 46 (1953): 458–70.

Walsh, Maura, and Dáibhí Ó Cróinín. *Cummian's Letter* De controversia paschali *and the* De ratione conputandi. Toronto: Pontifical Institute of Medieval Studies, 1988.

Warntjes, Immo. "The *Argumenta* of Dionysius Exiguus and their Early Recensions." In Warntjes and Ó Cróinín, *Computus*, 40–111.

——. "Irische Komputistik zwischen Isidor von Sevilla und Beda Venerabilis: Ursprung, karolingische Rezeption und Forschungsperspektiven." *Viator* 42 (forthcoming).

——. "The Munich Computus and the 84 (14)-year Easter Reckoning." *Proceedings of the Royal Irish Academy* 107C (2007): 31–85.

——. *The Munich Computus: Text and Translation; Irish Computistics between Isidore of Seville and the Venerable Bede and its Reception in Carolingian Times.* Sudhoffs Archiv: Beihefte 59. Stuttgart: Steiner, 2010.

Warntjes, Immo, and Dáibhí Ó Cróinín, eds. *Computus and its Cultural Context in the Latin West, AD 300–1200*. Studia Traditionis Theologiae 5. Turnhout: Brepols, 2010.

——, eds. *The Easter Controversy of Late Antiquity and the Early Middle Ages: Its Manuscripts, Texts, and Tables*. Studia Traditionis Theologiae 10. Turnhout: Brepols, (forthcoming).

Wattenbach, Wilhelm. "Über Hermann von Marienfeld aus Münster." *Sitzungsberichte der königlich preussischen Akademie der Wissenschaften zu Berlin*, phil.-hist. Kl., 9 (1884): 93–109.

Watts, Pauline M. "Prophecy and Discovery: On the Spiritual Origins of Christopher Columbus's 'Enterprise of the Indies'." *American Historical Review* 90 (1985): 73–102.

Weber, Stefan. *Iren auf dem Kontinent: Das Leben des Marianus Scottus von Regensburg und die Anfänge der irischen "Schottenklöster"*. Heidelberg: Mattes, 2010.

Weikmann, Hans Martin. *Heimo von Bamberg: De decursu temporum*. MGH Quellen zur Geistesgeschichte des Mittelalters 19. Hannover: Hahn, 2004.

Weinberg, Joanna. "Invention and Convention: Jewish and Christian Critique of the Jewish Fixed Calendar." *Jewish History* 14 (2000): 317–30.

Weinberger, Leon J. "On the Provenance of Benjamin B. Samuel Quštani." *Jewish Quarterly Review*, n.s., 68 (1977): 46–60.

Weinstock, Horst. "Roger Bacon und das 'hebräische' Alphabet." *Aschkenas* 2 (1992): 15–48.

Welborn, Mary Catherine. "Lotharingia as a Center of Arabic and Scientific Influence in the Eleventh Century." *Isis* 16 (1931): 188–99.

Welker, Lorenz. "Ottaviano Petrucci and the Political-Cultural Elite of his Time: The 1513 Print of Paul of Middelburg's *Summa Paulina de Recta Paschae Celebratione*." In *Venezia 1501: Petrucci e la stampa musicale*, edited by Giulio Cattin and Patrizia Dalla Vecchia, 107–15. Venice: Edizioni Fondazione Levi, 2005.

Wetzel, Georg. *Die Chronicen des Baeda Venerabilis* (Halle: Plötz, 1878).

Whalen, Brett Edward. *Dominion of God: Christendom and Apocalypse in the Middle Ages*. Cambridge, MA: Harvard University Press, 2009.

——. "Rethinking the Schism of 1054: Authority, Heresy, and the Latin Rite." *Traditio* 62 (2007): 1–24.

Whitrow, Gerald J. *Time in History: Views of Time from Prehistory to the Present Day*. Oxford: Oxford University Press, 1988.

Wiesenbach, Joachim. *Sigebert von Gembloux: Liber decennalis*. MGH Quellen zur Geistesgeschichte des Mittelalters 12. Weimar: Böhlau, 1986.

Willets, P. J. "A Reconstructed Astronomical MS from Christchurch Library Canterbury." *The British Museum Quarterly* 30 (1965): 22–30.

William of Auxerre. *Summa Aurea*. Edited by Jean Ribaillier. 5 vols. Spicilegium Bonaventurianum 16–20. Paris: CNRS, 1980–87.

William of Militona. *Quaestiones de sacramentis*. Edited by Caelestinus Piana and Gedeon Gál. 2 vols. Bibliotheca Franciscana Scholastica Medii Aevi 22–23. Quaracchi: Collegium S. Bonaventurae, 1961.

Yallop, Bernard D. *A Method for Predicting the First Sighting of the New Crescent Moon*. Cambridge: HM Nautical Almanac Office, 1997. http://www.icoproject.org/pdf/yallop_1997.pdf.

Yerushalmi, Yosef Hayim. *Zakhor: Jewish History and Jewish Memory*. Seattle: University of Washington Press, 1982.

Yuval, Israel Jacob. "Easter and Passover as Early Jewish-Christian Dialogue." In Bradshaw and Hoffman, *Passover*, 1:98–124.

——. "Passover in the Middle Ages." In Bradshaw and Hoffman, *Passover*, 1:127–60.

——. *Two Nations in Your Womb: Perceptions of Jews and Christians in Late Antiquity and the Middle Ages*. Translated by Barbara Harshav and Jonathan Chipman. Berkeley: University of California Press, 2006.

Zambelli, Paola. "Da Giulio II a Paolo III: Come l'astrologo provocatore Luca Gaurico divenne vescovo." In *La città dei segreti: Magia, astrologia e cultura esoterica a Roma (XV–XVIII)*, edited by Fabio Troncarelli, 299–323. Milan: Angeli, 1985.

——. "Many Ends for the World: Luca Gaurico Instigator of the Debate in Italy and in Germany." In *"Astrologi hallucinati": Stars and the End of the World in Luther's Time*, edited by Paola Zambelli, 239–63. Berlin: de Gruyter, 1986.

Zeitlin, Solomon. "The Judaean Calendar During the Second Commonwealth and the Scrolls." *Jewish Quarterly Review*, n.s., 57 (1966): 28–45.

——. "'The Secret of *BADHU*': A Specimen of 'Jewish Camouflage'." *American Journal of Theology* 24 (1920): 502–11.

Zerubavel, Eviatar. "Easter and Passover: On Calendars and Group Identity." *American Sociological Review* 47 (1982): 284–89.

Zinner, Ernst. *Regiomontanus: His Life and Work*. Translated by Ezra Brown. Amsterdam: North-Holland, 1990.

Zuckermann, Benedict. *Materialien zur Entwicklung der altjüdischen Zeitrechnung im Talmud*. Breslau: Preuss & Junger, 1882.

Zurbonsen, Friedrich. *Hermannus Zoestius und seine historisch-politischen Schriften: Nach handschriftlichen Quellen des fünfzehnten Jahrhunderts*. Beilage zum Programm des Königl. Gymnasiums zu Warendorf. Warendorf, 1884.

——. "Hermann Zoestius von Marienfeld und seine Schriften." *Westdeutsche Zeitschrift* 18 (1899): 146–73.

INDEX OF BIBLICAL REFERENCES

The Old Testament

The New Testament

INDEX OF NAMES

www.ingramcontent.com/pod-product-compliance
Lightning Source LLC
LaVergne TN
LVHW010656300926
849161LV00005B/1141